Linear Algebra with its Applications

RIVER PUBLISHERS SERIES IN MATHEMATICAL, STATISTICAL AND COMPUTATIONAL MODELLING FOR ENGINEERING

Series Editors:

HARISH GARG
Thapar Institute of Engineering & Technology, India

MANGEY RAM
Graphic Era University, India

Applied mathematical techniques along with statistical and computational data analysis has become vital skills across the physical sciences. The purpose of this book series is to present novel applications of numerical and computational modelling and data analysis across the applied sciences. We encourage applied mathematicians, statisticians, data scientists and computing engineers working in a comprehensive range of research fields to showcase different techniques and skills, such as differential equations, finite element method, algorithms, discrete mathematics, numerical simulation, machine learning, probability and statistics, fuzzy theory, etc.

Books published in the series include professional research monographs, edited volumes, conference proceedings, handbooks and textbooks, which provide new insights for researchers, specialists in industry, and graduate students.

Topics included in this series are as follows:

- Discrete mathematics and computation
- Fault diagnosis and fault tolerance
- Finite element method (FEM) modeling/simulation
- Fuzzy and possibility theory
- Fuzzy logic and neuro-fuzzy systems for relevant engineering applications
- Game Theory
- Mathematical concepts and applications
- Modelling in engineering applications
- Numerical simulations
- Optimization and algorithms
- Queueing systems
- Resilience
- Stochastic modelling and statistical inference
- Stochastic Processes
- Structural Mechanics
- Theoretical and applied mechanics

For a list of other books in this series, visit www.riverpublishers.com

Linear Algebra with its Applications

Ramakanta Meher

Sardar Vallabhbhai National Institute of Technology,
Surat, Gujarat, India

River Publishers

Routledge
Taylor & Francis Group
NEW YORK AND LONDON

Published 2025 by River Publishers
River Publishers
Alsbjergvej 10, 9260 Gistrup, Denmark
www.riverpublishers.com

Distributed exclusively by Routledge
605 Third Avenue, New York, NY 10017, USA
4 Park Square, Milton Park, Abingdon, Oxon OX14 4RN

Linear Algebra with its Applications / by Ramakanta Meher.

Routledge is an imprint of the Taylor & Francis Group, an informa business

ISBN 978-87-7004-157-7 (hardback)
ISBN 978-87-7004-781-4 (paperback)
ISBN 978-87-7004-771-5 (online)
ISBN 978-87-7004-770-8 (master ebook)

Contents

Preface

Linear algebra plays a vital role in studying different types of real-world problems to study the behavior of the issues. A definition of linear algebra that might be a part of algebra is concerned with equations of the first degree. Thus, at the fundamental level, it involves the discussion of matrices and determinants and the solutions of systems of linear equations, which have a wide application in further discussion of this subject. Linear algebra is a subject that has found the broadest range of applications in all branches of mathematics, physical and social sciences, and engineering. It has a more significant application in information sciences and control theory.

This book begins with a detailed discussion of matrix operation, its properties, and its applications in finding the solution of linear equations and determinants.

This textbook entitled *Linear Algebra with Its Applications* is intended to study matrices, vector spaces, eigenvalue, eigenvectors, linear transformation methods, inner product spaces, diagonalizations, applications to conics and quadrics, canonical forms, and least squares problems. This book contains 11 chapters.

Chapter 1 discusses the properties of matrices and matrix operations needed to study solutions of systems of linear equations and determinants. *Chapter 2* discusses the system of linear equations and its solution using the Gaussian elimination method, the Gauss–Jordan elimination method, and LU decomposition methods, and the definitions of determinants with their properties and Crammer's rule. In contrast, *Chapter 3* starts with a discussion of vector spaces in n-dimensional vector spaces that include the properties of a vector space and subspaces, linear combinations and spanning a vector space, finitely generated vector spaces, linear dependence and independence, basis and dimensions, rank, sum and intersection of subspaces, direct sums of subspaces, and more than two subspaces. *Chapter 4* discusses the properties of eigenvalues and eigenvectors and some properties of inner product spaces, including the Gram–Schmidt orthogonalization process and QR-factorization. *Chapter 5* discusses linear transformation, which includes

Range and Kernel of transformation, one-to-one and invertible transformations, ordination representation of vectors, change of basis, isomorphism, transformations in computer graphics, and fractal pictures of nature. *Chapter 6* discusses inner product spaces that include the Cauchy–Schwarz inequality, orthogonal complements, orthogonal sets and bases, projection of one vector onto another vector, orthogonal matrix theorem, the Gram–Schmidt orthogonalization process, projection of a vector onto a subspace, distance of a point from a subspace, and QR-factorization. *Chapter 7* discusses the matrix representation of linear transformations along with the importance of matrix representation, visualization of the matrix representation, and the relation between matrix representations.

In contrast, *Chapter 8* covers the diagonalizations that include minimal polynomials, the Cayley–Hamilton theorem, power of a matrix, diagonal matrix representation of a linear operator, diagonalization of matrices, diagonalization of symmetric matrices, and orthogonal diagonalization. *Chapter 9* discusses the application to conics and quadrics that covers quadratic forms, conics, quadrics, definite quadratic form, bilinear form, matrix representation of bilinear forms, symmetric and skew-symmetric bilinear form, symmetric bilinear forms and quadratic forms, eigenvalues of congruent matrices, Sylvester's law of inertia, skew-symmetric bilinear form, and the application to the reduction of quadrics. *Chapter 10* discusses the canonical forms that include triangularizable matrices, block triangular matrices, block diagonalization, Hermitian matrices, unitary matrix, Schur's theorem, spectral theorem, normal matrices, nilpotent operators, Jordan canonical form, rational canonical form, minimum polynomial and Jordan canonical form, Jordan normal form, properties of Jordan matrix, and minimum polynomial of Jordan normal form While the last chapter, *Chapter 11*, discusses the least square problems that cover approximation of functions, Fourier approximation, least square solutions, least square curves, eigenvalues by iteration and connectivity of networks, the power method for an $n \times n$ matrix, difficulties in the solution of the system of equations, the condition number of a matrix, and the coding theory.

This book is based on syllabi of linear algebra prescribed for undergraduate and postgraduate mathematics students in different institutions and universities in India and abroad. This book will be helpful for competitive examinations as well.

I welcome constructive criticisms, views, and suggestions from the reviewers and readers for further improvement.

List of Figures

List of Tables

List of Abbreviations

A.M	Algebraic multiplicity
G.M	Geometric multiplicity
Iff	If and only if
IFS	Iterated function system
LCM	Least common multiple
LD	Linearly dependent
LI	Linearly independent
LU	Lower-Upper
NTSC	National Television System Committee
REF	Reduced echelon form
RREF	Row-reduced echelon form
vph	Vehicles per hour

1

Matrices

This chapter discusses the classification, properties of matrices, and matrix operations needed to study the solution of linear equations and elementary matrices. It introduces the operations of addition and multiplications for matrices and defines the algebraic properties of these operations. The powers of matrices and inverses of matrices are also described. These tools lead to different methods for solving linear systems and insights into their behavior. It lays the foundation for using matrices to define functions called linear transformations or vector spaces. The reader will identify how matrices are used in various applications. They are applied in archaeology to determine the chronological order of artifacts, cryptography to ensure security, and demography to predict population movement. The inverse of a matrix is used in a model for analyzing the interdependence of economics. Wassily Leontief received a Nobel Prize for his work in this field. This model is now a standard tool for investigating economic structures ranging from cities and corporations to states and countries. Throughout these discussions, we shall be conscious of numerical implications in finding the need for efficiency and accuracy in implementing matrix models.

1.1 Introduction to Linear Equations: The Beginning of Algebra

To find the history of linear algebra, it is essential that, first, we determine what linear algebra is.

Linear algebra is the branch of mathematics that generally deals with linear equations such as $a_1x_1 + a_2x_2 + \ldots + a_nx_n = b$, linear maps such as $(x_1, x_2, \ldots, x_n) \mapsto a_1x_1 + a_2x_2 + \ldots + a_nx_n$, and their representation in vector space and through matrix representations.

Linear algebra applies to almost all areas of mathematics. For instance, linear fundamental algebra defines the necessary objects such as lines, planes, and rotations through geometrical representation. It is also used in most sciences and engineering because it allows many natural phenomena to model and efficiently compute such models. It is also the study of a particular algebraic structure called a vector space. Secondly, it is the study of linear sets of equations and their transformation properties. Finally, it is the branch of mathematics that investigates the properties of finite-dimensional vector space and linear mapping between such spaces and plays a central role in modern mathematics, essential in engineering and physical, social, and behavioral science.

This chapter shall introduce one of the vital parts of linear algebra, i.e., a matrix or a rectangular array of numbers and the standard matrix operations that are generally used in dealing with a linear system of equations. Matrices often come across in many engineering, physical, mathematics, and social sciences when data is given in the tabular form.

1.2 Matrices

An $m \times n$ matrix A is a rectangular order of real or complex numbers with $m-$rows and $n-$columns.

We shall write a_{ij} for the number that appears in the ith row and the jth column of A. This is called the (i, j) entry of A.

The extended form of the matrix A can be written as

$$\begin{bmatrix} a_{11} & a_{12} & \dots & a_{1n} \\ a_{21} & a_{22} & \dots & a_{2n} \\ \dots & \dots & \vdots & \dots \\ a_{n1} & a_{n2} & \dots & a_{nn} \end{bmatrix} \text{ or as } [a_{ij}]_{m\times n} \cdot$$

Here, the subscripts m and n tell us the corresponding number of rows and columns of the matrix A.

1.3 Power of the Matrix

Let A be an $n \times n$ matrix. Then the mth power of A is defined by the equations $A^0 = \boldsymbol{I_n}$ and $A^{m+1} = A^m A$, where m is a non-negative integer.

Note: We do not need pursuit to determine the negative powers at this junction.

Example 1.1:

If $A = \begin{bmatrix} 0 & 1 \\ -1 & 0 \end{bmatrix}$, then the power of matrix can be expressed as

$A^2 = \begin{bmatrix} -1 & 0 \\ 0 & -1 \end{bmatrix}$, $A^3 = \begin{bmatrix} 0 & -1 \\ 1 & 0 \end{bmatrix}$, and $A^4 = \begin{bmatrix} 1 & 0 \\ 0 & 1 \end{bmatrix} = I_2$.

Note: The higher power of A does not take precedence over new matrices.

1.3.1 Symmetric and skew-symmetric matrices

A matrix A is called *symmetric*, when $A = A^T$, i.e., it equals its transpose.

On the other hand, if A^T equals $-A$, then the matrix A is said to be *skew-symmetric*.

For example, the matrices $\begin{bmatrix} a & b \\ b & c \end{bmatrix}$ and $\begin{bmatrix} 0 & a \\ -a & 0 \end{bmatrix}$ respectively are the *symmetric* and *skew-symmetric* matrices.

Note:

(i) Symmetric and skew-symmetric matrices must be square matrices.
(ii) Symmetric matrices can be reduced to a diagonal matrix in a real sense.

1.4 The Inverse of a Square Matrix

An $n \times n$ square matrix A is said to be *invertible* if $|\boldsymbol{A}| \neq 0$ and there is an $n \times n$ matrix B such that $AB = \boldsymbol{I_n} = BA$.

If a matrix A is invertible, then B is called an inverse of A.

A matrix that is *not invertible* is called a *singular* matrix, while an invertible matrix is a *non-singular matrix*.

Theorem 1.1:

An $n \times n$ square matrix has at most one inverse.

Proof:

Let us consider an $n \times n$ square matrix A that has two inverses B_1 and B_2.

Then $AB_1 = AB_2 = I = B_1A = B_2A$.

The objective of the proof is to examine the product $(B_1A)\,B_2$.

Since $B_1A = I$, $(B_1A)\,B_2$ equals $IB_2 = B_2$.

On the other hand, by the associative law,

$$B_1(AB_2) \text{ equals } B_1I = B_1$$

. Therefore, $B_1 = B_2$.

Note: A^{-1} is the unique inverse of an invertible matrix A.

Theorem 1.2:

(i) If the matrix A is invertible, then A^{-1} is also invertible and $\left(A^{-1}\right)^{-1} = A$.

(ii) If A and B are invertible matrices of the same size, then their product AB is also invertible and

$$(AB)^{-1} = B^{-1}A^{-1}.$$

Proof:

(i) Indeed, we have $A.A^{-1} = I = A^{-1}A$, which show that A is an inverse of A^{-1}.
Since A^{-1} cannot have more than one inverse, its inverse must be A.

(ii) To prove the assertion, we need only to check that $B^{-1}A^{-1}$ is an inverse of AB.

Now $(AB)\left(B^{-1}A^{-1}\right) = A\left(BB^{-1}\right)A^{-1}$.
Upon using the associative law, it equals $AIA^{-1} = AA^{-1} = I$.
Similarly, $\left(B^{-1}A^{-1}\right)(AB) = I$.
Since the inverses are unique, $(AB)^{-1} = B^{-1}A^{-1}$.

Lemma 1.1:

Let A be an $m \times n$ matrix and let B be an $n \times p$ matrix.

$$\text{Then } (AB)^T = B^T A^T, \tag{1.1}$$

and if α and β are scalars,

$$\text{then } (\alpha A + \beta B)^T = \alpha A^T + \beta B^T. \tag{1.2}$$

Proof:

From definition:

$$\begin{aligned}\left((AB)^T\right)_{ij} &= (AB)_{ji} \\ &= \sum_k A_{jk}B_{ki} \\ &= \sum_k \left(B^T\right)_{ik}\left(A^T\right)_{kj} \\ &= \left(B^T A^T\right)_{ij}\end{aligned}$$

$$\begin{aligned}\left((\alpha A+\beta B)^T\right)_{ij} &= (\alpha A+\beta B)_{ji} \\ &= (\alpha A)_{ji}+(\beta B)_{ji} \\ &= \alpha(A)_{ji}+\beta(B)_{ji} \qquad = \alpha\left(A^T\right)_{ij}+\beta\left(B^T\right)_{ij} \\ &= \alpha A^T+\beta B^T\end{aligned}$$

Definition 1.1:

Let A be a $m \times n$ matrix. Then A^T denotes the $n \times m$ matrix, which is defined as follows: $\left(A^T\right)_{ij} = A_{ji}$.

The transpose of a matrix has the following essential property.

There is a particular matrix called I and defined by $I_{ij} = \delta_{ij}$.

Here δ_{ij} is the *Kronecker symbol* defined by

$$\delta_{ij} = \begin{cases} 1, & \text{If } i = j \\ 0, & \text{If } i \neq j \end{cases}.$$

It is said to be an identity matrix because it is a multiplicative identity in the following sense.

Lemma 1.2:

Suppose A is an $m \times n$ matrix and $\boldsymbol{I_n}$ is an $n \times n$ identity matrix. Then $A\boldsymbol{I_n} = A$. Next, if I_m is an $m \times m$ identity matrix, then it follows that $I_m A = A$.

Proof:

$$(A\boldsymbol{I_n})_{ij} = \sum_k A_{ik}\delta_{kj} = A_{ij}$$

and so $A\boldsymbol{I_n} = A$.

The other case is left as an exercise.

Theorem 1.3:

Let us consider an $n \times n$ matrix A. Then

(1) $A^r.A^s = A^{r+s}$.
(2) $(A^r)^s = A^{rs}$.
(3) $A^0 = \boldsymbol{I_n}$.
(4) $A^r.A^s = \underbrace{A \cdots A}_{r\,times}\,\underbrace{A \cdots A}_{s\,times} = \underbrace{A \cdots A}_{r+s\,times} = A^{r+s}$.

Here r and s are non-negative integers.

Theorem 1.4:

Let A be an $m \times n$ matrix and 0_{mn} be the zero matrices. Let B be an $n \times n$ square matrix, and 0_n and $\boldsymbol{I_n}$ be the zero and identity matrices. Then

(1) $A + 0_{mn} = 0_{mn} + A = A$.
(2) $B0_n = 0_n B = 0_n$.
(3) $B\boldsymbol{I_n} = \boldsymbol{I_n} B = B$.

Example 1.2:

Let $A = \begin{bmatrix} 2 & 1 & -3 \\ 4 & 5 & 8 \end{bmatrix}$ and $B = \begin{bmatrix} 2 & 1 \\ -3 & 4 \end{bmatrix}$.

We see that

$$A + 0_{23} = \begin{bmatrix} 2 & 1 & -3 \\ 4 & 5 & 8 \end{bmatrix} + \begin{bmatrix} 0 & 0 & 0 \\ 0 & 0 & 0 \end{bmatrix} = \begin{bmatrix} 2 & 1 & -3 \\ 4 & 5 & 8 \end{bmatrix}$$

$$B0_2 = \begin{bmatrix} 2 & 1 \\ -3 & 4 \end{bmatrix} \begin{bmatrix} 0 & 0 \\ 0 & 0 \end{bmatrix} = \begin{bmatrix} 0 & 0 \\ 0 & 0 \end{bmatrix} = 0_2$$

$$BI_2 = \begin{bmatrix} 2 & 1 \\ -3 & 4 \end{bmatrix} \begin{bmatrix} 1 & 0 \\ 0 & 1 \end{bmatrix} = \begin{bmatrix} 2 & 1 \\ -3 & 4 \end{bmatrix} = B.$$

Similarly, $0_{23} + A = A$, $0_2 B = 0_2$, and $I_2 B = B$.

1.5 Idempotent and Nilpotent Matrices

Definition 1.2:

An $n \times n$ square matrix A is said to be *idempotent*, if $A^2 = A$. Moreover, a square matrix A is said to be *nilpotent* of order p if there is a positive integer such that $A^p = 0$.

The least integer p such that $A^p = 0$ is called the degree of *nilpotency* of the matrix.

1.5.1 Elementary matrices

We now introduce a beneficial class of matrices called *elementary matrices*. An *elementary matrix* can be obtained from the *identity matrix* $\boldsymbol{I_n}$ through a single elementary row operation.

Illustration: Consider the following three-row operations $T_1, T_2, \ldots T_3$ on I_3 (one representing each kind of row operation). They lead to the three elementary matrices E_1, E_2, and E_3.

Elementary row operation	Corresponding elementary matrix
T_1: Interchange rows 2 and 3 of I_3	$E_1 = \begin{bmatrix} 1 & 0 & 0 \\ 0 & 0 & 1 \\ 0 & 1 & 0 \end{bmatrix}$
T_2: Multiply row 2 of I_3 by 5	$E_2 = \begin{bmatrix} 1 & 0 & 0 \\ 0 & 5 & 0 \\ 0 & 0 & 1 \end{bmatrix}$
T_3: Add two times rows of I_3 to row 2	$E_3 = \begin{bmatrix} 1 & 0 & 0 \\ 2 & 1 & 0 \\ 0 & 0 & 1 \end{bmatrix}$

Consider $I_3 = \begin{bmatrix} 1 & 0 & 0 \\ 0 & 1 & 0 \\ 0 & 0 & 1 \end{bmatrix}$.

Remark:

Suppose we want to perform a row operation T on a $m \times n$ matrix A. Let E be the elementary matrix obtained from $\boldsymbol{I_n}$ through the operation T. This row operation can be performed by multiplying A by E.

Example 1.3:

Let $A = \begin{bmatrix} a_1 & a_2 & a_3 \\ b_1 & b_2 & b_3 \\ c_1 & c_2 & c_3 \end{bmatrix}$ be a 3×3matrix.

Consider the three-row operation as stated above.

Let us show that the corresponding elementary matrices can indeed be used to perform these operations.

Interchange rows 2 and 3 of I_3:

$$\begin{bmatrix} 1 & 0 & 0 \\ 0 & 0 & 1 \\ 0 & 1 & 0 \end{bmatrix} \begin{bmatrix} a_1 & a_2 & a_3 \\ b_1 & b_2 & b_3 \\ c_1 & c_2 & c_3 \end{bmatrix} = \begin{bmatrix} a_1 & a_2 & a_3 \\ c_1 & c_2 & c_3 \\ b_1 & b_2 & b_3 \end{bmatrix}.$$

Multiply row 2 by 5 of I_3:

$$\begin{bmatrix} 1 & 0 & 0 \\ 0 & 5 & 0 \\ 0 & 0 & 1 \end{bmatrix} \begin{bmatrix} a_1 & a_2 & a_3 \\ b_1 & b_2 & b_3 \\ c_1 & c_2 & c_3 \end{bmatrix} = \begin{bmatrix} a_1 & a_2 & a_3 \\ 5b_1 & 5b_2 & 5b_3 \\ c_1 & c_2 & c_3 \end{bmatrix}.$$

Add twice of row 1 to row 2 of I_3:

$$\begin{bmatrix} 1 & 0 & 0 \\ 2 & 1 & 0 \\ 0 & 0 & 1 \end{bmatrix} \begin{bmatrix} a_1 & a_2 & a_3 \\ b_1 & b_2 & b_3 \\ c_1 & c_2 & c_3 \end{bmatrix} = \begin{bmatrix} a_1 & a_2 & a_3 \\ b_1 + 2a_1 & b_2 + 2a_2 & b_3 + 2a_3 \\ c_1 & c_2 & c_3 \end{bmatrix}.$$

Remark:

Every elementary matrix is square and invertible.

Theorem 1.5:

If A and B are row equivalent matrices and A is invertible, then B is invertible.

1.6 Elementary Matrices

An *elementary* matrix can be obtained from an $n \times n$ identity matrix $\boldsymbol{I}$ in one of three ways.

(1) Interchange i and j rows, i.e., $i \leftrightarrow j$, where $i \neq j$.
(2) Insert a scalar α as the (i, j) entry of the matrix, where $i \neq j$.
(3) Put a non-zero scalar α in the (i, i) position.

Definition 1.3:

An *elementary matrix* E can be obtained from an $n \times n$ *identity matrix* I through a *single elementary row operation.*

Let us perform a row operation T on an $n \times n$ matrix A. Let the elementary matrix E be obtained from the operation T. These row operations can be achieved by multiplying the matrix A by the elementary matrix E.

Note: Every elementary matrix E is square and invertible.

Theorem 1.6:

If A and B are row equivalent matrices and invertible, then the matrix B is invertible.

Proof:

Assume that the matrices A and B are row equivalent. Then there exists a sequential row operation $T_1, T_2, \cdots, T_n$ such that

$$B = T_n.T_{n-1} \cdots T_1 (A).$$

Let $E_1, E_2, \cdots, E_n$ be the elementary matrices of these operations.
Thus, $B = E_n.E_{n-1} \cdots E_1 A$.
The matrices $A, E_1, E_2, \cdots, E_n$ are all invertible.
Repeatedly applying the property of matrix inverse of a product to the following expression, we get

$$\begin{aligned} A^{-1}{E_1}^{-1}{E_2}^{-1}\ldots{E_n}^{-1} &= (E_1A)^{-1}\,{E_2}^{-1}\ldots{E_n}^{-1} \\ &= (E_2E_1A)^{-1}\,E_3-1\ldots E_n^{-1} \\ &= (E_nE_{n-1}\ldots E_2E_1A)^{-1} = B^{-1} \end{aligned}$$

Thus, the matrix B is invertible, and its inverse is given by

$$B^{-1} = A^{-1}{E_1}^{-1}{E_2}^{-1}\cdots{E_n}^{-1}.$$

Theorem 1.7:

Let A be an $n \times n$ matrix; then there exist elementary $n \times n$ matrices $E_1, E_2, \cdots, E_k$ such that the matrix $E_kE_{k-1}\cdots E_1A$ is in reduced echelon form.

Example 1.4:

Let us consider the matrix

$$A = \begin{bmatrix} 0 & 1 & 2 \\ 2 & 1 & 0 \end{bmatrix}.$$

Upon applying the row operations $R_1 \leftrightarrow R_2,\quad \left(\frac{1}{2}\right)R_1,\quad R_1 - \left(\frac{1}{2}\right)R_2$ successively, the matrix A can be put in reduced row echelon form B as

$$A \rightarrow \begin{bmatrix} 2 & 1 & 0 \\ 0 & 1 & 2 \end{bmatrix} \rightarrow \begin{bmatrix} 1 & \frac{1}{2} & 0 \\ 0 & 1 & 2 \end{bmatrix} \rightarrow \begin{bmatrix} 1 & 0 & -1 \\ 0 & 1 & 2 \end{bmatrix} = B.$$

Hence, $E_3E_2E_1A = B$,

where $E_1 = \begin{bmatrix} 0 & 1 \\ 1 & 0 \end{bmatrix}$, $E_2 = \begin{bmatrix} \frac{1}{2} & 0 \\ 0 & 1 \end{bmatrix}$, $E_3 = \begin{bmatrix} 1 & -\frac{1}{2} \\ 0 & 1 \end{bmatrix}$.

Theorem 1.8:

The following statements are equivalent for an $n \times n$ square matrix A, i.e.,

(1) The matrix A is invertible.
(2) The system $AX = 0$ has only a trivial solution.
(3) The matrix $\boldsymbol{I_n}$ is the reduced row echelon form of the matrix A.
(4) The matrix A is a product of elementary matrices.

Proof:

Here we shall establish the logical implication, i.e.,
$(1) \rightarrow (2)$, $(2) \rightarrow (3)$, $(3) \rightarrow (4)$, and $(4) \rightarrow (1)$, which serve to establish the equivalence of the above four statements.

If the matrix A is invertible, i.e., if (1) holds, then A^{-1} exists.

Thus, upon multiplying A^{-1} on both sides of the equation $AX = 0$, i.e.,

$$A^{-1}AX = A^{-1}.0,$$

we find $X = A^{-1}.0 = 0$, which is the trivial solution of the linear system.

Thus, statement (2) holds.

If the homogenous system $AX = 0$ has only a trivial solution, i.e., (2) holds, then it implies that the number of pivots of the matrix A in reduced row echelon form is n.

Since the matrix A is square, the matrix's reduced row echelon form must imply that (3) holds.

If the matrix $\boldsymbol{I_n}$ is the reduced row echelon form of the matrix A, i.e., (3) holds, then *Theorem 1.7* shows that there are k elementary matrices $E_1, E_2, \cdots, E_k$ such that

$$E_k E_{k-1} \cdots E_1 A = \boldsymbol{I_n}.$$

Since the elementary matrices $E_1, E_2, \cdots, E_k$ are invertible, so is $E_k E_{k-1}, \cdots, E_1$ and thus $A = (E_k E_{k-1} \cdots E_1)^{-1} = E_1{}^{-1} E_2{}^{-1} \cdots E_k{}^{-1}$.

Thus, (4) is true.

Finally, since a product of elementary matrices is always *invertible*, statement (iv) implies (1).

1.7 Finding the Inverse of a Matrix

An important application of this idea is that it can be used as an efficient method to determine the inverse of a matrix that is invertible.

This section describes an efficient method to compute the inverse of an invertible matrix. For example, assume that the $n \times n$ square matrix A is invertible. Then there exist elementary matrices $E_1, E_2 \cdots E_k$ of order $n \times n$ such that $E_k E_{k-1} \cdots E_1 A = \boldsymbol{I_n}$.

Therefore,

$$\begin{aligned} A^{-1} &= \boldsymbol{I_n} A^{-1} \\ &= (E_k E_{k-1} \cdots E_1 A)\, A^{-1} \\ &= (E_k E_{k-1} \cdots E_1)\, \boldsymbol{I_n} \end{aligned}.$$

This implies that the row operations that reduce the matrix A to its reduced row echelon form will automatically transform $\boldsymbol{I_n}$ to A^{-1} and it is the crucial observation that enables us to compute A^{-1}.

The procedure for computing the inverse, i.e., A^{-1} starts with the partitioned matrix $\left[A \vdots \boldsymbol{I_n}\right]$, puts it in reduced row echelon form. Thus, if the matrix A is invertible, then the reduced row echelon form will be $\left[\boldsymbol{I_n} \vdots A^{-1}\right]$.

Remark:

If the matrix A is not invertible, then it will be absurd to reach a reduced row echelon form of the matrix, which implies that the procedure will disclose the non-invertibility of a matrix A.

Example 1.5:

Determine the inverse of the matrix

$$A = \begin{bmatrix} 2 & -1 & 0 \\ -1 & 2 & -1 \\ 0 & -1 & 2 \end{bmatrix}.$$

Solution:

Let us consider the partitioned matrix $\left[A \vdots I_3\right]$. Upon using the elementary row operations as described above, the partitioned matrix $\left[A \vdots I_3\right]$ in reduced row echelon form,

$$\begin{bmatrix} 2 & -1 & 0 & \vdots & 1 & 0 & 0 \\ -1 & 2 & -1 & \vdots & 0 & 1 & 0 \\ 0 & -1 & 2 & \vdots & 0 & 0 & 1 \end{bmatrix} \Rightarrow \begin{bmatrix} 2 & -\frac{1}{2} & 0 & \vdots & \frac{1}{2} & 0 & 0 \\ -1 & 2 & -1 & \vdots & 0 & 1 & 0 \\ 0 & -1 & 2 & \vdots & 0 & 0 & 1 \end{bmatrix}$$

$$\Rightarrow \begin{bmatrix} 1 & -\frac{1}{2} & 0 & \vdots & \frac{1}{2} & 0 & 0 \\ 0 & \frac{3}{2} & -1 & \vdots & \frac{1}{2} & 1 & 0 \\ 0 & -1 & 2 & \vdots & 0 & 0 & 1 \end{bmatrix} \Rightarrow \begin{bmatrix} 1 & -\frac{1}{2} & 0 & \vdots & \frac{1}{2} & 0 & 0 \\ 0 & 1 & -\frac{2}{3} & \vdots & \frac{1}{3} & \frac{2}{3} & 0 \\ 0 & -1 & 2 & \vdots & 0 & 0 & 1 \end{bmatrix}$$

$$\Rightarrow \begin{bmatrix} 1 & 0 & -\frac{1}{3} & \vdots & \frac{2}{3} & \frac{1}{3} & 0 \\ 0 & 1 & -\frac{2}{3} & \vdots & \frac{1}{3} & \frac{2}{3} & 0 \\ 0 & -1 & \frac{4}{3} & \vdots & \frac{1}{4} & \frac{1}{2} & \frac{3}{4} \end{bmatrix} \Rightarrow \begin{bmatrix} 1 & 0 & 0 & \vdots & \frac{3}{4} & \frac{1}{2} & \frac{1}{4} \\ 0 & 1 & 0 & \vdots & \frac{1}{2} & 1 & \frac{1}{2} \\ 0 & 0 & 1 & \vdots & \frac{1}{4} & \frac{1}{2} & \frac{3}{4} \end{bmatrix}.$$

This is in reduced row echelon form.

Thus, the inverse of the matrix is

$$A^{-1} = \begin{bmatrix} \frac{3}{4} & \frac{1}{2} & \frac{1}{4} \\ \frac{1}{2} & 1 & \frac{1}{2} \\ \frac{1}{4} & \frac{1}{2} & \frac{3}{4} \end{bmatrix}.$$

This can be verified by checking $AA^{-1} = I_3 = A^{-1}A$.

1.8 Applications

1.8.1 Color model

A color model in the context of graphics is a method of implementing colors. Numerous models are used in practice, such as the RGB model (red, green, and blue) used in computer monitors and on a television screen. An RGB computer signal can be converted to a YIQ television signal using what is known as an $NTSC$ encoder ($NTSC$ stands for National Television System Committee). The conversion is accomplished using the matrix transformations

$$\begin{bmatrix} Y \\ I \\ Q \end{bmatrix} = \begin{bmatrix} .299 & .587 & .114 \\ .596 & -.275 & -.321 \\ .212 & -.523 & .311 \end{bmatrix} \begin{bmatrix} R \\ G \\ B \end{bmatrix}.$$

Let us look at the RGB model for Microsoft word. The default text color is black. Let us find the RGB value for black, change the text color to purple, and find the RGB values. On the right of the toolbar of Microsoft word, observe $\underset{-\triangledown}{A}$. The bar under the A is black, indicating the current text color. Point the cursor at this bar. It shows the font color (RGB (0,0,0)). The RGB setting for black is (0,0,0). To change the color, select the sequence "$\triangledown \rightarrow$ More colours$\rightarrow$ Customs. A spectrum of colors is displayed. Select a purple hue. The corresponding RGB values are seen to be $R = 213, g = 77, B = 187$. The bar under A has changed to purple, and any text entered at the keyboard is purple. The range of values for each of *R*, *G*, and *B* is 0–255, the set of numbers represented by a byte on a computer (note that $2^8 = 256$). You are asked to use the matrix transformation to find the range of *Y*, *I*, and *Q* values in the following exercises.

If we enter the RGB values for black, namely $R = 0, G = 0, B = 0$, into the preceding transformation, we find that $Y = 0, I = 0, Q = 0$. Black has the same RGB and YIQ values. The RGB values $R = 213, g = 77, B =$

187 for purple become $Y = 130.204, I = 45.746, Q = 63.042$. These are the YIQ values that would be used to duplicate this purple color on a television screen.

A signal is converted from a television screen to a computer monitor using the inverse of the above matrix.

$$\begin{bmatrix} R \\ G \\ B \end{bmatrix} = \begin{bmatrix} .299 & .587 & .114 \\ .596 & -.275 & -.321 \\ .212 & -.523 & .311 \end{bmatrix} \begin{bmatrix} Y \\ I \\ Q \end{bmatrix}.$$

That is,

$$\begin{bmatrix} R \\ G \\ B \end{bmatrix} = \begin{bmatrix} 1 & .956 & .620 \\ 1 & -.272 & -.647 \\ 1 & -1.108 & 1.705 \end{bmatrix} \begin{bmatrix} Y \\ I \\ Q \end{bmatrix}.$$

1.8.2 Cryptography

In the previous application, we talked about two different ways that colors are coded. We now turn our attention to coding messages. Cryptography is the process of coding and decoding messages. The word comes from the Greek "*Kryptos*" meaning "hidden." The technique can be traced back to the ancient Greeks. Today governments use sophisticated methods of coding and decoding messages. One code that is extremely difficult to break uses a large invertible matrix to encode a message. The receiver of the message decodes it using the inverse of the matrix. This first matrix is called the *encoding matrix*, and its inverse is called the *decoding matrix*.

We illustrate the method for a 3×3 matrix.

Let the message be BUY IBM STOCK

and the encoding matrix be $\begin{bmatrix} -3 & -3 & -4 \\ 0 & 1 & 1 \\ 4 & 3 & 4 \end{bmatrix}$.

We assign a number to each letter of the alphabet. For convenience, let us associate each letter with its position in the alphabet. A is 1, B is 2, and so on. Let a space between words be denoted by the number 27. The digital form of the message is

$$\begin{array}{ccccccccccccc} B & U & Y & - & I & B & M & - & S & T & O & C & K \\ 2 & 21 & 25 & 27 & 9 & 2 & 13 & 27 & 19 & 20 & 15 & 3 & 11 \end{array}.$$

Since we will use a 3×3 matrix to encode the message, we break the digital message up into a sequence of 3×1 column matrices as follows:

$$\begin{bmatrix} 2 \\ 21 \\ 25 \end{bmatrix}, \begin{bmatrix} 27 \\ 9 \\ 2 \end{bmatrix}, \begin{bmatrix} 13 \\ 27 \\ 19 \end{bmatrix}, \begin{bmatrix} 20 \\ 15 \\ 3 \end{bmatrix}, \begin{bmatrix} 11 \\ 27 \\ 27 \end{bmatrix}.$$

Observe that adding two spaces at the end of the message to complete the last matrix was necessary. We now put the message into code by multiplying each of the above column matrices by the encoding matrix. This can be conveniently done by writing the given column matrices as a matrix column and pre-multiplying that matrix by the encoding matrix. We get

$$\begin{bmatrix} -3 & -3 & -4 \\ 0 & 1 & 1 \\ 4 & 3 & 4 \end{bmatrix} \begin{bmatrix} 2 & 27 & 13 & 20 & 11 \\ 21 & 9 & 27 & 15 & 27 \\ 25 & 2 & 19 & 3 & 27 \end{bmatrix}$$

$$= \begin{bmatrix} -169 & -116 & -196 & -117 & -222 \\ 46 & 11 & 46 & 18 & 54 \\ 171 & 143 & 209 & 137 & 233 \end{bmatrix}.$$

The columns of this matrix give the encoded message. The message is transmitted in the following linear form:

$$-169, 46, 171, -116, 11, 143, -196, 46, 209, -117, 18, 137, -222, 54, 233.$$

To decode the message, the receiver writes this string as a sequence of 3×1 column matrices and repeats the technique using the inverse of the encoding matrix. The inverse of this encoding matrix, the decoding matrix is

$$\begin{bmatrix} 1 & 0 & 1 \\ 4 & 4 & 3 \\ -4 & -3 & -3 \end{bmatrix}.$$

Thus, to decode the message

$$\begin{bmatrix} 1 & 0 & 1 \\ 4 & 4 & 3 \\ -4 & -3 & -3 \end{bmatrix} \begin{bmatrix} -169 & -116 & -196 & -117 & -222 \\ 46 & 11 & 46 & 18 & 54 \\ 171 & 143 & 209 & 137 & 233 \end{bmatrix}$$

$$= \begin{bmatrix} 2 & 27 & 13 & 20 & 11 \\ 21 & 9 & 27 & 15 & 27 \\ 25 & 2 & 19 & 3 & 27 \end{bmatrix}.$$

The columns of this matrix written in linear form give the original message.

$$\begin{matrix} 2 & 21 & 25 & 27 & 9 & 2 & 13 & 27 & 19 & 20 & 15 & 3 & 11 \\ B & U & Y & - & I & B & M & - & S & T & O & C & K \end{matrix}.$$

Exercises

1. $A = \begin{bmatrix} \lambda_1 & 0 & \cdots & 0 \\ 0 & \lambda_2 & \cdots & 0 \\ \vdots & \vdots & \vdots & \vdots \\ 0 & 0 & \cdots & \lambda_n \end{bmatrix}$, $B = \begin{bmatrix} \mu_1 & 0 & \cdots & 0 \\ 0 & \mu_2 & \cdots & 0 \\ \vdots & \vdots & \vdots & \vdots \\ 0 & 0 & \cdots & \mu_n \end{bmatrix}$.

   Compute (a) AB (b) BA (c) A^{-1}, if it exists.

An $n \times n$ matrix A is said to be *nilpotent* if $A^n = 0$ for some positive integer n. The smallest positive integer n for which $A^n = 0$ is called the *degree of nilpotence* of A.

2. Check whether the following matrices are nilpotent. In the case of nilpotent matrices, find the degree of nilpotence.

   (a) $\begin{bmatrix} 0 & 0 & 0 & 0 \\ \alpha & 0 & 0 & 0 \\ \beta_1 & \beta_2 & 0 & 0 \\ \gamma_1 & \gamma_2 & \gamma_3 & 0 \end{bmatrix}$ (b) $\begin{bmatrix} 0 & 1 & 2 & -1 \\ 0 & 0 & 1 & 2 \\ 0 & 0 & 0 & 1 \\ 0 & 0 & 0 & 0 \end{bmatrix}$.

   (c) $\begin{bmatrix} 1 & 5 & -2 \\ 1 & 2 & -1 \\ 3 & 6 & -3 \end{bmatrix}$ (d) $\begin{bmatrix} 1 & -3 & -4 \\ -1 & 3 & -4 \\ 1 & -3 & -4 \end{bmatrix}$.

3. If A and B are square matrices of the same order, then prove that

   (a) $A^2 - B^2 = (A - B)(A + B)\, AB = BA$

   (b) $A^2 \pm 2AB + B^2 = (A \pm B)^2$

4. Denote the matrix $A = \begin{bmatrix} \alpha_{11} & \alpha_{12} & \alpha_{13} \\ \alpha_{21} & \alpha_{22} & \alpha_{23} \\ \alpha_{31} & \alpha_{32} & \alpha_{33} \end{bmatrix}$.

   as $A = \begin{bmatrix} A_{11} & A_{12} \\ A_{21} & A_{22} \end{bmatrix}$,

   where

A_{11}= the matrix $[\alpha_{11}]$. $A_{12} = [\alpha_{11}, \alpha_{12}]$,

$$A_{21} = \begin{bmatrix} \alpha_{21} \\ \alpha_{31} \end{bmatrix}, A_{22} = \begin{bmatrix} \alpha_{22} & \alpha_{23} \\ \alpha_{32} & \alpha_{33} \end{bmatrix}.$$

Similarly, define B, B_{11},....by replacing α_{ij} by β_{ij}. Then prove that

$$A + B = \begin{bmatrix} A_{11} + B_{11} & A_{12} + B_{12} \\ A_{21} + B_{21} & A_{22} + B_{22} \end{bmatrix}$$

and $AB = \begin{bmatrix} A_{11}B_{11} + A_{12}B_{12} & A_{11}B_{12} + A_{12}B_{22} \\ A_{21}B_{11} + A_{22}B_{21} & A_{21}B_{12} + A_{22}B_{22} \end{bmatrix}.$

5. If α, β are any scalars, then prove that $A^2 - (\alpha + \beta) A + \alpha\beta I = (A - \alpha I)(A - \beta I)$, where A is any square matrix of order n and $I = \boldsymbol{I_n}$.
6. If α, β are scalars such that $A = \alpha B + \beta I$, then prove that $AB = BA$.
7. A square matrix A is said to be involutory if $A^2 = I$. Prove that the matrices $\begin{bmatrix} 1 & \alpha \\ 0 & -1 \end{bmatrix}$ and $\begin{bmatrix} 1 & 0 \\ \alpha & -1 \end{bmatrix}$ are involutory for all scalars α. Determine all 2×2involutory matrices.
8. Let $p(x) = \alpha_0 + \alpha_1 x + \cdots + \alpha_n x^n$ be a polynomial of degree n, and A be a square matrix of order m. Then the *matrix polynomial* $p(A)$ is defined as

$p(A) = \alpha_0 I + \alpha_1 A + \cdots + \alpha_n A^n$, where $I = I_m$.
If $f(x) = 7x^2 - 3x + 5$, $g(x) = 3x^3 - 2x^2 + 5x - 1$
$A = \begin{bmatrix} 1 & 2 \\ -1 & 3 \end{bmatrix}$, $B = \begin{bmatrix} -1 & 1 \\ 1 & -2 \end{bmatrix}$, evaluate

(a) $f(A)$ (b) $g(A)$

(c) $f(B)$ (d) $g(B)$

(e) $f(2A + 3B)$ (f) $g(3A - 7B)$.

9. Using matrix methods, solve the following system of linear equations:

(a) $x + 2y = 3$
$y = 1$

(b) $\alpha x + \beta y = a$
$\beta x = b$

(c) $x - 2z = 3$
$2x + y = 2$
$x + 2z = 3$

10. If A and B are two non-singular matrices of the same order, then prove that AB is non-singular and $(AB)^{-1} = B^{-1}A^{-1}$.

11. Prove that the following matrices are non-singular and find their inverses:

(a) $\begin{bmatrix} 1 & 2 \\ -1 & 0 \end{bmatrix}$ (b) $\begin{bmatrix} 1 & 0 & 0 \\ 2 & 1 & 0 \\ 3 & 4 & 2 \end{bmatrix}$

(c) $\begin{bmatrix} 1 & -1 & 2 \\ 3 & 0 & 1 \\ 0 & 1 & -1 \end{bmatrix}$ (d) $\begin{bmatrix} 1 & -1 & 2 & 3 \\ 0 & 2 & 1 & 1 \\ 1 & 3 & 1 & -1 \\ 1 & 1 & -1 & 0 \end{bmatrix}$

(e) $\begin{bmatrix} 2 & 2 & 1 & 1 \\ 0 & 2 & 1 & 1 \\ 0 & 0 & 1 & 1 \\ 0 & 0 & 1 & 0 \end{bmatrix}$

12. Find the values of α and β for which the following matrix is invertible. Find the inverse when it exists.

$$\begin{bmatrix} \alpha & \beta & 0 \\ 0 & \alpha & \beta \\ \beta & 0 & \alpha \end{bmatrix}.$$

2

System of Linear Equations and Determinants

This chapter discusses the system of linear equations and its solution using the Gaussian elimination method, Gauss-Jordan elimination method and LU decomposition methods, the definitions of determinants, their properties, and Crammer's rule. Linear algebra is a branch of mathematics that plays a central role in modern mathematics and is essential to engineers and physical, social, and behavioral scientists. When mathematics is used to solve a problem, it often becomes necessary to find a solution to a so-called system of linear equations to study methods for solving such equations. This chapter introduces methods for solving linear equations and looks at some of the solutions' properties. It is essential to know the solutions to a given system of equations and why they are the solutions. If the system describes some real-life situation, then understanding the behavior of the solutions can lead to a better understanding of the circumstances. Associated with every square matrix is a number called its determinant. The determinant of a matrix is a tool used in many branches of mathematics, science, and engineering. At the same time, the method of Gauss-Jordan elimination enables us to find the inverse of a specific matrix. For example, it does not give us an algebraic formula for the inverse of an arbitrary matrix, a formula that can be used in theoretical work. Determinants provide us with such procedures. Furthermore, criteria for when specific systems of linear equations have unique, none, or many solutions can be stated in terms of determinants. In this chapter, the determinant is defined, and its properties are developed.

2.1 Introduction

Let us consider a general system of m-linear equations in n-variables

$$\begin{aligned} &c_{11}x_1 + c_{12}x_2 + \cdots + c_{1n}x_{\mathrm{n}} = b_1 \\ &c_{21}x_1 + c_{22}x_2 + \cdots + c_{2n}x_{\mathrm{n}} = b_2 \\ &\vdots \end{aligned} \quad . \qquad (2.1)$$

$$c_{m1}x_1 + c_{m2}x_2 + \cdots + c_{mn}x_n = b_m$$

Upon using the matrix notation, eqn (2.1) can be expressed as follows:

$$\begin{bmatrix} c_{11}x_1 + c_{12}x_2 + \cdots + c_{1n}x_n \\ c_{21}x_1 + c_{22}x_2 + \cdots + c_{2n}x_n \\ \vdots \\ c_{m1}x_1 + c_{m2}x_2 + \cdots + c_{mn}x_n \end{bmatrix} = \begin{bmatrix} b_1 \\ b_2 \\ \vdots \\ b_m \end{bmatrix}. \tag{2.2}$$

If the left-hand side of eqn (2.2) can be written as a product of the matrix coefficients A and a column matrix of unknown variables X, i.e., AX and the right-hand side of the column matrix of constants is B, then the system of the linear equation can be expressed as

$$\begin{bmatrix} c_{11} & c_{12} & \cdots & c_{1n} \\ c_{21} & c_{22} & \cdots & c_{2n} \\ \vdots & \vdots & \ddots & \vdots \\ c_{m1} & c_{m2} & \cdots & c_{mn} \end{bmatrix} \begin{bmatrix} x_1 \\ x_2 \\ \vdots \\ x_n \end{bmatrix} = \begin{bmatrix} b_1 \\ b_2 \\ \vdots \\ b_m \end{bmatrix}. \tag{2.3}$$

Thus, we can write the above system of equations (2.3) in matrix form as

$$AX = B.$$

2.2 Elementary Transformation and Row Operations

Elementary transformation can be used to change a system of linear equations into another system of linear equations that has the same solution. *Elementary transformation* is used to solve the system of linear equations by eliminating the unknown variables.

In a matrix, the operations involved in *elementary transformation* are called *elementary row operations.*

The following table distinguishes the differences between elementary transformations and row operations.

Elementary transformation	**Row operations**
Interchange of two equations	Interchange of two rows of a matrix
A non-zero constant is multiplying on both sides of an equation	Multiply the elements of the row by a non-zero constant
A multiple of one equation adds to another equation.	A multiple of the elements of one row adds to the corresponding elements of another row

The equations related to elementary transformations are called equivalent systems, whereas the matrices associated with *elementary row operations* are called *row equivalent matrices*.

Remark: The elementary transformations preserve the solutions of a system of linear equations since the order of the equations does not alter the solution.

Example 2.1:

Solve the system

$$\begin{aligned} x_1 + x_2 + x_3 &= 2 \\ 2x + 3x_2 + x_3 &= 3 \\ x - x_2 - 2x_3 &= -6 \end{aligned}$$

Solution:

Elementary transformation:

Step-I:

Eliminate x from the second and third equations, i.e.,
Eqn (2) + $(-2) \times$ Eqn (1)
Eqn (3) + $(-1) \times$ Eqn (1)

$$\begin{aligned} x_1 + x_2 + x_3 &= 2 \\ x_2 - x_3 &= -1. \\ 2x_2 - 3x_3 &= -8 \end{aligned}$$

Step-II:

Eliminate y from the first and third equations, i.e., Eqn (1) + $(-1) \times$ Eqn (2)

$$\begin{aligned} x_1 + 2x_3 &= 2 \\ x_2 - x_3 &= -1. \\ -5x_3 &= -10 \end{aligned}$$

Eliminate z from the first and second equations

$$\begin{aligned} x_1 &= -1 \\ x_2 &= 1 \quad . \\ x_3 &= 2 \end{aligned}$$

Matrix method:

$$\begin{bmatrix} 1 & 1 & 1 & 2 \\ 2 & 3 & 1 & 3 \\ 1 & -1 & -2 & -6 \end{bmatrix}$$

$$\approx \begin{matrix} R_2 + (-2)\,R_1 \\ R_3 + (-1)\,R_1 \end{matrix} \begin{bmatrix} 1 & 1 & 1 & 2 \\ 0 & 1 & -1 & -1 \\ 0 & -2 & -3 & -8 \end{bmatrix}$$

$$\approx \begin{matrix} R_1 + (-1)\,R_2 \\ R_3 + (2)\,R_2 \end{matrix} \begin{bmatrix} 1 & 1 & 1 & 2 \\ 0 & 1 & -1 & -1 \\ 0 & 0 & -5 & -10 \end{bmatrix}$$

$$\approx \left(-\frac{1}{5}\right) R_3 \begin{bmatrix} 1 & 0 & 2 & 3 \\ 0 & 1 & -1 & -1 \\ 0 & 0 & 1 & 2 \end{bmatrix}$$

$$\approx \begin{matrix} R_1 + (-2)\,R_3 \\ R_2 + R_3 \end{matrix} \begin{bmatrix} 1 & 0 & 0 & -1 \\ 0 & 1 & 0 & 1 \\ 0 & 0 & 1 & 2 \end{bmatrix}$$

$$\Rightarrow \begin{matrix} x = -1 \\ y = 1 \\ z = 2 \end{matrix} .$$

Example 2.2:

Solve the system

$$\begin{aligned} x_1 - 2x_2 + 4x_3 &= 12 \\ 2x_1 - x_2 + 5x_3 &= 18. \\ -x_1 + 3x_2 - 3x_3 &= -8 \end{aligned}$$

Solution:

$$\begin{bmatrix} 1 & -2 & 4 & 12 \\ 2 & -1 & 5 & 18 \\ -1 & 3 & -3 & -8 \end{bmatrix} \approx \begin{matrix} R_2 + (-2)\,R_1 \\ R_3 + R_1 \end{matrix} \begin{bmatrix} 1 & -2 & 4 & 12 \\ 0 & 3 & -3 & -6 \\ 0 & 1 & 1 & 4 \end{bmatrix}$$

$$\approx \left(\frac{1}{3}\right) R_2 \begin{bmatrix} 1 & -2 & 4 & 12 \\ 0 & 1 & -1 & -2 \\ 0 & 1 & 1 & 4 \end{bmatrix} \approx \begin{matrix} R_1 + (2)\,R_2 \\ R_3 + (-1)\,R_2 \end{matrix} \begin{bmatrix} 1 & 0 & 2 & 8 \\ 0 & 1 & -1 & -2 \\ 0 & 0 & 2 & 6 \end{bmatrix}$$

$$\approx \frac{1}{2}R_3 \begin{bmatrix} 1 & 0 & 2 & 8 \\ 0 & 1 & -1 & -2 \\ 0 & 0 & 1 & 3 \end{bmatrix} \approx \begin{matrix} R_1 + (2)\,R_3 \\ R_2 + R_3 \end{matrix} \begin{bmatrix} 1 & 0 & 0 & 2 \\ 0 & 1 & 0 & 1 \\ 0 & 0 & 1 & 3 \end{bmatrix}$$

$x_1 = 2\,, x_2 = 1\,, x_3 = 3$

2.3 Row Echelon Form and Reduced Row Echelon Form

Definition 2.1:

A matrix A is said to be in *row echelon form* if:

1. Any zero-element rows are present in the matrix, which are grouped at the bottom of the matrix.
2. The first non-zero element of each row is 1, which is called the leading 1.
3. All the elements below a leading 1 are zero. The leading 1 of each row after the first is positioned to the right side of the leading 1 of the previous row.

Example 2.3:

$$\begin{bmatrix} 1 & -1 & 2 \\ 0 & 1 & 2 \\ 0 & 0 & 1 \end{bmatrix} \begin{bmatrix} 1 & 3 & -6 & 4 \\ 0 & 0 & 1 & 3 \\ 0 & 0 & 0 & 0 \end{bmatrix} \begin{bmatrix} 1 & 4 & 6 & 2 & 5 & 2 \\ 0 & 0 & 1 & 2 & 3 & 4 \\ 0 & 0 & 0 & 0 & 1 & 6 \end{bmatrix}.$$

Now we will discuss the reduced echelon form in a more general form.

Definition 2.2:

A matrix is in a row-reduced echelon form, if:

1. Any zero-element rows are present in the matrix, which are grouped at the bottom of the matrix.
2. The first non-zero elements of each row is 1. This element is called leading 1.
3. The leading 1 of each row after the first is positioned to the right of the leading 1 of the previous row.
4. The other elements in a column that contains a leading 1 are zero.

Example 2.4:

$$\begin{bmatrix} 1 & 0 & 8 \\ 0 & 1 & 2 \\ 0 & 0 & 0 \end{bmatrix} \begin{bmatrix} 1 & 0 & 0 & 7 \\ 0 & 1 & 0 & 3 \\ 0 & 0 & 1 & 9 \end{bmatrix} \begin{bmatrix} 1 & 4 & 0 & 0 \\ 0 & 0 & 1 & 0 \\ 0 & 0 & 0 & 1 \end{bmatrix}$$

$$\begin{bmatrix} 1 & 2 & 0 & 3 & 0 & 4 \\ 0 & 0 & 1 & 2 & 0 & 7 \\ 0 & 0 & 0 & 0 & 1 & 6 \\ 0 & 0 & 0 & 0 & 0 & 0 \end{bmatrix} \begin{bmatrix} 1 & 2 & 0 & 4 \\ 0 & 0 & 0 & 0 \\ 0 & 0 & 1 & 3 \end{bmatrix} \begin{bmatrix} 1 & 2 & 0 & 3 & 0 \\ 0 & 0 & 3 & 4 & 0 \\ 0 & 0 & 0 & 0 & 1 \end{bmatrix}$$

$$\begin{bmatrix} 1 & 7 & 0 & 8 \\ 0 & 1 & 0 & 3 \\ 0 & 0 & 1 & 2 \\ 0 & 0 & 0 & 0 \end{bmatrix}.$$

The difference between a *row-reduced echelon form* (*RREF*) and a *reduced echelon form* (REF) is that the element above and below a leading 1 is zero in a *row-reduced echelon form* (*RREF*) while only the element below the leading 1 need to be zero in a *reduced echelon form* (REF).

Remark: The *row-reduced echelon form* of a matrix is *unique*, and the Gauss-Jordan elimination method is a meaningful systematic way to arrive at the row-reduced echelon form. In contrast, the Gauss-Jordan elimination method can be used to arrive at the *reduced echelon form.*

2.4 Solving Linear Systems via Gaussian Elimination

This section discusses an essential part of linear algebra, i.e., linear system of equation, and determines the solution properties of the linear system of equations.

Non-homogenous linear system: Let us consider a system of m linear equations in n unknowns $x_1, x_2, \cdots, x_\text{n}$:

$$\begin{array}{l} c_{11}x_1 + c_{12}x_2 + \cdots + c_{1n}x_\text{n} = b_1 \\ c_{21}x_1 + c_{22}x_2 + \cdots + c_{2n}x_\text{n} = b_2 \\ \cdots \quad \cdots \quad \cdots \\ c_{m1}x_1 + c_{m2}x_2 + \cdots + c_{mn}x_\text{n} = b_\text{n} \end{array} . \tag{2.4}$$

The column vector defines the solution of the linear system, i.e.,

$$\begin{bmatrix} x_1 \\ x_2 \\ \vdots \\ x_\text{n} \end{bmatrix}.$$

That satisfies all the equations of the system (2.4) and is called the *general solution* of the linear system (2.4). A linear system having a solution is consistent, whereas *no solution* is said to be *inconsistent*.

2.4.1 The Homogenous Case

Theorem 2.1:
A *homogenous system* of linear equations in n-variables

$$\begin{aligned} c_{11}x_1 + c_{12}x_2 + \cdots + c_{1n}x_{\mathrm{n}} &= 0 \\ c_{21}x_1 + c_{22}x_2 + \cdots + c_{2n}x_{\mathrm{n}} &= 0 \\ \cdots \quad \cdots & \\ c_{m1}x_1 + c_{m2}x_2 + \cdots + c_{mn}x_{\mathrm{n}} &= 0 \end{aligned}$$

always has the solution $x_1 = 0, x_2 = 0, \cdots, x_{\mathrm{n}} = 0$. This solution is called the *trivial solution*.

Example 2.5:

$$\begin{aligned} x + 2y - 5z &= 0 \\ -2x - 3y + 6z &= 0 \end{aligned} .$$

This system has the solution $x = 0, y = 0, z = 0$, i.e., which is a trivial one.

Theorem 2.2:
A homogenous system of linear equations that has more variables than equations has many solutions. One of these solutions is the trivial solution.

Example 2.6:
Let us consider the system

$$\begin{bmatrix} 1 & 2 & -5 & 0 \\ -2 & -3 & 6 & 0 \end{bmatrix} \approx \begin{bmatrix} 1 & 0 & 3 & 0 \\ 0 & 1 & -4 & 0 \end{bmatrix}$$

$$x + 3z = 0$$

$$y - 4z = 0$$

$$\Rightarrow \begin{aligned} x &= -3z \\ y &= 4z \end{aligned}$$

$$\Rightarrow z = r\,, x = -3r\,, y = 4r.$$

$\Rightarrow$ For $r = 0$, it implies that $x = y = z = 0$.

2.4.2 The non-homogenous case

If A is the coefficient matrix of a system $AX = B$ having n-equations in n-variables, then it has a unique solution, and it is row equivalent to I_{n}.

$$\Rightarrow [A : B] \approx [I_{\mathrm{n}} : X]$$

$$[A : B_1 B_2 \cdots B_{\mathrm{n}}] \approx [I_{\mathrm{n}} : X_1 X_2 \cdots X_{\mathrm{n}}]$$

It is leading to the solutions $X_1, X_2, \cdots, X_{\mathrm{n}}$.

2.5 Criteria for Consistency and Uniqueness

2.5.1 Gaussian elimination

This section introduces another elimination method called the *Gaussian elimination method.*

Gaussian elimination:

(1) First, write down the augmented matrix $[A : B]$ of the system of linear equations $AX = B$.
(2) Find an echelon form of the augmented matrix $[A : B]$ using elementary row operations. This is done by creating leading 1s and then zeros below each leading 1, column by column, starting with the first column.
(3) Represent the system of linear equations corresponding to the echelon form.
(4) Use back substitution to arrive at the solution.

Remark:

An essential feature of Gauss elimination is that it constitutes a practical algorithm for solving linear systems, which can easily be implemented in standard programming languages.

Example 2.7:

Solve the system

$$x_1 + 2x_2 + 3x_3 + 2x_4 = -1$$
$$-x_1 - 2x_2 - 2x_3 + x_4 = 2$$
$$2x_1 + 4x_2 + 8x_3 + 12x_4 = 4$$

using the Gauss elimination method.

Solution:

$$\approx \begin{matrix} R_2 + R_1 \\ R_3 + (-2)\,R_1 \end{matrix} \begin{bmatrix} 1 & 2 & 3 & 2 & -1 \\ 0 & 0 & 1 & 3 & 1 \\ 0 & 0 & 2 & 8 & 6 \end{bmatrix}$$

$$\approx \; R_3 + (-2)\,R_2 \begin{bmatrix} 1 & 2 & 3 & 2 & -1 \\ 0 & 0 & 1 & 3 & 1 \\ 0 & 0 & 0 & 2 & 4 \end{bmatrix}$$

$$\approx \frac{1}{2}R_3 \begin{bmatrix} 1 & 2 & 3 & 2 & -1 \\ 0 & 0 & 1 & 3 & 1 \\ 0 & 0 & 0 & 1 & 2 \end{bmatrix}.$$

The corresponding system of equation is

$$\begin{aligned} &x_1 + 2x_2 + 3x_3 + 2x_4 = -1 \\ &x_3 + 3x_4 = 1 \\ &x_4 = 2. \end{aligned}$$

By back substitutions, we get $x_3 = 1 - 6 = -5$

$$x_1 + 2x_2 = 10 \Rightarrow x_1 = -2x_2 + 10.$$

Let $x_2 = r$. Then, the systems have many solutions.
The solutions are

$$x_1 = -2r + 10,\; x_2 = r,\; x_3 = -5,\; x_4 = 2.$$

Example 2.8:

$$\begin{aligned} &x_1 + 2x_2 + 3x_3 + 2x_4 = -1 \\ &-x_1 - 2x_2 - 2x_3 + x_4 = 2 \\ &2x_1 + 4x_2 + 8x_3 + 12x_4 = 4. \end{aligned}$$

The back substitutions can also be performed using matrices.
The final matrix is then the reduced echelon form of the systems.

2.5.2 Gauss-Jordan elimination

The Gauss-Jordan method of solving a linear equation system using matrices involves creating specific matrices. These numbers are developed systematically, column by column.

The Gauss-Jordan elimination is used to solve the n-equations in n-variables with a unique solution.

That is, $[A : B] \approx [I_n : X]$ can be used for a system that gives a unique solution.

Now we will discuss the method in its more general form where the number of equations can differ from the number of unknowns. There can be a unique solution, many solutions, or no solution to this system of equations.

Gauss-Jordan elimination:

(1) First, write down the augmented matrix $[A : B]$ of the system of linear equations $AX = B$.
(2) Derive the reduced echelon form of the augmented matrix $[A : B]$ using elementary row operation. This is done by creating leading 1s and then zeros above and below each leading 1, column by column, starting with the first column.
(3) Represent the system of equations corresponding to the reduced echelon form. This reduced system gives the solution to the system.

Example 2.9:

Determine the reduced echelon form of the following matrix:

$$\begin{bmatrix} 0 & 0 & 2 & -2 & 2 \\ 3 & 3 & -3 & 9 & 12 \\ 4 & 4 & -2 & 11 & 12 \end{bmatrix}$$

using the Gauss elimination method.

Solution:

Step-I:

$$\begin{bmatrix} 0 & 0 & 2 & -2 & 2 \\ 3 & 3 & -3 & 9 & 12 \\ 4 & 4 & -2 & 11 & 12 \end{bmatrix} \approx R_1 \leftrightarrow R_2 \begin{bmatrix} (3) & 3 & -3 & 9 & 12 \\ 0 & 0 & 2 & -2 & 2 \\ 4 & 4 & -2 & 11 & 12 \end{bmatrix}.$$

This non-zero element is called a pivot.

Step-II:

Create a 1 in the pivot location by multiplying the pivot row by $\frac{1}{\text{Pivot}}$

$$\approx \frac{1}{3}R_1 \begin{bmatrix} 1 & 1 & -1 & 3 & 4 \\ 0 & 0 & 2 & -2 & 2 \\ 4 & 4 & -2 & 11 & 12 \end{bmatrix}.$$

Step-III:

Create zero elsewhere in the pivot column by adding suitable multiples of the pivot row to all other rows of the matrix

$$\approx R_3 + (-4)\,R_1 \begin{bmatrix} 1 & 1 & -1 & 3 & 4 \\ 0 & 0 & 2 & -2 & 2 \\ 0 & 0 & 2 & -1 & -4 \end{bmatrix}.$$

Step-IV:

Cover the pivot row and all rows above it.

Repeat Step-I and Step-II for the remaining submatrix.

Repeat Step-III for the whole matrix. Continue this until the reduced echelon form (REF) is reached.

$$\begin{bmatrix} 1 & 1 & -1 & 3 & 4 \\ 0 & 0 & 2 & -2 & 2 \\ 0 & 0 & 2 & -1 & -4 \end{bmatrix} = \begin{bmatrix} 1 & 1 & -1 & 3 & 4 \\ 0 & 0 & (2_{\text{pivot}}) & -2 & 2 \\ 0 & 0 & 2 & -1 & -4 \end{bmatrix}$$

$$\approx \frac{1}{2}R_2 \begin{bmatrix} 1 & 1 & -1 & 3 & 4 \\ 0 & 0 & 1 & -1 & 1 \\ 0 & 0 & 2 & -1 & -4 \end{bmatrix}$$

$$\approx \begin{matrix} R_1 + R_2 \\ R_3 + (-2)\,R_2 \end{matrix} \begin{bmatrix} 1 & 1 & 0 & 2 & 5 \\ 0 & 0 & 1 & -1 & 1 \\ 0 & 0 & 0 & (1_{\text{Pivot}}) & -6 \end{bmatrix}$$

$$\approx \begin{matrix} R_1 + (-2)\,R_3 \\ R_2 + R_3 \end{matrix} \begin{bmatrix} 1 & 1 & 0 & 0 & 17 \\ 0 & 0 & 1 & 0 & -5 \\ 0 & 0 & 0 & 1 & -6 \end{bmatrix},$$

which is in reduced echelon form (REF).

Theorem 2.3:

If the number of pivots in echelon form is less than the number of unknowns in a homogenous system, then the linear system has a *non-trivial* solution.

Note:

If the number of pivots is r, then the $n - r$ non-pivotal unknowns can be given arbitrary values, and there will be a *non-trivial solution* whenever $n - r > 0$.

On the other hand, if $n = r$, none of the unknowns can be given arbitrary values; it provides a *unique solution*, namely the trivial one.

Corollary 2.1:

A homogenous linear system of m-equations in n unknowns always have a non-trivial solution, if $m < n$.

If r is the number of pivots, then $r \leq m < n$.

Example 2.10:

For which value of t, the following homogenous system has a non-trivial solution?

$$\begin{aligned} &6x_1 - x_2 + x_3 = 0 \\ &tx_1 + x_3 = 0 \\ &x_2 + tx_3 = 0 \end{aligned} .$$

Solution:

Let us proceed to put the linear system in echelon form by applying to it successively the operations

$\frac{1}{6} \times \text{eqn.}\,(1)$, $\text{eqn}\,(2) - t \times \text{eqn}\,(1)$ $\text{eqn}\,(2) \leftrightarrow \text{eqn}\,(3)$ and $\text{eqn}\,(3) - \frac{t}{6} \times \text{eqn}\,(2)$.

$$\begin{aligned} &x_1 - \left(\tfrac{1}{6}\right) x_2 + \left(\tfrac{1}{6}\right) x_3 = 0 \\ &x_2 + tx_3 = 0 \\ &\left(1 - \tfrac{t}{6} - \tfrac{t^2}{6}\right) x_3 = 0 \end{aligned}$$

.

The number of pivots will be less than 3, i.e., $r = 3$.

Now as $x_3 \neq 0$, implies, $1 - \frac{t}{6} - \frac{t^2}{6} = 0$, i.e., $t = 2$ or $t = -3$.

are the only values of t, *which gives the number of unknowns* for which the linear system has a *non-trivial* solution.

Example 2.11:

Consider a linear system

$$\begin{aligned} x_1 + 3x_2 + 3x_3 + 2x_4 &= 1 \\ 2x_1 + 6x_2 + 9x_3 + 5x_4 &= 5 \\ -x_1 - 3x_2 + 3x_3 \qquad &= 5 \end{aligned} .$$

The augmented matrix here is

$$\begin{bmatrix} 1 & 3 & 3 & 2 & \vdots & 1 \\ 2 & 6 & 9 & 5 & \vdots & 5 \\ -1 & -3 & 3 & 0 & \vdots & 5 \end{bmatrix}.$$

Now we can convert it into row echelon form

$$\begin{bmatrix} 1 & 3 & 3 & 2 & \vdots & 1 \\ 0 & 0 & 1 & \frac{1}{3} & \vdots & 1 \\ 0 & 0 & 0 & 0 & \vdots & 0 \end{bmatrix}.$$

Since the bottom right-hand entry of the matrix is 0, the linear system is consistent.

Thus, the linear system corresponding to the last matrix is

$$\begin{aligned} & x_1 + 3x_2 + 3x_3 + 2x_4 = 1 \\ & x_3 + \tfrac{1}{3}x_4 = 1 \\ & 0 = 0 \end{aligned}.$$

Thus, the general solution given by back substitution is

$$x_1 = -2 - s - 3r,\ x_2 = r\ ,\ x_3 = 1 - \frac{s}{3},\ x_4 = s,$$

where r and s are arbitrary scalars.

Theorem 2.4:

Let us consider a linear system $AX = B$ having n-unknowns with augmented matrix $[A : B]$. Then we have the following:

(1) The system $AX = B$ is *consistent* if and only if the matrices A and $[A : B]$ have the same number of pivots in *row echelon form*.

(2) Suppose the system $AX = B$ is *consistent* and r denotes the number of pivots of matrix A in row echelon form. Then the $n - r$ unknowns that correspond to columns of matrix A not containing a *pivot* can be given arbitrary values. Thus, the system has a *unique solution*.

Proof:

Let the linear system be consistent. The row echelon form of the augmented matrix must have only zero entries in the last column below the final pivot, but this is just the condition for A and $[A : B]$ to have the same number of pivots.

Finally, suppose the linear system is consistent. In that case, the unknowns corresponding to columns that do not contain pivots may be given arbitrary values, and the remaining unknowns can be found by back substitution.

Comparison of Gaussian and Gaussian-Jordan elimination:

The method of Gaussian elimination is, in general, more efficient than the Gaussian-Jordan elimination, in that it involves fewer operations of addition and multiplications. It is during the back substitution that Gaussian elimination picks up this advantage.

2.6 Method of LU Decomposition

Definition 2.3:

Let us consider a square matrix A that can be factored into an upper triangular matrix product and a lower triangular matrix. This factorization is called LU decomposition of the matrix A.

Remark:

Not every matrix has an LU decomposition, and when it exists, it is not unique. The method that now introduces can solve a system of linear equations if the matrix A has an LU decomposition.

Method of LU decomposition:

Let $AX = B$ be a system of n-equation in n-variables where the matrix A has LU decomposition,

$$\text{i.e., } A = LU.$$

The system thus can be written as

$$LUX = B.$$

The method involves writing this system as two subsystems, one of which is lower triangular and the other upper triangular

$$\begin{aligned} UX &= Y \\ LY &= B. \end{aligned}$$

Observe that substituting for Y from the first equation into the second gives the original system

$$LUX = B.$$

In practice, we first solve $LY = B$ for Y and then solve $UX = Y$ to get the solution X.

2.7 Construction of an LU Decomposition of a Matrix

(1) Use row operations to arrive at an upper triangular matrix U.
(2) The operations must involve the addition of multiples of rows to rows. In general, if row interchanges are required to arrive at U, then an LU form does not exist.
(3) The diagonal elements of the lower triangular matrix L are 1s. The non-zero elements of L correspond to the row operations.
(4) The row operation $R_k + cR_j$ implies that $l_{kj} = -c$.

2.7.1 Solution of matrix equation $AX = B$ using LU decomposition

(1) Find the LU decomposition of the matrix A. (If the matrix A has no LU decomposition, then the method is not applicable.)
(2) Use forward substitution to solve $LY = B$.
(3) Use back substitution to solve $UX = Y$.

Example 2.12:

Solve the following system of equations

$$\begin{aligned} x_1 - 3x_2 + 4x_3 &= 12 \\ -x_1 + 5x_2 - 3x_3 &= -12 \\ 4x_1 - 8x_2 + 23x_3 &= 58 \end{aligned}$$

using LU decomposition.

Solution:

$$\begin{bmatrix} 1 & -3 & 4 \\ -1 & 5 & -3 \\ 4 & -8 & 23 \end{bmatrix} \approx \begin{matrix} R_2 + R_1 \\ R_3 - 4R_1 \end{matrix} \begin{bmatrix} 1 & -3 & 4 \\ 0 & 2 & 1 \\ 0 & 4 & 7 \end{bmatrix}$$

$$\approx R_3 - 2R_2 \begin{bmatrix} 1 & -3 & 4 \\ 0 & 2 & 1 \\ 0 & 0 & 5 \end{bmatrix}.$$

These row operations lead to the following LU decomposition of A:

$$A = \begin{bmatrix} 1 & 0 & 0 \\ -1 & 1 & 0 \\ 4 & 2 & 1 \end{bmatrix} \begin{bmatrix} 1 & -3 & 4 \\ 0 & 2 & 1 \\ 0 & 0 & 5 \end{bmatrix}.$$

We again solve the given system $LUX = B$ by solving the two subsystems $LY = B$ and $UX = Y$.

$$LY = B \Rightarrow \begin{bmatrix} 1 & 0 & 0 \\ -1 & 1 & 0 \\ 4 & 2 & 1 \end{bmatrix} \begin{bmatrix} y_1 \\ y_2 \\ y_3 \end{bmatrix} = \begin{bmatrix} 12 \\ -12 \\ 58 \end{bmatrix}.$$

This lower triangular system has the solution

$$x_1 = 1, x_2 = -1, x_3 = 2.$$

Thus, the solution to the given system is

$$x_1 = 1, x_2 = -1, x_3 = 2.$$

Theorem 2.5:

Let A be an $m \times n$ matrix and c be a non-zero scalar.

(1) If a matrix B is obtained from A by multiplying the elements of a row (column) by α, then $|B| = \alpha\,|A|$.
(2) If a matrix B is obtained from A by interchanging two rows (column), then $|B| = -\,|A|$.
(3) If a matrix B is obtained from A by adding multiple rows (column) to another row (column), then $|B| = |A|$.

Proof:

Let matrix B be obtained by multiplying the *k*th row of A by c.
The *k*th row of B is thus $\alpha c_{k1} \alpha c_{k2} \cdots \alpha c_{kn}$.
Expand $|B|$ in terms of this row,

$$\begin{aligned} |B| &= \alpha c_{k1} C_{k1} + \alpha c_{k2} C_{k2} + \cdots + \alpha c_{kn} C_{kn} \\ &= \alpha \left(c_{k1} C_{k1} + c_{k2} C_{k2} + \cdots + c_{kn} C_{kn} \right) \\ &= \alpha\,|A| \end{aligned} \quad .$$

The corresponding result for columns is obtained by expanding the matrix B in terms of the *k*th column.

Definition 2.4:

A square matrix A is said to be *singular* if $|A| = 0$. A is non-singular if $|A| \neq 0$. The following theorem gives information about some of the circumstances under which we can expect a matrix to be singular.

Theorem 2.6:

Let A be a square matrix. A is singular if:

(1) All the elements of the row (column) are zero.
(2) Two rows (columns) are equal.
(3) Two rows (columns) are proportional.

(Note that (2) is a particular case of (2.7.1), but we list it to give it specific emphasis.)

Proof:

(1) Let all the elements of the *k*th row of A be zero. Expand $|A|$ in terms of the *k*th row:

$$\begin{aligned} |A| &= c_{k1}C_{k1} + c_{k2}C_{k2} + \cdots + c_{kn}C_{kn} \\ &= 0\,C_{k1} + 0\,C_{k2} + \cdots + 0\,C_{kn} \\ &= 0 \end{aligned}.$$

The corresponding results can be seen to hold for columns by expanding the determinant in terms of the columns of zeros.

(2) Interchange the equal rows of A to get a matrix B, which is equal to A; thus, $|B| = |A|$.

We know that interchanging two rows of a matrix negates the determinants.

Thus, $|B| = -|A|$.

The two results $|B| = |A|$ and $|B| = -|A|$ combine to give $|A| = -|A|$.

Thus, $2\,|A| = 0$, implying that $|A| = 0$.

The proof for columns is similar.

Example 2.13:

Show that the following matrices are singular.

$$(1)\ A = \begin{bmatrix} 2 & 0 & -7 \\ 3 & 0 & 1 \\ -4 & 0 & 9 \end{bmatrix}, (2)\ B = \begin{bmatrix} 2 & -1 & 3 \\ 1 & 2 & 4 \\ 2 & 4 & 8 \end{bmatrix}.$$

Solution:

All the elements in columns C_2 of A are zero.

Thus, $|A| = 0$.

Observe that every element in row R_3 of B is twice the corresponding element in row R_2.

We write $(R_3) = 2\,(R_2)$.

Since R_2 and R_3 are proportional,

$$|B| = 0.$$

The following theorem tells us how determinants interact with various matrix operations.

Theorem 2.7:

Let A and B be $n \times n$ matrices and α be a non-zero scalar.

(1) Determinant of a scalar multiples: $|\alpha A| = \alpha^n |A|$.
(2) Determinant of a product: $|AB| = |A|\,|B|$.
(3) Determinant of a transpose: $\left|A^T\right| = |A|$.
(4) Determinant of an inverse: $\left|A^{-1}\right| = \frac{1}{|A|}$.

Proof of (1):

Each row of αA is a row of A multiplied by α.
Upon applying Theorem 2.5(1), to each of the n-rows of αA, it becomes

$$|\alpha A| = \alpha^n |A|$$

Proof of (4):

Since $AA^{-1} = I_{\mathrm{n}}$.

$$\text{Thus, } \left|AA^{-1}\right| = |I_{\mathrm{n}}| \text{ Implies } |A|\left|A^{-1}\right| = 1$$

$$\text{Upon using (2), It finds } \left|A^{-1}\right| = \frac{1}{|A|}.$$

Example 2.14:

If A and B are square matrices of the same size, with A being singular, then prove that AB is also singular. Is the converse true?

Solution:

The matrix A is singular.
Thus, $|A| = 0$. Applying the properties of determinants, we get

$$|AB| = |A|\,|B| = 0\,.$$

Therefore, the matrix AB is singular.
We now investigate the converse:
Does AB being singular mean that A is singular?

We get $\begin{array}{l} |AB| = 0 \Rightarrow |A|\,|B| = 0 \\ \Rightarrow |A| = 0 \text{ or } \ |B| = 0 \end{array}$.

Since the product AB is singular, it implies that either A or B is singular (we do not exclude the possibility of both being singulars). The converse is not valid.

Example 2.15:

Let S be the set of 2×2 singular matrices. Prove that S is not closed under addition but is closed under scalar multiplication.

Solution:

To prove that S is not closed under addition, one example will suffice.

Let $A = \begin{bmatrix} 1 & 1 \\ 1 & 1 \end{bmatrix}$ and $B = \begin{bmatrix} 1 & 2 \\ 2 & 4 \end{bmatrix}$.

Then $|A| = 0$ and $|B| = 0$ imply that both matrices A and B are singular.

Their sum is $A + B = \begin{bmatrix} 2 & 3 \\ 3 & 5 \end{bmatrix}$.

Observe that $|A + B| = 1 \neq 0$, and $A + B$ is non-singular.

Thus, S is not closed under addition.

To prove that S is closed under scalar multiplication, we have to consider the general case.

Let α be a scalar and C be a singular matrix.

Thus, $|C| = 0$ and then $|\alpha C| = \alpha^2 |C| = 0$. The matrix αC is also singular.

Therefore, S is closed under scalar multiplication.

This example illustrates the closure properties under addition and scalar multiplication and with independent conditions.

It is possible to have one condition hold without the other holding.

Row operations can be used in a systematic manner similar to the Gauss-Jordan elimination to compute determinants. We lead up this method with a discussion of the determinants of upper triangular matrices.

2.8 Determinants and Matrix Inverses

We have introduced the concept of the determinant, discussed various ways of computing determinants, and looked at the algebraic properties of determinants. We shall now see how a determinant can give information about the inverse of a matrix and solutions to equation systems.

First, it introduces some necessary definitions for developing a formula for the inverse of a non-singular matrix.

Definition 2.5:

Let A be an $n \times n$ matrix and C_{ij} be the co-factor of c_{ij}. The matrix whose (i, j) th elements is C_{ij} is called the matrix of co-factors of A.

The transpose of this matrix A is called *the adjoint of* A and is denoted by $\mathrm{adj}\,(A)$.

$$\begin{bmatrix} C_{11} & C_{12} & \cdots & C_{1n} \\ C_{21} & C_{22} & \cdots & C_{2n} \\ \cdots & \cdots & \cdots & \cdots \\ C_{\text{n}1} & C_{\text{n}2} & \cdots & C_{\text{nn}} \end{bmatrix} \text{ Matrix of co-factors.}$$

$$\begin{bmatrix} C_{11} & C_{21} & \cdots & C_{\text{n}1} \\ C_{12} & C_{22} & \cdots & C_{\text{n}2} \\ \cdots & \cdots & \cdots & \cdots \\ C_{1n} & C_{2n} & \cdots & C_{\text{nn}} \end{bmatrix} \text{ Adjoint matrix.}$$

The following theorem provides a formula to determine the inverse of a non-singular matrix.

Theorem 2.8:

Let A be a square matrix with $|A| \neq 0$. Then the matrix A is invertible with

$$A^{-1} = \frac{1}{|A|}\text{adj}\,(A)\,.$$

Proof:

Consider the matrix product $A.\text{adj}\,(A)$.

The (i, j)th element of this product is

(i, j)th element = (row i of A) $\times$ column j of adj (A))

$$= [c_{\text{i}1}, c_{\text{i}2}, \cdots, c_{\text{in}}] \begin{bmatrix} C_{j1} \\ C_{j2} \\ \vdots \\ C_{jn} \end{bmatrix}.$$
$$= c_{\text{i}1}C_{j1} + c_{\text{i}2}C_{j2} + \cdots + c_{\text{in}}C_{jn}$$

If $i = j$, this is the expansion of $|A|$ in terms of the ith row.

If $i \neq j$, then it expands the determinants of a matrix in which the jth row of matrix A has been replaced by the ith row of A, a matrix having two identical rows.

Therefore (i, j) th *element* $= \begin{cases} |A| & \text{, if } i = j \\ 0 & \text{, if } i \neq j \end{cases}$.

The product $A.\text{adj}\,(A)$ is thus a diagonal matrix with the diagonal elements all being $|A|$. Factor out all the diagonal elements to get $|A|\,I_{\text{n}}$.

Thus, $A.\text{adj}\,(A) = |A|\,I_{\text{n}}$.

Since $|A| \neq 0$, we can rewrite this equation as $A.\left(\frac{1}{|A|}.\,\text{adj}\,(A)\right) = I_{\text{n}}$.

Similarly, it can be shown that

$$\left(\frac{1}{|A|}.\operatorname{adj}(A)\right).A = I_{\mathrm{n}}.$$

Thus, $A^{-1} = \frac{1}{|A|}.\operatorname{adj}(A)$, thus proving the theorem.

The necessity of this result lies in that it gives us a formula for the inverse of an arbitrary non-singular matrix that can be used in theoretical work. The Gauss-Jordan algorithm presented earlier is much more efficient than that formula for computing the inverse of a specific matrix. However, the Gauss-Jordan method cannot be used to describe the inverse of an arbitrary matrix.

The following theorem complements the previous one. It tells us that the non-singular matrices are the only matrices that have inverses.

Theorem 2.9:

A square matrix A is invertible, iff $|A| \neq 0$.

Proof:

Assume that A is invertible.
Thus, $AA^{-1} = I_{\mathrm{n}}$.
This implies that $\left|AA^{-1}\right| = |I_{\mathrm{n}}|$.
Properties of determinant give $|A|\left|A^{-1}\right| = 1$.
Thus, $|A| \neq 0$.
Conversely, Theorem 2.8 tells us that if $|A| \neq 0$.
Then matrix A is invertible.
The inverse of A, i.e., A^{-1}, exists if and only if $|A| \neq 0$.

Example 2.16:

Determine which of the following matrices are invertible:

$$(1)\quad A = \begin{bmatrix} 1 & -1 \\ 3 & 2 \end{bmatrix}. \qquad (2)\quad B = \begin{bmatrix} 4 & 2 \\ 2 & 1 \end{bmatrix}$$

$$(3)\quad C = \begin{bmatrix} 2 & 4 & -3 \\ 4 & 12 & -7 \\ -1 & 0 & 1 \end{bmatrix} \qquad (4)\quad D = \begin{bmatrix} 1 & 2 & -1 \\ -1 & 1 & 2 \\ 2 & 8 & 0 \end{bmatrix}.$$

Solution:

Compute the determinant of each matrix and apply the previous theorem. We get

(1) $|A| = 5 \neq 0$ implies that A is invertible.
(2) $|B| = 0$ implies that B is singular. Thus, the inverse does not exist.
(3) $|C| = 0$ implies that C is singular. Therefore, the inverse does not exist.
(4) $|D| = 2 \neq 0$ implies that D is invertible.

Example 2.17:

Determine the inverse of the matrix

$$A = \begin{bmatrix} 2 & 0 & 3 \\ -1 & 4 & -2 \\ 1 & -3 & 5 \end{bmatrix}.$$

Solution:

The determinant, i.e., $|A|$ is computed and found to be 25. Thus, the inverse of A exists.

We find that

$$\text{adj}\,(A) = \begin{bmatrix} 14 & -9 & -12 \\ 3 & 7 & 1 \\ -1 & 6 & 8 \end{bmatrix}.$$

The inverse of matrix A gives

$$A^{-1} = \frac{1}{25}\text{adj}\,(A) = \begin{bmatrix} \frac{14}{25} & \frac{-9}{25} & \frac{-12}{25} \\ \frac{3}{25} & \frac{7}{25} & \frac{1}{25} \\ \frac{-1}{25} & \frac{6}{25} & \frac{8}{25} \end{bmatrix}.$$

2.9 Determinants and Systems of Linear Equations

Next, we discuss the relationship between the existence and uniqueness of the solution to a system of n-linear equations in n-variables and the determinants of the system's coefficients.

Theorem 2.10:

Let $AX = B$ be a system of n-linearly independent in n-variables. If $|A| \neq 0$, then there is a unique solution. If $|A| = 0$, there may be many or no solutions.

Proof:

If $|A| \neq 0$, we know that A^{-1} exists and that there is then a unique solution given by $X = A^{-1}B$.

If $|A| = 0$, then the determinant of every matrix including the reduced echelon form of A is zero. Which implies the reduced echelon form of A is not I_n. Thus the solution to the system $AX = B$ is not unique.

The following systems of equations show that there may be many or no solutions.

$$\begin{aligned} x_1 - 2x_2 + 3x_3 &= 1 \\ 3x_1 - 4x_2 + 5x_3 &= 3 \\ 2x_1 - 3x_2 + 4x_3 &= 2 \end{aligned} \qquad \begin{aligned} x_1 + 2x_2 + 3x_3 &= 3 \\ 2x_1 + x_2 + 3x_3 &= 3 \\ x_1 + x_2 + 2x_3 &= 0 \end{aligned}$$

Many solutions No solutions.

$$x_1 = r\,, x_2 = 2r\,, x_3 = r.$$

Example 2.18:

Check the uniqueness of solutions of the following system of equations:

$$\begin{aligned} 3x_1 + 3x_2 - 2x_3 &= 2 \\ 4x_1 + x_2 + 3x_3 &= -5 \\ 7x_1 + 4x_2 + x_3 &= 9 \end{aligned} \quad .$$

Solution:

The determinant of the matrix of coefficients gives

$$\begin{vmatrix} 3 & 3 & -2 \\ 4 & 1 & 3 \\ 7 & 4 & 1 \end{vmatrix} = 0.$$

Thus, the system does not have a unique solution.

We now introduce a Crammer rule result to solve a system of n-linear equations in n-variables with a unique solution.

This rule is of theoretical importance in that it gives us a formula for the solution of a system of equations.

2.10 Crammer's Rule

Theorem 2.11 (Crammer's rule):

Let $AX = B$ be a system of n-linear equations in n-variables such that $|A| \neq 0$. Then the system has a solution given by

$$x_1 = \frac{|A_1|}{|A|}\,, x_2 = \frac{|A_2|}{|A|}\,, \cdots, x_n = \frac{|A_n|}{|A|},$$

which is unique and A_i is the matrix obtained by replacing the column i of A with B.

Proof:

Since $|A| \neq 0$, the solution to the system $AX = B$ is unique and is given by

$$X = A^{-1}B = \frac{1}{|A|}\text{adj}\,(A)\,B$$

x_i. The ith element of X is given by

$$\begin{aligned} x_i &= \tfrac{1}{|A|}\,[\text{row i of adj}\,(A)] \times B \\ &= \tfrac{1}{|A|}\,[c_{1i}, c_{2i}, \cdots, c_{ni}] \begin{bmatrix} b_1 \\ b_2 \\ \vdots \\ b_n \end{bmatrix} . \\ &= \tfrac{1}{|A|}\,(b_1c_{1i} + b_2c_{2i} + \cdots + b_nc_{ni}) \end{aligned}$$

The expression in parenthesis is the co-factor expansion of $|A_i|$ in terms of the ith column.

Thus, $x_i = \frac{|A_i|}{|A|}$.

Example 2.19:

Use Crammer's rule and solve the following system of equations:

$$\begin{aligned} x_1 + 3x_2 + x_3 &= -2 \\ 2x_1 + 5x_2 + x_3 &= -5 \\ x_1 + 2x_2 + 3x_3 &= 6 \end{aligned} .$$

Solution:

The matrix of coefficients A and column matrix of constants B are

$$A = \begin{bmatrix} 1 & 3 & 1 \\ 2 & 5 & 1 \\ 1 & 2 & 3 \end{bmatrix} \text{ and } B = \begin{bmatrix} -2 \\ -5 \\ 6 \end{bmatrix}.$$

It is found that $|A| = -3 \neq 0$.

Thus, Crammer's rule can be applied, and we get

$$A_1 = \begin{bmatrix} -2 & 3 & 1 \\ -5 & 5 & 1 \\ 6 & 2 & 3 \end{bmatrix}, A_2 = \begin{bmatrix} 1 & -2 & 1 \\ 2 & -5 & 1 \\ 1 & 6 & 3 \end{bmatrix}, A_3 = \begin{bmatrix} 1 & 3 & -2 \\ 2 & 5 & -5 \\ 1 & 2 & 6 \end{bmatrix},$$

giving

$$|A_1| = -3\,,\ |A_2| = 6\,,\ |A_3| = -9\ .$$

Crammer's rule now gives

$$x_1 = \frac{|A_1|}{|A|} = \frac{-3}{-3} = 1\,,\ x_2 = \frac{|A_2|}{|A|} = \frac{6}{-3} = -2,\ x_3 = \frac{|A_3|}{|A|} = \frac{-9}{-3} = 3.$$

The unique solution is

$$x_1 = 1\,,\ x_2 = -2\ ,\ x_3 = 3.$$

Example 2.20:

Find the values of λ for which the following system of the equation has non-trivial solutions. Find the solutions for each value of λ:

$$\begin{array}{l} (\lambda + 2)\,x_1 + (\lambda + 4)\,x_2 = 0 \\ 2x_1 + (\lambda + 1)\,x_2 = 0 \end{array}\ .$$

Solution:

This system is a homogenous system of linear equations.

The above system is a homogenous system of linear equations. Thus, it has a trivial solution. But according to Theorem 2.10, there is the possibility of other solution, if the determinants of the matrix of coefficients is zero.

Equating this determinant to zero, we get

$$\begin{vmatrix} \lambda + 2 & \lambda + 4 \\ 2 & \lambda + 1 \end{vmatrix} = 0$$

$$(\lambda + 2)\,(\lambda + 1) - 2\,(\lambda + 4) = 0$$

$$\lambda^2 + \lambda - 6 = 0$$

$$(\lambda - 2)\,(\lambda + 3)\ = 0$$

The determinant is zero if $\lambda = -3$ or $\lambda = 2$ results in the system.

$\lambda = -3$ results in the system

$$\begin{array}{l} x_1 + x_2 = -2 \\ 2x_1 - 2x_2 = 0 \end{array} \quad \Rightarrow x_1 = r\,,\ x_2 = r.$$

This system has many solutions.

$\lambda = 2$ results in the system

$$\begin{array}{l} 4x_1 + 6x_2 = 0 \\ 2x_1 + 3x_2 = 0 \end{array}\ .$$

This system has many solutions

$$x_1 = -\frac{3r}{2}, x_2 = r.$$

2.11 Curve Fitting, Electrical Network, and Traffic Flow

Systems of linear equations are used in such diverse fields as electrical engineering, economics, and traffic analysis. We now discuss applications in some of these fields.

2.11.1 Curve fitting

The following problem occurs in different branches of the sciences. A set of data points $(x_1, y_1), (x_2, y_2), \cdots, (x_n, y_n)$ is given, and it is necessary to find a polynomial whose graph passes through the points. The points are often measurements in an experiment. The x-coordinates are called *base points*. It can be shown that if the base points are all distinct, then a unique polynomial of degree $n-1$ (or less)

$$y = a_0 + a_1 x + \cdots + a_{n-2} x^{n-2} + a_{n-1} x^{n-1}$$

can be *fitted* to the points (see Figure 2.1).

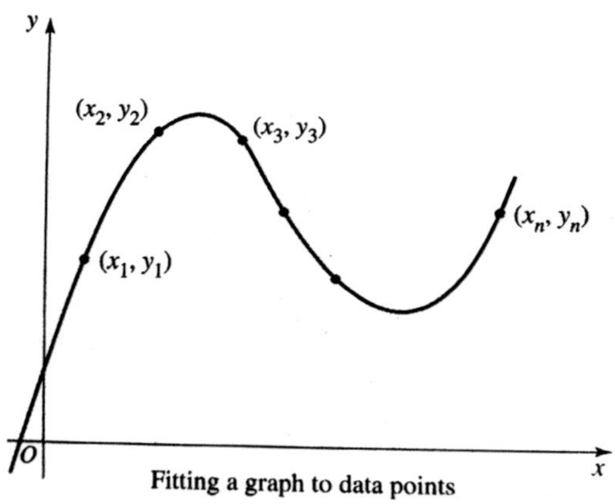

Figure 2.1 Curve fitting with a set of data points.

The coefficients $a_0, a_1, \cdots, a_{n-2}, a_{n-1}$ of the appropriate polynomial can be found by substituting the points into the polynomial equation and then solving a system of linear equations. (It is usual to write the polynomial in terms of ascending powers to find these coefficients. The columns of the matrix of coefficients of the system of equations then often follow a pattern. More will be said about this later.)

We now illustrate the procedure by fitting a polynomial of degree two, a parabola to a set of three such data points.

Example 2.21:

Determine the equation of the polynomial of degree two whose graph passes through the points $(1,6)\ (2,3)\ (3,2)$.

Solution:

Observe that in this example, we are given *three* points, and we want to find a polynomial of degree two (not less than the number of data points). Let the polynomial be

$$y = a_0 + a_1 x + a_2 x^2.$$

We give three points and shall use these three sets of information to determine the three unknowns a_0, a_1, and a_2, substituting

$$x = 1, y = 6; x = 2, y = 3; x = 3.y = 2.$$

In turn, the polynomial leads to the following system of three linear equations in a_0, a_1, and a_2.

$$\begin{array}{l} a_0 + a_1 + a_2 = 6 \\ a_0 + 2a_1 + 4a_2 = 3 \\ a_0 + 3a_1 + 9a_2 = 2 \end{array} .$$

Solve this system using the Gauss-Jordan elimination

$$\begin{bmatrix} 1 & 1 & 1 & 6 \\ 1 & 2 & 4 & 3 \\ 1 & 3 & 9 & 2 \end{bmatrix} \approx \begin{array}{c} R_2 + (-1)\, R_1 \\ R_3 + (-1)\, R_1 \end{array} \begin{bmatrix} 1 & 1 & 1 & 6 \\ 0 & 1 & 3 & -3 \\ 0 & 2 & 8 & -4 \end{bmatrix}$$

$$\approx \begin{array}{c} R_1 + (-1)\, R_2 \\ R_3 + (-2)\, R_2 \end{array} \begin{bmatrix} 1 & 0 & -2 & 9 \\ 0 & 1 & 3 & -3 \\ 0 & 0 & 2 & 2 \end{bmatrix}$$

$$\underset{\frac{1}{2}R_3}{\approx} \begin{bmatrix} 1 & 0 & -2 & 9 \\ 0 & 1 & 3 & -3 \\ 0 & 0 & 1 & 1 \end{bmatrix}$$

$$\approx \begin{array}{c} R_1 + (2)\, R_3 \\ R_2 + (-3)\, R_3 \end{array} \begin{bmatrix} 1 & 0 & 0 & 11 \\ 0 & 1 & 0 & -6 \\ 0 & 0 & 1 & 1 \end{bmatrix}.$$

We get $a_0 = 11$, $a_1 = -6$, $a_2 = 1$. The parabola that passes through these points is $y = 11 - 6x + x^2$ (see Figure 2.2).

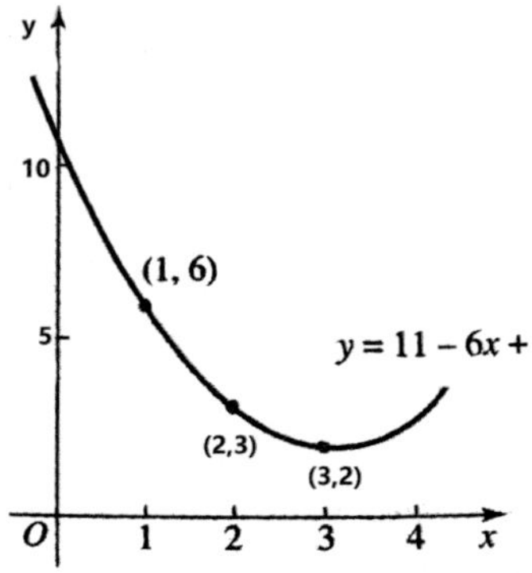

Figure 2.2 Fitted parabola with a set of data points.

2.12 Electrical Network Analysis

Systems of linear equations are used to determine the currents through various branches of electrical networks. The following two laws, which are based on experimental verification in the laboratory, lead to the equations.

Kirchhoff's law:

1. Junction:

All the current flowing into a junction must flow out of it.

2. Paths:

The sum of the *IR* terms (*I* denotes current and *R* denotes resistance) in any direction around a closed path is equal to the total voltage in the path in that direction.

Example 2.22:

Consider the electrical network of Figure 2.3. Let us determine the currents through each branch of this network.

Solution:

The batteries are of 8 and 16 volts. The following convention is used in electrical engineering to indicate the terminal of the battery out of which the current flows.

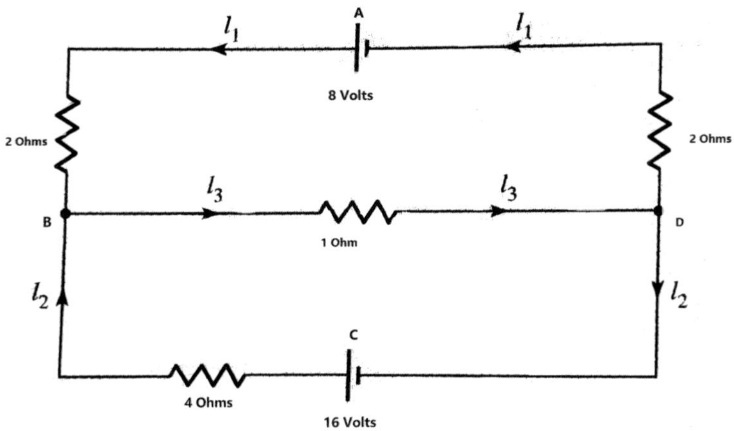

Figure 2.3 Electrical network diagram.

The resistances are one 1-ohm, one 4-ohms, and two 2-ohms. The current entering each battery will be the same as that leaving it.

Let the currents in the various branches of the above circuit be I_1, I_2, and I_3. Kirchhoff's law refers to junctions and closed paths. There are two junctions in this circuit, namely points B and D. There are three closed paths: ABDA, CBDC, and ABCDA. Apply the laws to the junctions and paths.

Junctions:

Junction B:

$$I_1 + I_2 = I_3.$$

Junction D:

$$I_3 = I_1 + I_2.$$

These two equations result in a single linear equation

$$I_1 + I_2 - I_3 = 0\,.$$

Paths:

Path ABDA:

$$2I_1 + 1I_3 + 2I_1 = 8.$$

Path CBDC:

$$4I_2 + 1I_3 = 16\,.$$

It is not necessary to look further at paths ABCDA. We now have a system of three linear equations in three unknowns I_1, I_2, and I_3. Path ABCDA, in fact, leads to an equation that is a combination of the last two equations; there is no new information.

The problem thus reduces to solving the following system of three linear equations in three variables:

$$\begin{aligned} &I_1 + I_2 - I_3 = 0 \\ &4I_1 + I_3 = 8 \\ &4I_2 + 1I_3 = 16 \end{aligned} \quad .$$

Using the method of the Gauss-Jordan elimination, we get

$$\begin{bmatrix} 1 & 1 & -1 & 0 \\ 4 & 0 & 1 & 8 \\ 0 & 4 & 1 & 16 \end{bmatrix} \underset{R_2 + (-4)R_1}{\approx} \begin{bmatrix} 1 & 1 & -1 & 0 \\ 0 & -4 & 5 & 8 \\ 0 & 4 & 1 & 16 \end{bmatrix}$$

$$\underset{\left(\frac{-1}{4}\right) R_2}{\approx} \begin{bmatrix} 1 & 1 & -1 & 0 \\ 0 & 1 & -\frac{5}{4} & -2 \\ 0 & 4 & 1 & 16 \end{bmatrix}$$

$$\approx \begin{matrix} R_1 + (-1)R_2 \\ R_3 + (-4)R_2 \end{matrix} \begin{bmatrix} 1 & 0 & \frac{1}{4} & 2 \\ 0 & 1 & -\frac{5}{4} & -2 \\ 0 & 0 & 6 & 24 \end{bmatrix}$$

$$\underset{\left(\frac{1}{6}\right) R_3}{\approx} \begin{bmatrix} 1 & 0 & \frac{1}{4} & 2 \\ 0 & 1 & -\frac{5}{4} & -2 \\ 0 & 0 & 1 & 4 \end{bmatrix}$$

$$\approx \begin{matrix} R_1 + \left(\frac{-1}{4}\right) R_3 \\ R_2 + \left(\frac{5}{4}\right) R_3 \end{matrix} \begin{bmatrix} 1 & 0 & 0 & 1 \\ 0 & 1 & 0 & 3 \\ 0 & 0 & 1 & 4 \end{bmatrix}.$$

The currents are $I_1 = 1, I_2 = 3, I_3 = 4$. The units are amps. The solution is unique, as is to be expected in this physical situation.

Example 2.23: in the Figure 2.4

Determine the currents through the various branches of the electrical network in the *figure*. This example illustrates how one has to be conscious of direction in applying law 2 for closed circuits.

Solution:

Junction:

Junction B:

$$I_1 + I_2 = I_3.$$

Junction D:

$$I_3 = I_1 + I_2.$$

Giving

$$I_1 + I_2 - I_3 = 0.$$

Paths:

Path ABCDA:

$$1I_1 + 2I_3 = 12.$$

Path ABDA:

$$1I_1 + 2(-I_2) = 12 + (-16).$$

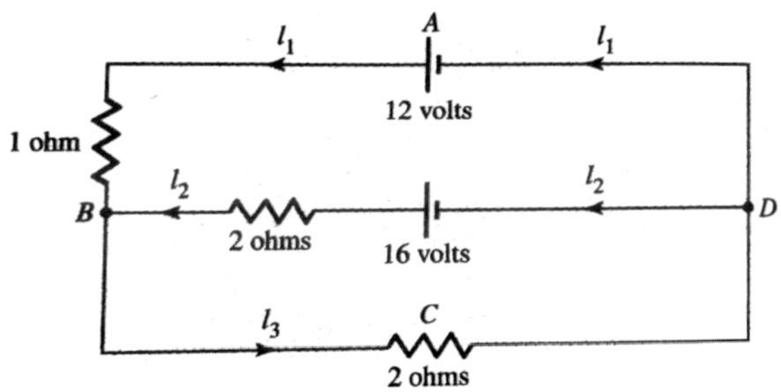

Figure 2.4 Electrical network diagram.

Observe that we have selected the direction ABDA around this last path. The current along the branch BD in this direction is $-I_2$, and the voltage is -16. We now have three equations in the three variables I_1, I_2, and I_3.

$$\begin{aligned} I_1 + I_2 - I_3 &= 0 \\ I_1 \quad + 2I_3 &= 12 \\ I_1 - 2I_2 \quad &= -4 \end{aligned} .$$

Solving these equations, we get
$I_1 = 2, I_2 = 3, I_3 = 5$amps.

In practice, electrical networks can involve many resistances and circuits; determining currents through branches involves solving large equations on a computer.

2.12.1 Traffic flow

Network analysis, as we saw in the previous discussion, plays an essential role in electrical engineering. In recent years, the concepts and tools of network analysis have been found to be helpful in many other fields, such as information theory and the study of the transportation system. The following analysis of traffic flow mentioned in the introduction illustrates how linear equations with many solutions can arise in practice.

Consider the typical road network of Figure 2.5. It represents an area of downtown Jacksonville, Florida. The streets are all one-way, with the arrows indicating the direction of traffic flow. The traffic is measured in vehicles per hour (vph). The figures in and out of the network given here are based on mid-week peak traffic hours, 7−9 a.m. and 4−6 p.m. Let us construct a mathematical model that can be used to analyze the flow with x_1, C, x_4 in the network.

Assume that the following traffic law applies.

All traffic entering an intersection must leave that intersection.

This conservation of flow constraint (compare it to Kirchoff's laws for electrical networks) leads to linear equations.

These are by intersection:

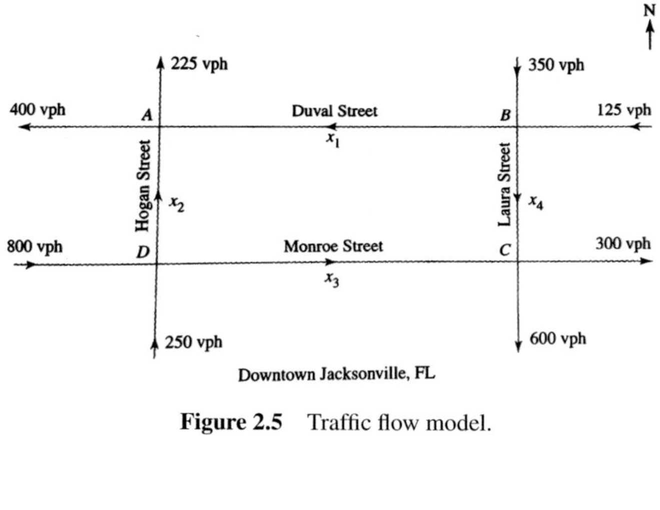

Figure 2.5 Traffic flow model.

A: Traffic in $= x_1 + x_2$; Traffic out $= 400 + 225$. Thus, $x_1 + x_2 = 625$.
B: Traffic in $= 350 + 125$; Traffic out $= x_1 + x_4$. Thus, $x_1 + x_4 = 475$.
C: Traffic in $= x_3 + x_4$; Traffic out $= 600 + 300$. Thus, $x_3 + x_4 = 900$.
D: Traffic in $= 800 + 250$; Traffic out $= x_2 + x_3$. Thus, $x_2 + x_3 = 1050$.

The following system of linear equations describes the constraints on the traffic:

$$\begin{aligned} x_1 + x_2 &= 625 \\ x_1 + x_4 &= 475 \\ x_3 + x_4 &= 900 \\ x_2 + x_3 &= 1050 \end{aligned} .$$

The method of Gauss-Jordan elimination is used to solve this system of equations. The augmented matrix and reduced echelon form of the preceding system are as follows:

$$\begin{bmatrix} 1 & 1 & 0 & 0 & 625 \\ 1 & 0 & 0 & 1 & 475 \\ 0 & 0 & 1 & 1 & 900 \\ 0 & 1 & 1 & 0 & 1050 \end{bmatrix} \approx \cdots \approx \begin{bmatrix} 1 & 0 & 0 & 1 & 475 \\ 0 & 1 & 0 & -1 & 150 \\ 0 & 0 & 1 & 1 & 900 \\ 0 & 0 & 0 & 0 & 0 \end{bmatrix}.$$

The system of equations that corresponds to this reduced echelon form is

$$\begin{aligned} x_1 + x_4 &= 475 \\ x_2 - x_4 &= 150 \\ x_3 + x_4 &= 900 \end{aligned} .$$

Expressing each leading variable in terms of the remaining variable, we get

$$\begin{aligned} x_1 &= -x_4 + 475 \\ x_2 &= x_4 + 150 \\ x_3 &= -x_4 + 900 \end{aligned} .$$

As was perhaps to be expected, the system of equations has many solutions. There are many traffic flows possible. One does have a certain amount of choice at an intersection. Let us now use this mathematical model to arrive at information. Suppose it becomes necessary to perform road work on the stretch DC of Monroe Street. It is desirable to have as small a flow x_3 as possible along this trench of road. The flows can be controlled along various branches using traffic lights. What is the minimum value of x_3 along DC that would not lead to traffic congestion? We use the preceding system of equations to answer this question.

All traffic flows must be non-negative (a negative flow would be interpreted as traffic moving in the wrong direction on a one-way street). The third equation tells us that "x_3" a minimum and will be as large as possible, as long as it does not go above 900. The most significant value of x_4 that can be without causing negative values of x_1 or x_2 is 475. Thus, the smallest value of x_3 is $-475 + 900$ or 425. Any road works on Monroe should allow for at least 425 vph.

In practice, networks are much faster than the one discussed here, leading to larger systems of linear equations that are handled on computers. Various values of variables can be fed in, and different scenarios created.

Exercises

1. If A is a square matrix of order n and α a scalar, then prove that $\det(\alpha A) = \alpha^n \det A$.
2. Let A be a skew-symmetric matrix of odd order. Then prove that $\det A = 0$.
3. For a triangular matrix A, prove that $\det A$ is the product of its diagonal entries.
4. Prove that

$$\begin{vmatrix} 1 & \alpha & \alpha^2 \\ 1 & \beta & \beta^2 \\ 1 & \delta & \delta^2 \end{vmatrix} = (\alpha - \beta) \begin{vmatrix} 0 & 1 & \alpha+\beta \\ 1 & \beta & \beta^2 \\ 1 & \delta & \delta^2 \end{vmatrix}$$
$$= (\alpha - \beta)(\beta - \delta)(\delta - \alpha) \begin{vmatrix} 0 & 1 & \alpha+\beta \\ 0 & 0 & 1 \\ 1 & \delta & \delta^2 \end{vmatrix}.$$

5. Without expanding, prove that

(a) $\begin{vmatrix} 1 & 1 & 3 \\ 2 & 9 & 1 \\ 4 & 11 & 7 \end{vmatrix} = 0.$

(b) $\begin{vmatrix} b+c & 1 & a \\ c+a & 1 & b \\ b+a & 1 & c \end{vmatrix} = 0.$

(c) $\begin{vmatrix} a & b & c \\ x & y & z \\ p & q & r \end{vmatrix} = \begin{vmatrix} y & b & q \\ x & a & p \\ z & c & r \end{vmatrix} = \begin{vmatrix} x & y & z \\ p & q & r \\ a & b & c \end{vmatrix}.$

(d) $\begin{vmatrix} x & y & z \\ x^2 & y^2 & z^2 \\ yz & zx & xy \end{vmatrix} = \begin{vmatrix} 1 & 1 & 1 \\ x^2 & y^2 & z^2 \\ x^3 & y^3 & z^3 \end{vmatrix}.$

(e) $\begin{vmatrix} a+b & b+c & c+a \\ p+q & q+r & r+p \\ x+y & y+z & z+x \end{vmatrix} = 2\begin{vmatrix} a & b & c \\ p & q & r \\ x & y & z \end{vmatrix}.$

(f) $\begin{vmatrix} x & x+3 & x+6 \\ x+1 & x+4 & x+7 \\ x+2 & x+5 & x+8 \end{vmatrix} = 0.$

(g) $\begin{vmatrix} 4 & 7 & 10 \\ 10 & 13 & 16 \\ 20 & 23 & 26 \end{vmatrix} = 0.$

(h) $\begin{vmatrix} 1 & a & a^2-bc \\ 1 & b & b^2-ca \\ 1 & c & c^2-ab \end{vmatrix} = 0.$

(i) $\begin{vmatrix} x-y & y-z & z-x \\ y-z & z-x & x-y \\ z-x & x-y & y-z \end{vmatrix} = 0.$

6. Without expanding, prove that

$$\begin{vmatrix} 1 & yz & y+z \\ 1 & zx & z+x \\ 1 & xy & x+y \end{vmatrix} = \begin{vmatrix} 1 & 1 & 1 \\ x & y & z \\ x^2 & y^2 & z^2 \end{vmatrix}.$$

7. Prove that

(a) $\begin{vmatrix} 1 & x & x^2 \\ 1 & y & y^2 \\ 1 & z & z^2 \end{vmatrix} = (x-y)(y-z)(z-x).$

(b) $\begin{vmatrix} 1 & x & yz \\ 1 & y & zx \\ 1 & z & xy \end{vmatrix} = (x-y)(y-z)(z-x).$

(c) $\begin{vmatrix} x+a & b & c & d \\ a & x+b & c & d \\ a & b & x+c & d \\ a & b & c & x+d \end{vmatrix} = x^3(x+a+b+c+d).$

8. If ω_1, ω_2, and ω_3 are the three cube roots of unity, then prove that

$$\begin{vmatrix} x_1 & x_2 & x_3 \\ x_3 & x_1 & x_2 \\ x_2 & x_3 & x_1 \end{vmatrix} = \prod_{\mathrm{i}=1}^{3}\left(x_1 + x_2\omega_{\mathrm{i}} + x_1\omega_{\mathrm{i}}^2\right).$$

This determinant is called a *circulant* of the third order. Write down a circulant of order n. Write also its value.

9. (a) Prove that the equation of a circle through three points $(x_1, y_1), (x_2, y_2)$, and (x_3, y_3) is given by

$$\begin{vmatrix} x^2+y^2 & x & y & 1 \\ x_1^2+y_1^2 & x_1 & y_1 & 1 \\ x_2^2+y_2^2 & x_2 & y_2 & 1 \\ x_3^2+y_3^2 & x_3 & y_3 & 1 \end{vmatrix} = 0.$$

(b) Determine the equations of the sphere passing through the points $(x_1, y_1, z_1), (x_2, y_2, z_2), (x_3, y_3, z_3)$, and (x_4, y_4, z_4) in the determinant form.

10. Solve the equation

$$\begin{vmatrix} x+a & b & c \\ c & x+b & a \\ a & b & x+c \end{vmatrix} = 0.$$

11. Without expanding, prove that

$$\begin{vmatrix} 1+a & 1 & 1 & 1 \\ 1 & 1+b & 1 & 1 \\ 1 & 1 & 1+c & 1 \\ 1 & 1 & 1 & 1+d \end{vmatrix} = abcd\left(1+\frac{1}{a}+\frac{1}{b}+\frac{1}{c}+\frac{1}{d}\right).$$

12. If A and B are square matrices of order n, then prove that

$$\det\left(A^T B\right) = \det\left(AB^T\right) = \det\left(A^T B^T\right) = \det(AB).$$

13. If A is a square matrix, then prove that $\det(A^n) = (\det A)^n$ for all positive integers n.
14. Prove that the determinant of an idempotent matrix is either 0 or 1.
15. Evaluate $\det A$, if A is a nilpotent matrix.
16. Prove that

(a) $$\begin{vmatrix} 1 & a & a^2 \\ 1 & b & b^2 \\ 1 & c & c^2 \end{vmatrix}^2 = \begin{vmatrix} 3 & a+b+c & a^2+b^2+c^2 \\ a+b+c & a^2+b^2+c^2 & a^3+b^3+c^3 \\ a^2+b^2+c^2 & a^3+b^3+c^3 & a^4+b^4+c^4 \end{vmatrix}.$$

(b) $$\begin{vmatrix} 0 & z & y \\ z & 0 & x \\ y & x & 0 \end{vmatrix}^2 = \begin{vmatrix} y^2+z^2 & xy & xz \\ xy & z^2+x^2 & yz \\ zx & yz & x^2+y^2 \end{vmatrix}.$$

(c) $$\begin{vmatrix} 2yz-x^2 & z^2 & y^2 \\ z^2 & 2zx-y^2 & x^2 \\ y^2 & x^2 & 2xy-z^2 \end{vmatrix} = \begin{vmatrix} x & y & z \\ y & z & x \\ z & x & y \end{vmatrix}^2.$$

17. Prove that $\begin{vmatrix} \alpha_{11} & \alpha_{12} & \alpha_{13} \\ \alpha_{21} & \alpha_{22} & \alpha_{23} \\ \alpha_{31} & \alpha_{32} & \alpha_{33} \end{vmatrix}^2 = \begin{vmatrix} A_{11} & A_{12} & A_{13} \\ A_{21} & A_{22} & A_{23} \\ A_{31} & A_{32} & A_{33} \end{vmatrix}$, where A_{ij} is the co-factor of α_{ij}.

18. Use Crammer's rule (if applicable) to find the solutions of the following system of linear equations:

(a) $\begin{array}{l} 3x + y = 1 \\ 5x + 2y = 3 \end{array}$.

(b) $\begin{array}{l} 2x - 3y = 7 \\ x + 4y = 1 \end{array}$.

(c) $\begin{array}{l} x + 2y + 3z = 3 \\ 2x - z = 4 \\ 4x + 2y + 2z = 5 \end{array}$.

(d) $\begin{array}{l} x - y + 2z = 1 \\ 2x + 2z = 3 \\ 3x + y + 3z = 7 \end{array}$.

(e) $\begin{array}{l} x + y + 2z = 3 \\ 2x + 2y + 2z = 7 \\ 3x + 4y + 3z = 2 \end{array}$.

19. Reduce the following matrices to row-reduced echelon form:

(a) $\begin{bmatrix} 1 & -1 & 1 \\ 3 & -1 & 2 \\ 3 & 1 & 1 \end{bmatrix}$ (b) $\begin{bmatrix} 3 & 2 & 3 & 1 \\ 4 & 3 & 5 & 2 \\ 2 & 1 & 1 & 0 \end{bmatrix}$

(c) $\begin{bmatrix} 1 & 2 & 3 & 4 & -1 \\ 4 & 1 & 5 & -6 & 10 \\ 2 & 0 & 2 & -2 & 2 \end{bmatrix}$ (d) $\begin{bmatrix} 1 & 3 & 2 & 0 \\ 2 & 0 & 1 & 1 \\ 1 & 2 & 3 & 0 \\ 3 & -1 & 0 & 1 \end{bmatrix}$

(e) $\begin{bmatrix} -1 & 1 & 1 & 0 \\ 1 & -1 & 1 & 1 \\ 0 & 1 & 2 & -1 \\ 1 & 1 & 0 & -2 \end{bmatrix}$ (f) $\begin{bmatrix} 1 & 2 & 0 & 0 \\ 1 & 1 & -1 & 2 \\ 0 & 2 & 1 & -1 \end{bmatrix}$

(g) $\begin{bmatrix} 3 & 1 & -1 \\ 1 & 2 & 3 \\ 4 & 0 & 1 \\ 1 & 5 & 3 \end{bmatrix}$ (h) $\begin{bmatrix} 0 & 6 & 6 & 1 \\ -8 & 7 & 2 & 3 \\ -3 & 2 & 1 & 1 \\ 1 & 1 & -1 & 0 \end{bmatrix}$

(i) $\begin{bmatrix} 1 & 0 & 7 & 9 \\ 5 & 2 & 2 & 10 \\ 3 & -2 & 3 & 11 \\ 2 & -1 & 3 & 8 \end{bmatrix}$.

20. Solve the following systems of linear equations by using the row-reduced method:

(a) $\begin{aligned} 2x - 3y &= 1 \\ 2x - y + z &= 2 \\ 3x + y - 2z &= 1 \end{aligned}$

(b) $\begin{aligned} y - 2z &= 3 \\ 3x + z &= 4 \\ x + y + z &= 1 \end{aligned}$

(c) $\begin{aligned} x - y + z &= 0 \\ 2x + y - 3z &= 1 \\ -x + y + 2z &= -1 \end{aligned}$

(d) $\begin{aligned} x - y + 3z &= 1 \\ 2x + y - z &= 2 \\ 3x - y + 2z &= 2 \end{aligned}$

(e) $\begin{aligned} x + y - 2z &= 3 \\ 3x + y - z &= 8 \\ 2x - y + z &= 0 \end{aligned}$.

21. Determine whether the following systems of linear equations are consistent. Discuss the solution completely in the case of consistent systems.

(a) $\begin{aligned} x_1 - x_2 + 2x_3 + 3x_4 &= 1 \\ 2x_1 + 2x_2 + 2x_4 &= 1 \\ 4x_1 + x_2 - x_3 - x_4 &= 1 \\ x_1 + 2x_2 + 3x_3 &= 1 \end{aligned}$

(b) $\begin{aligned} x_1 + 2x_2 + 4x_3 + x_4 &= 4 \\ 2x_1 - x_3 + 3x_4 &= 4 \\ x_1 - 2x_2 - x_3 &= 0 \\ 3x_1 + x_2 - x_3 - 5x_4 &= 7 \end{aligned}$

(c) $\begin{aligned} 2x_1 + x_3 - x_4 + x_5 &= 2 \\ x_1 + x_3 - x_4 + x_5 &= 1 \\ 12x_1 + 2x_2 + 8x_3 + 2x_5 &= 12 \end{aligned}$

(d) $\begin{aligned} x_1 + 2x_2 - x_3 - 2x_4 &= 0 \\ 2x_1 + 4x_2 + 2x_3 + 4x_4 &= 4 \\ 3x_1 + 6x_2 + 3x_3 + 6x_4 &= 6 \end{aligned}$

$$\begin{aligned} x_1 - x_3 &= 1 \\ 2x_1 + x_2 + x_3 &= 2 \end{aligned}$$

(e) $$\begin{aligned} x_2 - x_3 &= 3 \\ x_1 + x_2 + x_3 &= 4 \\ 2x_2 - x_3 &= 0 \end{aligned}$$

(f) $$\begin{aligned} x_1 + 2x_3 &= 1 \\ 2x_1 + x_2 + 2x_3 &= 1 \\ x_2 - 2x_3 &= 1 \\ x_1 + x_2 &= 1 \\ x_1 - x_2 + 4x_3 &= 1 \end{aligned}$$

(g) $$\begin{aligned} 2x_1 + x_2 + x_3 + x_4 &= 2 \\ 3x_1 - x_2 + x_3 - x_4 &= 2 \\ x_1 + 2x_2 - x_3 + x_4 &= 1 \\ 6x_1 + 2x_2 + x_3 + x_4 &= 5 \end{aligned}$$

(h) $$\begin{aligned} x_1 + 3x_2 - 3x_3 + 2x_4 &= 1 \\ 4x_1 + x_2 - 2x_3 + x_4 &= 1 \\ 6x_1 + 5x_2 + 10x_3 + 3x_4 &= 15 \\ x_1 + 2x_2 + 3x_3 + x_4 &= 6 \end{aligned}$$

(i) $$\begin{aligned} x_1 - 2x_2 - x_3 &= -1 \\ 2x_1 - x_3 - 3x_4 &= 1 \\ 3x_1 + x_2 - x_3 - 5x_4 &= 1 \\ 2x_1 + 3x_3 + x_4 &= 0 \end{aligned}$$

(j) $$\begin{aligned} 3x_1 + 6x_2 + 3x_3 + 6x_4 &= 5 \\ x_1 + 2x_2 - x_3 - 2x_4 &= -1 \\ 3x_1 + 6x_2 + x_3 + 2x_4 &= 3 \\ x_1 + 2x_2 + 2x_3 + 4x_4 &= 3 \end{aligned}$$

(k) $$\begin{aligned} x_1 - x_3 &= 2 \\ x_1 + x_2 + 2x_3 &= 4 \\ x_1 + x_2 - 2x_3 &= 4 \\ x_1 + x_2 + x_3 &= 4 \\ x_1 + 3x_2 - x_3 &= 8 \end{aligned}$$

(l) $$\begin{aligned} 2x_1 + x_2 + 2x_3 &= 1 \\ x_1 + x_2 &= 0 \\ x_1 - 2x_2 + 6x_3 &= 3 \\ x_1 - 2x_3 &= 1 \\ x_1 - x_2 + 4x_3 &= 2 \end{aligned}$$

(m) $$\begin{aligned} x_1 + x_2 - x_3 - 6x_4 + 6x_5 &= -19 \\ x_1 + 7x_4 - 7x_5 &= 28 \\ 2x_2 - 3x_3 + 18x_4 - 4x_5 &= 24 \end{aligned}$$

(n) $$\begin{aligned} x_1 - 3x_2 + x_3 - x_4 &= 7 \\ 2x_1 + 4x_2 - x_3 + 6x_4 &= -6 \\ 2x_1 + x_2 + x_4 &= 0 \end{aligned} \quad .$$

3

Vector Spaces

This chapter introduces the notion of an abstract vector space with its typical features. Vector spaces occur in various branches of mathematics and have many applications in science and engineering. In this chapter, we generalize the concept of vector space R^n and examine its underlying algebraic structure. Any set with this structure has the same mathematical properties as R^n and will be called a vector space. The results that were developed for the vector space R^n will also apply to such spaces. We shall, for example, find that specific spaces of matrices and functions have the same mathematical properties as the vector space R^n. The concepts of linear independence, spanning set, basis, linear transformations, and dimension will be extended to these spaces.

Standard features of vector spaces:

The typical features of a vector space $\mathbf{V}$ are:

(1) It consists of a non-empty set of objects, including zero, called vectors, and zero is called a "zero" vector.
(2) The addition of two vectors gives another vector.
(3) Multiplication of a vector by a scalar gives a vector.
(4) It requires a field to perform the operations.

3.1 Field

A field generally consists of scalars, usually R or C, and accordingly the vector space will be called *real* or *complex* vector space depending on whether the field is R or C.

3.2 Vector Spaces

The vector space R^n is a set of n vectors on which two operations, namely addition and scalar multiplication, have been defined.

A vector space is also said to be a *linear space*.

3.3 The Notion of a Vector Space

Definition 3.1:

A vector space **V** is a set of elements called vectors having addition and scalar multiplication operation on **V**.

The vector space **V** satisfies the following conditions:

Let $u, u_1, u_2, u_3 \in \boldsymbol{V}$ and $k_1 \& k_2$ are scalars over the field F.

Closure axiom:

(1) If the sum $u_1 + u_2$ exists and is an element of **V**, then **V** is said to be closed under the operation addition.
(2) If ku_1 is a component of **V**, then **V** is said to be closed under the operation scalar multiplication.

Addition axiom:

(3) $u_1 + u_2 = u_2 + u_1$ (Commutative).
(4) $u_1 + (u_2 + u_3) = (u_1 + u_2) + u_3$ (Associative).
(5) There exists an element u of **V** called the zero-vector denoted as 0_v such that $u + 0_v = u$.
(6) For every element $u \in$ **V**, there exists a part called the negative of u denoted by $-u$ such that $u + (-u) = 0_v$.

Scalar multiplication axiom:

(7) $k(u_1 + u_2) = ku_1 + ku_2$.
(8) $(k_1 + k_2) u_1 = k_1u_1 + k_2u_1$.
(9) $k_1(k_2u_1) = (k_1k_2) u_1$.
(10) $1u = u$.

The two most common sets of scalars used in vector spaces are real numbers and complex numbers.

Examples of vector space:

The vector space **V** over R or C is then called *real* and *complex* vector space.

Vector space of matrices M_{mn}:

The set of real $m \times n$ matrices M_{mn} is a vector space over R.

Vector spaces of functions:

(1) The set of all functions $f(x)$ form a vector space over R.
(2) All functions defined on real numbers with pointwise addition and scalar multiplication operations are a vector space.

(3) The complex vector space C^n over C.

Theorem 3.1:

Let u be a vector and 0_v the zero-vector defined on a vector space $\mathbf{V}$. Let $k \in F$ be any scalar and 0 the zero scalars. Then

(1) $0u = 0_v$.
(2) $c0_v = 0_v$.
(3) $(-1)\,u = -u$.
(4) If $ku = 0$, then either $k = 0$ or $u = 0$.

Proof:

$$\begin{aligned} 0u + 0u &= (0+0)\,u \qquad \text{(Axiom}-8) \\ &= 0u \end{aligned}$$

Add the $-$ve of $0u$, namely $-0u$, to both sides of this equation:

$$(0u + 0u) + (-0u) = 0u + (-0u)$$
$$\Rightarrow 0u + (0u + (-0u)) = 0, \quad \text{(Axiom:(4)\&(5))}$$
$$\Rightarrow 0u + 0_v = 0_v \quad \text{(Axiom:(6))}$$
$$\Rightarrow 0u = 0_v \quad \text{(Axiom:(5))}$$
$$(-1)u + u = (-1)u + 1u \quad \text{(Axiom:(10))}$$
$$= [(-1) + 1]u \quad \text{(Axiom:(8))}$$
$$= 0u = 0_v \quad \text{(Property of Scalar 0)}$$

Thus $(-1)\,u$ is the negative of u. (Axiom: vi)

3.4 Subspaces

Definition 3.2:

Let $\mathbf{U}$ be a non-empty subset of a vector space $\mathbf{V}$. If $\mathbf{U}$ satisfies the operations of addition and scalar multiplication of a vector space $\mathbf{V}$ over a field F, then $\mathbf{U}$ is said to be a subspace of $\mathbf{V}$.

That is, the non-empty subset $\mathbf{U}$ is a *subspace of a vector space* if it is closed under the operation of addition and scalar multiplication.

The non-empty subset $\mathbf{U}$ then acquires the other properties of vector space from the vector space $\mathbf{V}$.

Generally, a vector subspace is itself a vector space contained within a larger vector space.

Example 3.1:

Let us consider a linear homogenous system $\mathbf{AX} = \mathbf{0}$ in $n-$unknowns defined over some field $\mathbf{F}$.

Let S denote the set of all solutions in $n-$column vectors $\boldsymbol{X}$ of the linear system $\mathbf{AX} = \mathbf{0}$ over the field $\mathbf{F}$. Then S is a subset of R^n, and it indeed contains the zero vector.

Now since S satisfies all the operations of addition and scalar multiplication of a vector space $\mathbf{V}$ over a field $\mathbf{F}$, so S is said to be a vector subspace of R^n.

Vector addition:

If $\mathbf{X_1}$ and $\mathbf{X_2}$ are the solutions of the linear system $\mathbf{AX} = \mathbf{0}$ and k is any scalar, then

$$\mathbf{A}\left(\mathbf{X_1} + \mathbf{X_2}\right) = \mathbf{AX_1} + \mathbf{AX_2} = \mathbf{0}.$$

Thus, $\mathbf{X_1} + \mathbf{X_2}$ belongs to S.

Scalar multiplication:

Let $\mathbf{X_1} \in S$ and $k \in \mathbf{F}$ be any scalar.

Then

$$\mathbf{A}\left(\mathbf{k}\,\mathbf{X_1}\right) = \mathbf{k}\left(\mathbf{AX_1}\right) = \mathbf{0}.$$

Thus, $k\mathbf{X_1}$ belongs to S.

Hence, S is a subspace of the vector space $\mathbf{V} = \mathbf{R^n}$.

Remark:

This subspace S is called the *solution space* of the homogenous linear system $\mathbf{AX} = \mathbf{0}$, also known as the *null space* of the matrix $\mathbf{A}$.

Example 3.2:

Let us consider the subset $\mathbf{W}$ of R^3 consisting of vectors of the form (a, a, b), where the first two components are the same, i.e.,

$$\mathbf{W} = \{(a, a, b) : a, b \in R\}\,.$$

If we add two such vectors (a, a, b) and (c, c, d), we get a vector $(a + c, a + c, b + d)$ with identical first components.

If we multiply (a, a, b) by a scalar k, we get (ka, ka, kb), again a vector with identical first components.

$\Rightarrow$ $\mathbf{W}$ is closed under addition and scalar multiplication.

Hence, $\mathbf{W}$ is a subspace of R^3.

Example 3.3:

Consider the subset $\mathbf{W}$ of R^3 consisting of vectors of the form (a, a^2, b), where the second component is the square of the first, i.e.,

$$\mathbf{W} = \{(a, a^2, b) : a, b \in R\}.$$

On adding two such vectors (a, a^2, b) and (c, c^2, d), we get $(a + c, a^2 + c^2, b + d)$.

The second component of the vector is not a square of the first.

$\Rightarrow$ The vector is not in $\mathbf{W}$.

$\Rightarrow\mathbf{W}$ is not closed under addition.

Hence, $\mathbf{W}$ is not a subspace of R^3.

Example 3.4:

Prove that $\mathbf{D_{22}}$ the set of 2×2 diagonal matrices forms a subspace of the vector space $\mathbf{V} = M_{22}$.

Solution:

Vector addition:

Let

$$u_1 = \begin{bmatrix} a & 0 \\ 0 & b \end{bmatrix}, u_2 = \begin{bmatrix} p & 0 \\ 0 & q \end{bmatrix}.$$

We get

$$\begin{aligned} u_1 + u_2 &= \begin{bmatrix} a & 0 \\ 0 & b \end{bmatrix} + \begin{bmatrix} p & 0 \\ 0 & q \end{bmatrix} \\ &= \begin{bmatrix} a+p & 0 \\ 0 & b+q \end{bmatrix} \in U \end{aligned}.$$

Since $u_1 + u_2 \in \mathbf{U}$ is a diagonal matrix, $\mathbf{U}$ is closed under addition.

Scalar multiplication:

Let k be a scalar.

We get $ku_1 = k \begin{bmatrix} a_1 & 0 \\ 0 & b_1 \end{bmatrix} = \begin{bmatrix} ka_1 & 0 \\ 0 & kb_1 \end{bmatrix}$.

This implies that $ku_1 \in \mathbf{U}$, since ku_1 is a 2×2 diagonal matrix.

Thus, $\mathbf{U}$ is closed under scalar multiplication.

Hence, $\mathbf{U}$ is a subspace of $\mathbf{M_{22}}$.

Example 3.5:

Let $\mathbf{P_n}(\mathbf{x})$ denote the set of all real polynomial functions having $\deg \leq n$. Then $\mathbf{P_n}(\mathbf{x})$ forms a vector space under the operation of addition and multiplication on polynomial $\mathbf{P_n}(\mathbf{x})$ in a pointwise manner.

Theorem 3.2:

Let $\mathbf{U} \subset \mathbf{V}$ be a subspace of a vector space $\mathbf{V}$. Then $\mathbf{U}$ contains the zero vector $\mathbf{0_v}$ of $\mathbf{V}$.

Proof:

Let $\mathbf{U} \subset \mathbf{V}$ and $u \in \mathbf{U}$ be an arbitrary vector in $\mathbf{U}$ and $\mathbf{0_v} \in \mathbf{V}$ be the zero vector of $\mathbf{V}$.

Let $0 \in \mathbf{F}$ be the zero scalars.

As we know, $\mathbf{0u} = \mathbf{0_v}$ (as $\mathbf{U}$ is closed under scalar multiplication).

So, it implies that $\mathbf{0_v}$ is in $\mathbf{U}$.

Remark:

This theorem tells us, for example, that all subspaces of R^3 containing the zero vector, i.e., (0,0,0), which means that all subspaces of a three-dimensional space pass through the origin. This theorem can sometimes be used as a quick check to show that any subsets cannot be subspaces.

That is, if a given subset does not contain the zero vector $\mathbf{0_v}$, then it cannot be a subspace of a vector space $\mathbf{V}$.

Example 3.6:

Let $\mathbf{U} = \{(a, a, a + 2) : a \in R\mathbb{R}\}$ and $\mathbf{U} \subset \mathbf{V}$. Then show that $\mathbf{U}$ is not a subspace of R^3.

Solution:

First, we check about the presence of zero vector in $\mathbf{U}$, i.e., whether $(0, 0, 0)$ is in $\mathbf{U}$ or not.

For this, we have to check whether there a value of a for which $(a, a, a + 2)$ is $(0, 0, 0)$.

On equating $\mathbf{U}$ to $\mathbf{0_v} = (0, 0, 0)$, we get

$$(a, a, a + 2) = (0, 0, 0)\,.$$

On equating the corresponding components, it gives

$a = 0$ and $a + 2 = 0 \Rightarrow a = -2$.

This implies that this system of the equation has no solution.

Thus, $\mathbf{0_v} = (0, 0, 0)$ is not an element of $\mathbf{U}$. Hence, $\mathbf{U}$ is not a subspace of $\mathbf{V} = R^3$.

3.5 Linear Combinations

Definition 3.3:

Let $v_1, v_2, \cdots, v_n$ be vectors in a vector space $\mathbf{V}$. Then the vector $v \in \mathbf{V}$ is a linear combination of the vectors $v_1, v_2, \cdots, v_n$; if there exist scalars $c_1, c_2, \cdots, c_n$, such that $v = c_1 v_1, + c_2 v_2 + \cdots + c_n, v_n$.

Example 3.7:

The vector $(7, 3, 2)$ is a linear combination of the vectors $(1, 3, 0)$ and $(2, -3, 1)$ because it can be written as a linear combination of the vectors $(1, 3, 0)$ and $(2, -3, 1)$, i.e., $(7, 3, 2) = 3\,(1, 3, 0) + 2\,(2, -3, 1)$.

Example 3.8:

The vector $(3, 4, 2)$ is not a linear combination of the vectors $(1, 1, 0)$ and $(2, 3, 0)$ because there are no values of c_1 and c_2 for which the vector $(3, 4, 2)$ can be expressed as a linear combination of $(1, 1, 0)$ and $(2, 3, 0)$, i.e., $(3, 4, 2) \neq c_1\,(1, 1, 0) + c_2\,(2, 3, 0)$ is true.

3.6 Spanning a Vector Space

Definition 3.4:

Let $v_1, v_2, \cdots, v_m$ be a set of m vectors in a vector space $\mathbf{V}$. If every vector $v \in \mathbf{V}$ can be expressed as a linear combination of the set of vectors $v_1, v_2, \cdots, v_m$, then these vectors $v_1, v_2, \cdots, v_m$ span the vector space $\mathbf{V}$.

Example 3.9:

The vectors $(1, 0, 0)$, $(0, 1, 0)$, and $(0, 0, 1)$ span R^3. Since any arbitrary vector (a, b, c) of R^3 can be expressed as a linear combination of these vectors $(1, 0, 0)$, $(0, 1, 0)$, and $(0, 0, 1)$, i.e.,

$$(a, b, c) = a\,(1, 0, 0) + b\,(0, 1, 0) + c\,(0, 0, 1)\,.$$

It implies the set of vectors $(1, 0, 0)$, $(0, 1, 0)$, and $(1, 1, 1)$ also spans R^3, as (a, b, c) can be expressed as a linear combination of the set of vectors $(1, 0, 0)$, $(0, 1, 0)$, and $(1, 1, 1)$, i.e.,

$$(a, b, c) = (a - c)\,(1, 0, 0) + (b - c)\,(0, 1, 0) + c\,(1, 1, 1)\,.$$

But the vectors $(1, 0, 0)$, $(0, 2, 0)$, and $(3, 4, 0)$ do not span R^3 because any vector $(a, b, c) \in R^3$ cannot be written as a linear combination of these vectors $(1, 0, 0)$, $(0, 2, 0)$, and $(3, 4, 0)$.

Similarly, the vectors $(1,1,0)$ and $(0,0,1)$ span the subspace R^3 that consists of vectors of the form (a,a,b) because we can write $(a,a,b) = a\,(1,1,0) + b\,(0,0,1)$.

Example 3.10:

Show that the vectors $(1,2,0)\,,(0,1,-1)$, and $(1,1,2)$ span R^3.

Solution:

Next, we determine whether any arbitrary vector of R^3 can be expressed as a linear combination of the given vectors $(1,2,0)\,,(0,1,-1)$, and $(1,1,2)$.

Let (x,y,z) be an arbitrary element of R^3.

We have to determine whether we can write

$$\begin{array}{l} (x,y,z) = c_1\,(1,2,0)\, + c_2\,(0,1,-1) + c_3\,(1,1,2) \\ \Rightarrow (x,y,z) = (c_1 + c_3, 2c_1 + c_2 + c_3, -c_2 + 2c_3) \end{array}.$$

Thus,

$$c_1 + c_3 = x, 2c_1 + c_2 + c_3 = y, -c_2 + 2c_3 = z.$$

Using Gauss-Jordan elimination, it is found that

$$c_1 = 3x - y - z, c_2 = -4x + 2y + zc_3 = -2x + y + z.$$

Hence, the vectors $(1,2,0)\,,(0,1,-1)$, and $(1,1,2)$ span R^3.

3.7 Generating a Vector Space

Theorem 3.3:

Let $v_1, v_2, \cdots, v_m$ be a set of vectors in a vector space $\mathbf{V}$. Let $\mathbf{U}$ be the set consisting of all linear combination of the vectors $v_1, v_2, \cdots, v_m$. Then $\mathbf{U}$ is a subspace of $\mathbf{V}$ spanned by these vectors $v_1, v_2, \cdots, v_m$.

That is, the set $\mathbf{U}$ is said to be the vector subspace of a vector space $\mathbf{V}$ generated by $v_1, v_2, \cdots, v_m$, and it is denoted as $Span\,[v_1, v_2, \cdots, v_m]$.

Proof:

Let $u_1 = a_1v_1 + a_2v_2 + \cdots + a_mv_m$ and $u_2 = b_1v_1 + b_2v_2 + \cdots + b_mv_m$ be any arbitrary elements of $\mathbf{U}$. Then

$$\begin{array}{l} u_1 + u_2 = (a_1v_1 + a_2v_2 + \cdots + a_mv_m) + (b_1v_1 + b_2v_2 + \cdots + b_mv_m) \\ \quad = (a_1 + b_1)\,v_1 + (a_2 + b_2)\,v_2 + \cdots + (a_m + b_m)\,v_m \end{array}.$$

It implies that $u_1 + u_2$ is a linear combination of $v_1, v_2, \cdots, v_m$.

It implies that $u_1 + u_2 \in \mathbf{U}$ is in $\mathbf{U}$.

Thus, $\mathbf{U}$ is closed under vector addition.

Let k be an arbitrary scalar. Then

$$\begin{aligned} ku_1 &= k\left(a_1 v_1 + a_2 v_2 + \cdots + a_m v_m\right) \\ &= ka_1 v_1 + ka_2 v_2 + \cdots + ka_m v_m \end{aligned},$$

is a linear combination of $v_1, v_2, \cdots, v_m$.

It implies that ku_1 is in $\mathbf{U}$.

Thus, $\mathbf{U}$ is closed under scalar multiplication.

It implies that $\mathbf{U}$ is a subspace of $\mathbf{V}$.

By the definition of $\mathbf{U}$, since every vector in the vector space $\mathbf{U}$ can be written as a linear combination of $v_1, v_2, \cdots, v_m$,

thus, $v_1, v_2, \cdots, v_m$ is in $Span$ $\mathbf{U}$.

Example 3.11:

Let v_1 and v_2 be two vectors in the vector space R^3.

The subspace $Span\left[v_1, v_2\right]$ generated by v_1 and v_2 is the set of all vectors of the form $c_1 v_1 + c_2 v_2$, as shown in Figure 3.1.

In general, this space is the plane defined by the vectors v_1 and v_2. Now if v_1 and v_2 are collinear, then the vector space will be the line defined by these vectors.

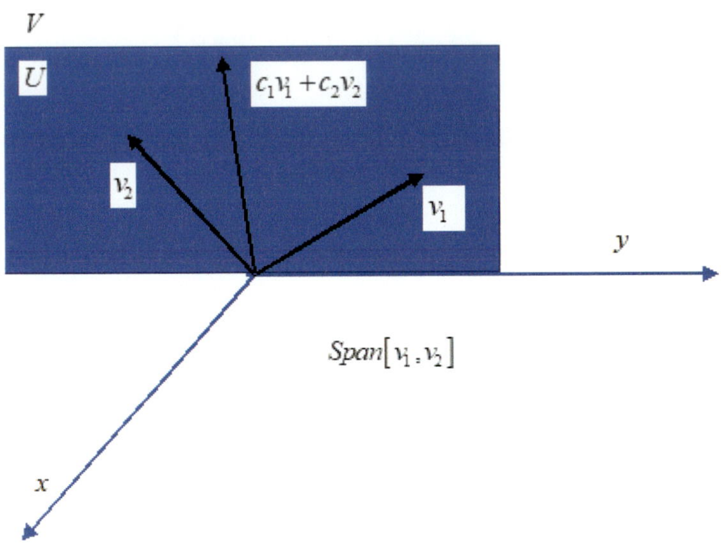

Figure 3.1 Spanning of vectors.

Example 3.12:

Let v_1 and v_2 be any two vectors that span a subspace of a vector space $\mathbf{U}$ of a vector space $\mathbf{V}$. Let k_1 and k_2 be non-zero scalars. Then show that k_1v_1 and k_2v_2 also span $\mathbf{U}$.

Solution:

Let $v \in \mathbf{U}$ be a vector in $\mathbf{U}$.
Since $v_1, v_2 \in Span\,\mathbf{U}$, there exist scalars a_1 and a_2 such that $v = a_1v_1 + a_2v_2$.
However, we can write

$$v = \frac{a_1}{k_1}(k_1v_1) + \frac{a_2}{k_2}(k_2v_2).$$

Thus, the vectors k_1v_1 and k_2v_2 are also in $Span\,\mathbf{U}$.

3.8 Finitely Generated Vector Spaces

A vector space $\mathbf{V}$ is *finitely generated* if there is a finite subset $\{v_1, v_2, \cdots, v_k\}$ of $\mathbf{V}$ such that $v = \langle v_1, v_2, \cdots, v_k \rangle$, i.e., if every vector $v \in \mathbf{V}$ can be expressed as a *linear combination* of vectors $v_1, v_2, \cdots, v_k$ and so has the form $c_1v_1 + c_2v_2 + \cdots + c_kv_k$, for some scalar c_i.

On the other hand, the vector space $\mathbf{V}$ is infinitely generated if there is no finite subset present in $\mathbf{V}$ that generates the vector space $\mathbf{V}$.

Example 3.13:

Show that the Euclidean space R^n forms a finitely generated vector space, i.e., R^n is finitely generated.

Proof:

Let $X_1, X_2, ..., X_n$ be the column vectors of the identity matrix $\mathbf{I_n}$ in R^n.

If $A = \begin{bmatrix} a_1 \\ a_2 \\ \vdots \\ a_n \end{bmatrix}$ is any vector in R^n, then $A = a_1X_1 + a_2X_2 + \cdots + a_nX_n$.

This implies that the column vectors $X_1, X_2, \cdots, X_n$ generate the vector space R^n, and, therefore, this vector space R^n is *finitely generated.*

Contrarily, one does not have to look far for *infinitely generated vector spaces*.

Example 3.14:

Show that the vector space $\mathbf{P}(X) = \{p(x) : x \in R\mathbb{R}\}$ of all real polynomials is *infinitely generated.*

Proof:

It can be proved by the method of proof by contradiction. Let us suppose that the vector space $\mathbf{P}(X)$ is finitely generated with polynomials $p_1(x), p_2(x), \cdots, p_k(x)$, and let us look for a contradiction.

Let us assume that $p_i(x) \neq 0,\ for\ all\ i = 1, 2, \cdots, k.$

Let m be the largest of their degrees. Then the degree of any linear combination of $p_1(x), p_2(x), \cdots, p_k(x)$ certainly cannot exceed m. But this means that x^{m+1} is not such a linear combination.

Therefore, $p_1(x), p_2(x), \cdots, p_k(x)$ do not generate $\mathbf{P}(X)$, and we have grasped a contradiction.

Hence proved.

3.9 Linear Dependence and Independence

Definition 3.5:

1. The set of vectors $\{v_1, v_2, \cdots, v_m\}$ in a vector space $\mathbf{V}$ is *linearly dependent* if there exist scalars $c_1, c_2, \cdots, c_m$, not all zero such that $c_1v_1 + c_2v_2 + \cdots + c_mv_m = 0_v$.
2. The set of vectors $\{v_1, v_2, \cdots, v_m\}$ in a vector space $\mathbf{V}$ is said to be linearly independent if $c_1v_1 + c_2v_2 + \cdots + c_mv_m = 0_v$ implies $c_1 = c_2 = \cdots = c_m = 0$, i.e., all $c_i^{\prime}s = 0$.

Example 3.15:

The set $\{(1,0,0), (0,1,0), (0,0,1)\}$ is said to be *linearly independent* in R^3, since

$$c_1(1,0,0) + c_2(0,1,0) + c_3(0,0,1) = (0,0,0)$$

can only be satisfied if $c_1 = 0, c_2 = 0$ and $c_3 = 0$.

Example 3.16:

Consider the set $\{(1,2,3), (5,1,0), (2,0,0)\}$.

The identity $c_1(1,2,3) + c_2(5,1,0) + c_3(2,0,0) = (0,0,0)$ leads to $c_1 = 0, c_2 = 0$ and $c_3 = 0$.

It implies that the vectors $(1,2,3), (5,1,0)$, and $(2,0,0)$ are linearly independent in R^3.

Example 3.17:

Consider the set $\{(4,1,0),(2,1,3),(0,1,2)\}$. It can be seen that

$$1.(4,1,0)-2(2,1,3)+3(0,1,2)=(0,0,0).$$

Thus, the vectors $(4,1,0)$, $(2,1,3)$, and $(0,1,2)$ are linearly dependent in R^3.

Theorem 3.4:

A set consisting of more than one vector in a vector space $\mathbf{V}$ is linearly dependent if and only if it is possible to express one of the vectors as a linear combination of other vectors in $\mathbf{V}$.

Proof:

Let us consider the set $\{v_1, v_2, ..., v_m\}$ to be linearly dependent. Therefore, there exist scalars $c_1, c_2, \cdots, c_m$, not all zeros, such that

$$c_1v_1 + c_2v_2 + \cdots + c_mv_m = 0_v.$$

Let us assume $c_1 \neq 0$.

Then the proceeding identity can be rewritten as

$$v_1 = \left(-\frac{c_2}{c_1}\right)v_2 + \cdots + \left(-\frac{c_m}{c_1}\right)v_m.$$

Thus, v_1 is a linear combination of $v_2, \cdots, v_m$.

Conversely:

Assume that v_1 is a linear combination of $v_2, \cdots, v_m$; therefore, there exist scalars $d_2, d_3, \cdots, d_m$ such that

$$v_1 = d_2v_2 + \cdots + d_mv_m.$$

Rewrite this equation as

$$1.v_1 + (-d_2)v_2 + \cdots + (-d_m)v_m = 0_v.$$

Thus, the set $\{v_1, v_2, + \cdots + v_m\}$ is linearly dependent.

Linear dependence of $\{v_1, v_2\}$:

The set $\{v_1, v_2\}$ is *linearly dependent* if and only if it is possible to write any one vector as a scalar multiple of the other vector.

Let $v_2 = cv_1$; then it implies that v_1, v_2 are collinear.

On the other hand, the set $\{v_1, v_2\}$ is linearly independent if it is not possible to express one vector as a multiple of the other, as shown in Figure 3.2.

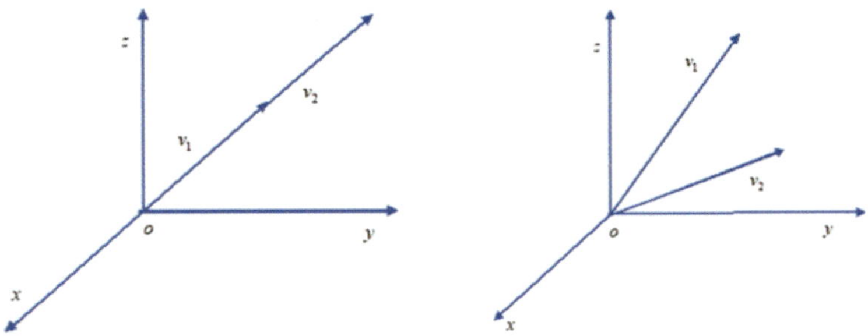

$\{v_1, v_2\}$ Linearly dependent; vectors lie on a line. **$\{v_1, v_2\}$ Linearly independent; vectors do not lie on a line**

Linearly dependent and independent of $\{v_1, v_2\}$ in R^2.

Figure 3.2 Dependency of Vectors.

Theorem 3.5:

Any set of vectors in a vector space $\mathbf{V}$ that contains the zero vector 0_v is linearly dependent.

Proof:

Let us consider the set $\{0_v, v_2, \cdots, v_m\}$, which contains the zero vector.

Let us examine the identity by considering the linear combinations of $0_v, v_2, \cdots, v_m$, i.e.,

$$c_1 0_v + c_2 v_2 + \cdots + c_m v_m = 0_v,$$

which shows that the identity is valid for

$c_2 = \cdots = c_m = 0$ and $c_1 \neq 0$ (i.e., not all zero).

Thus, the set of vectors $0_v, v_2, \cdots, v_m$ are linearly dependent.

Theorem 3.6:

Let us consider the set $\{v_1, v_2, \cdots, v_m\}$ to be linearly dependent in a vector space $\mathbf{V}$. Any collection of vectors in $\mathbf{V}$ that consists of $v_1, v_2, \cdots, v_m$ will also be linearly dependent.

Proof:

Let the set $\{v_1, v_2, \cdots, v_m\}$ be linearly dependent; so there exist scalars $c_1, c_2, \cdots, c_m$, not all zero such that $c_1 v_1 + c_2 v_2 + \cdots + c_m v_m = 0_v$.

Consider the extended set of n vectors $\{v_1, v_2, \cdots, v_m, v_{m+1}, \cdots, v_n\}$ that contains the given vectors $v_1, v_2, \cdots, v_m$.

With the choices of scalars, not all zero, namely $c_1, c_2, \cdots, c_m, 0, 0..0$, such that

$$c_1 v_1 + c_2 v_2 + \cdots + c_m v_m + 0.v_{m+1} + \cdots + 0.v_n = 0_v,$$

which implies $\{v_1, v_2, \cdots, v_m, v_{m+1}, \cdots, v_n\}$ are linearly dependent.

Example 3.18:

Show that the vectors $\begin{bmatrix} -1 \\ 2 \end{bmatrix}, \begin{bmatrix} 1 \\ 2 \end{bmatrix}, \begin{bmatrix} 2 \\ -4 \end{bmatrix}$ are linearly dependent on R^2.

Solution:

Consider the linear combination of the vectors

$$c_1 \begin{bmatrix} -1 \\ 2 \end{bmatrix} + c_2 \begin{bmatrix} 1 \\ 2 \end{bmatrix} + c_3 \begin{bmatrix} 2 \\ -4 \end{bmatrix} = \begin{bmatrix} 0 \\ 0 \end{bmatrix},$$

where c_1, c_2, and, c_3 are any scalars.

This is equivalent to the homogenous linear system

$$\begin{aligned} -c_1 + c_2 + 2c_3 &= 0 \\ 2c_1 + 2c_2 - 4c_3 &= 0. \end{aligned}$$

As the number of unknowns is higher than the number of equations, this system has a *non-trivial* solution.

Hence, the vectors are linearly dependent.

The following theorem discusses more the linear dependency of column vectors in a matrix equation.

Theorem 3.7:

Let $C_1, C_2, \cdots, C_m$ be vectors in the vector space R^n. Put $\mathbf{A} = \left[C_1 \vdots C_2 \vdots \cdots \vdots C_m \right]$, as $n \times m$ matrix. Then the vectors $C_1, C_2, \cdots, C_m$ are *linearly dependent* if and only if the number of pivots of $\mathbf{A}$ in row echelon form is fewer than m.

Proof:

Let us consider the identity

$$k_1 C_1 + k_2 C_2 + \cdots + k_m C_m = 0,$$

where $k_1, k_2, \cdots, k_m$ are scalars.

On equating the entries of the vectors on the left side of the equation to zero, it finds that this equation is similar to the homogenous linear system

$$\mathbf{A}\begin{bmatrix} k_1 \\ k_2 \\ \vdots \\ k_m \end{bmatrix} = 0.$$

Thus, the condition of the non-trivial solution $k_1, k_2, \cdots, k_m$ is that the number of pivots is fewer than m, i.e., $n < m$. Hence, this is the case for the set of column vectors to be linearly dependent as the number of row vectors n is less than the number of column vectors, i.e., m.

3.10 Properties of Bases

Theorem 3.8:

Let the vectors $v_1, v_2, \cdots, v_n$ span a vector space $\mathbf{V}$. Then each vector $v \in \mathbf{V}$ can be expressed uniquely as a linear combination of these vectors $v_1, v_2, \cdots, v_n$ if and only if the vectors $v_1, v_2, \cdots, v_n$ are linearly independent.

Proof:

Let us consider that the vectors $v_1, v_2, \cdots, v_n$ are linearly independent. Let $v \in \mathbf{V}$ be a vector in $\mathbf{V}$.

Since the vectors $v_1, v_2, \cdots, v_n$ span the vector space, we can express a vector $v \in \mathbf{V}$ as a linear combination of $v_1, v_2, \cdots, v_n$.

Let us consider that v can be described as a linear combination of these vectors in more than one way.

Suppose that we can write

$v = c_1v_1 + c_2v_2 + \cdots + c_nv_n$ and $v = d_1v_1 + d_2v_2 + \cdots + d_nv_n$.

Then

$$c_1v_1 + c_2v_2 + \cdots + c_nv_n = d_1v_1 + d_2v_2 + \cdots + d_nv_n.$$

It implies that

$$(c_1 - d_1)\, v_1 + (c_2 - d_2)\, v_2 + \cdots + (c_n - d_n)\, v_n = 0_v.$$

Since $v_1, v_2, \cdots, v_n$ are linearly independent, it implies that

$$c_1 - d_1 = 0\,, c_2 - d_2 = 0, \cdots, c_n - d_n = 0,$$

which implies that

$$c_1 = d_1, c_2 = d_2, \cdots, c_n = d_n.$$

Thus, it can be concluded that there is only one way of expressing v as a linear combination of the vectors $v_1, v_2, \cdots, v_n$.

Conversely:

Let v be a vector in $\mathbf{V}$. Let us assume that v can be written in only one way as a linear combination of the vectors $v_1, v_2, \cdots, v_n$.

Note that

$$0v_1 + 0v_2 + \cdots + 0v_n = 0_v.$$

This must be the only way that 0_v can be written as a linear combination of $v_1, v_2, \cdots, v_n$.

Thus, $c_1v_1 + c_2v_2 + \cdots + c_nv_n = 0_v$ can only be satisfied when $c_1 = 0, c_2 = 0, \cdots, c_n = 0$.

Hence, it implies that $v_1, v_2, \cdots, v_n$ are linearly independent.

3.11 Basis and Dimensions

Definition 3.6:

A finite set of vectors $\{v_1, v_2, \cdots, v_n\}$ is a basis for a vector space $\mathbf{V}$ if the set $\{v_1, v_2, \cdots, v_n\}$ spans $\mathbf{V}$ and is linearly independent.

That is, each vector $v \in \mathbf{V}$ can be expressed particularly as a linear combination of the vectors $v_1, v_2, \cdots, v_n$ in a basis.

Example 3.19:

The set $\{(1,0,0), (0,1,0), (0,0,1)\}$ spans R^3 and is linearly independent. It implies that it is a standard basis for R^3.

Example 3.20:

The set $\{(1,2,0), (1,1,-2), (0,1,-1)\}$ also spans R^3 and is linearly independent and implies that it is a basis for R^3.

Example 3.21:

The set $\{x^2 + 1, 3x - 1, -4x + 1\}$ spans P_2 and is linearly independent. It is a basis for P_2.

Theorem 3.9:

Let $\{v_1, v_2, ..., v_n\}$ be a basis for a vector space $\mathbf{V}$. If $\{w_1, w_2, \cdots, w_m\}$ is a set of more than n-vectors in $\mathbf{V}$, then this set is linearly dependent.

Proof:

Consider the identity

$$c_1 w_1 + c_2 w_2 + \cdots + c_n w_m = 0_v. \tag{3.1}$$

We shall show that values of $c_1, c_2, \cdots, c_m$ not all zeros exist, satisfying the identity. Thus, it is proved that the vectors are linearly dependent.

The set $\{v_1, v_2, \cdots, v_n\}$ is a basis for $\mathbf{V}$. Thus, each of the vectors $w_1, w_2, ..., w_m$ can be expressed as a linear combination of $v_1, v_2, ..., v_n$.

Let

$$\begin{array}{l} w_1 = a_{11}v_1 + a_{12}v_2 + \cdots + a_{1n}v_n \\ w_2 = a_{21}v_1 + a_{22}v_2 + \cdots + a_{2n}v_n \\ \cdots \\ w_m = a_{m1}v_1 + a_{m2}v_2 + \cdots + a_{mn}v_n \end{array}.$$

Substituting for $w_1, w_2, \cdots, w_m$ in eqn (3.1), we get

$$\begin{array}{r} c_1 (a_{11}v_1 + a_{12}v_2 + ... + a_{1n}v_n) + c_2 (a_{21}v_1 + a_{22}v_2 + \cdots + a_{2n}v_n) \\ + ... + c_n (a_{m1}v_1 + a_{m2}v_2 + \cdots + a_{mn}v_n) = 0_v \end{array}.$$

Rearranging, we get

$$\begin{array}{r} v_1 (c_1a_{11} + c_2a_{21} + ... + c_na_{m1}) + v_2 (c_1a_{12} + c_2a_{22} + \cdots + c_na_{m2}) \\ + \cdots + v_n (c_1a_{1n} + c_2a_{2n} + \cdots + c_na_{mn}) = 0_v \end{array}.$$

Since $v_1, v_2, ...v_n$ are linearly independent, this identity can only be satisfied if all the coefficients $v_1, v_2, ...v_n$ are zero. Thus,

$$\begin{array}{l} c_1a_{11} + c_2a_{21} + \cdots + c_na_{m1} = 0 \\ c_1a_{12} + c_2a_{22} + \cdots + c_na_{m2} = 0 \\ \cdots \\ c_1a_{1n} + c_2a_{2n} + ... + c_na_{mn} = 0 \end{array}.$$

Thus, finding $c_i's$ that satisfy eqn (3.1) reduces the solution to this system of $n-$equations in $m-$variables.

Since $m > n$, the number of variables is greater than the number of the equation. We know that such a system of the homogenous equation has many solutions. There are, therefore, non-zero values of $c_i's$ that satisfies eqn (3.1).

Thus, the set $\{w_1, w_2, \cdots, w_m\}$ is linearly dependent.

Theorem 3.10:

Let $\mathbf{V}$ be a vector space. All bases for a vector space $\mathbf{V}$ have the same number of vectors.

Proof:

Suppose $\{v_1, v_2, \cdots, v_n\}$ and $\{w_1, w_2, \cdots, w_m\}$ are two bases for a vector space $\mathbf{V}$. Then, if we assume $\{v_1, v_2, \cdots, v_n\}$ as a basis for $\mathbf{V}$ and $\{w_1, w_2, \cdots, w_m\}$ as a set of linearly independent vectors in $\mathbf{V}$, then *Theorem 3.9* tells us that $m \leq n$.

Conversely:

Let us assume $\{w_1, w_2, \cdots, w_m\}$ as a basis for $\mathbf{V}$ and $\{v_1, v_2, \cdots, v_n\}$ as a set of linearly independent vectors then $n \leq m$.

Thus, $n = m$, which proves that both the bases of $\mathbf{V}$ consist of the same number of vectors.

Definition 3.7:

The dimension of a vector space defines the number of basis vectors present in a vector space $\mathbf{V}$, i.e. if a vector space $\mathbf{V}$ has a basis consisting of $n-$vectors, then the dimension of the vector space $\mathbf{V}$ is said to be n.

Note: $\dim(\mathbf{V})$ for the dimension of $\mathbf{V}$.

Example 3.22:

Consider the set of vectors $\{(1, -2, 3), (2, 3, 1)\}$ in R^3.

The vectors $(1, -2, 3)$ and $(2, 3, 1)$ generate a subspace $\mathbf{U}$ of R^3 consisting of all vectors of the form

$$v = k_1(1, -2, 3) + k_2(2, 3, 1).$$

Thus, the vectors $(1, -2, 3)$ and $(2, 3, 1)$ span this subspace $\mathbf{U}$ of R^3.

Similarly, the second vector $(2, 3, 1)$ is not a scalar multiple of the first vector $(1, -2, 3)$; so these vectors are linearly independent.

Thus, $\{(1, -2, 3), (2, 3, 1)\}$ forms a basis for $\mathbf{V}$.

Hence, $\dim(U) = 2$.

Theorem 3.11:

(1) The origin $(0, 0, 0)$ is a subspace of R^3, and hence the dimension of this subspace is zero.

(2) The one-dimensional subspaces of the vector space R^3 are the lines through the origin.

(3) The two-dimensional subspaces of the vector space R^3 are the planes through the origin.

Proof:

(1) Let $\mathbf{U}$ be the set consisting of a single element $\{(0, 0, 0)\}$, i.e., the zero vector of $\mathbf{V} = R^3$.

Let $k \in F$ be an arbitrary scalar.
Since $(0, 0, 0) + (0, 0, 0) = (0, 0, 0)$
and $k\,(0, 0, 0) = (0, 0, 0)$,
it implies that the set **U** is closed under the operation of addition and scalar multiplication.
Thus, **U** is a subspace of R^3.
Hence, the dimension of this subspace **U** is defined to be zero.

(2) Let $\{u\}$ be a basis for a one-dimensional subspace **U** of R^3.
Each vector u in **U** is thus of the form cu, for some scalar c.
These vectors form a line that passes through the origin.

(3) Let $\{u_1, u_2\}$ be a basis for a two-dimensional subspace **U** of R^3.
Since $\{u_1, u_2\}$ is a basis of $\mathbf{V} = R^3$, every vector in **U** is of the form $c_1u_1 + c_2u_2$ and thus **U** is a plane through the origin $(0, 0, 0)$, as shown in Figure 3.3

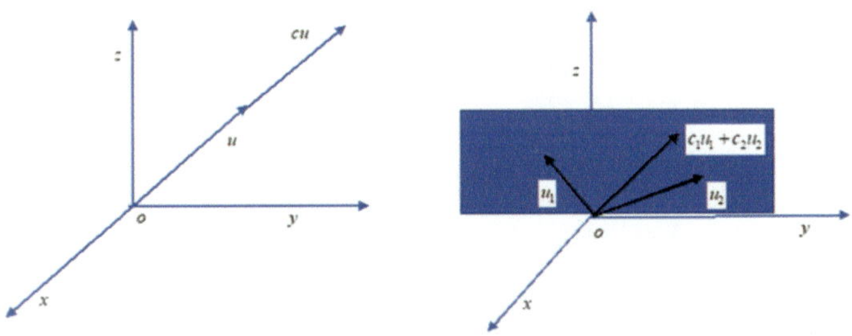

One-dimensional subspace of R^3 with a basis $\{u\}$ is a line through the origin.

The two-dimensional subspace of R^3 with a basis $\{u_1, u_2\}$ is a plane through the origin.

Figure 3.3 One- and two-dimensional subspace of R^3.

Theorem 3.12:

Let **V** be an n-dimensional vector space.

(1) If $\mathbf{U} = \{u_1, u_2, \cdots, u_n\}$ is a set of n-linearly independent vectors in **V**, then **U** is a basis for vector space **V**.

(2) If $\mathbf{U} = \{u_1, u_2, \cdots, u_n\}$ is a set of n-vectors that span the vector space **V**, then **U** is a basis for vector space **V**.

Theorem 3.13:

Let us consider an n-dimensional vector space **V**. Let $\{v_1, v_2, \cdots, v_m\}$ be a set of m linearly independent vectors in the vector space **V**, where $m < n$. Then there exist vectors $v_{m+1}, v_{m+2}, \cdots, v_n$ such that $\{v_1, v_2, \cdots, v_m, v_{m+1}, v_{m+2}, \cdots, v_n\}$ is a basis of $\mathbf{V}$.

Proof:

Since $m < n$, $\{v_1, v_2, \cdots, v_m\}$ cannot be a basis of $\mathbf{V}$.

Thus, there exists a vector v_{m+1} in $\mathbf{V}$, which does not lie in the subspace generated by $v_1, v_2, \cdots, v_m$.

The set $\{v_1, v_2, \cdots, v_m, v_{m+1}\}$ will be linearly independent.

Now if $m + 1 = n$, then $\{v_1, v_2, \cdots, v_m, v_{m+1}\}$ is a basis of $\mathbf{V}$.

If $m+1 < n$, there will be a vector v_{m+2} that does not lie in the subspace generated by $\{v_1, v_2, \cdots, v_m, v_{m+1}, v_{m+2}\}$.

If $m + 2 = n$, then $\{v_1, v_2, \cdots, v_m, v_{m+1}, v_{m+2}\}$ is a basis for $\mathbf{V}$.

One thus continues adding vectors until a basis is found.

3.12 Rank

Definition 3.8:

Let $\mathbf{A}$ be a $m \times n$ matrix. The rows of the matrix $\mathbf{A}$ may be outlined as row vectors $R_1, R_2, \cdots, R_m$ and the columns as column vectors $C_1, C_2, \cdots, C_n$. Each row vector $R_i, (i = 1, 2, \cdots, n)$ will have $n-$components, and each column vector will have $m-$components. The row vectors $R_i, (i = 1, 2, \cdots, n)$ will span a subspace of the vector space $\mathbf{V} = R^n$ called the *row space of* $\mathbf{A}$, while the column vectors $C_j, (j = 1, 2, \cdots, m)$ will span a subspace called the column space $\mathbf{A}$.

Example 3.23:

Consider the matrix

$$A = \begin{bmatrix} 1 & -2 & 1 & 0 \\ 3 & -4 & 1 & 6 \\ 5 & 3 & -1 & 2 \end{bmatrix}.$$

Solution:

Let us consider the row vectors of the matrix $\mathbf{A}$ are

$$\begin{aligned} R_1 &= (1, -2, 1, 0) \\ R_2 &= (3, -4, 1, 6) \\ R_3 &= (5, 3, -1, 2) \end{aligned}.$$

The row vectors R_1, R_2 and R_3 span a subspace $\mathbf{U}$ of $\mathbf{V} = R^4$ called the row space of the matrix $\mathbf{A}$. Similarly, the column vectors C_1, C_2, C_3 and C_4

of the matrix **A** are

$$C_1 = \begin{bmatrix} 1 \\ 3 \\ 5 \end{bmatrix}, \ C_2 = \begin{bmatrix} -2 \\ -4 \\ 3 \end{bmatrix}, \ C_3 = \begin{bmatrix} 1 \\ 1 \\ -1 \end{bmatrix}, \ C_4 = \begin{bmatrix} 0 \\ 6 \\ 2 \end{bmatrix}.$$

These column vectors C_1, C_2, C_3 and C_4 span a subspace **U** of $\mathbf{V} = R^4$ called the *column space* of the matrix **A**.

Theorem 3.14:

The row space and the column space of a matrix **A** have the same dimension.

Proof:

Let the row vectors of **A** be $R_1, R_2, \cdots, R_m$ and the ith vector be $R_i = (a_{i1}, a_{i2}, \cdots, a_{in})$.

Let r be the dimension of the row space, and the vectors $v_1, v_2, \cdots, v_r$ form a basis for the row space.

Let the jth vector of this set be $v_j = (d_{j1}, d_{j2}, \cdots, d_{jn})$.

Each of the row vectors of the matrix **A** is a linear combination of $v_1, v_2, \cdots, v_r$.

Let

$$\begin{aligned} R_1 &= c_{11}v_1 + c_{12}v_2 + \cdots + c_{1r}v_r \\ R_2 &= c_{21}v_1 + c_{22}v_2 + \cdots + c_{2r}v_r \\ &\cdots \\ R_m &= c_{m1}v_1 + c_{m2}v_2 + \cdots + c_{mr}v_r \end{aligned}.$$

Equating the ith components of the vectors on the left and right, we get

$$\begin{aligned} a_{1i} &= c_{11}d_{1i} + c_{12}d_{2i} + \cdots + c_{1r}d_{ri} \\ a_{2i} &= c_{21}d_{1i} + c_{22}d_{2i} + \cdots + c_{2r}d_{ri} \\ &\cdots \\ a_{mi} &= c_{m1}d_{1i} + c_{m2}d_{2i} + \cdots + c_{mr}d_{ri} \end{aligned}.$$

This may be written as

$$\begin{bmatrix} a_{1i} \\ a_{2i} \\ \vdots \\ a_{mi} \end{bmatrix} = d_{1i}\begin{bmatrix} c_{11} \\ c_{21} \\ \vdots \\ c_{m1} \end{bmatrix} + d_{2i}\begin{bmatrix} c_{12} \\ c_{22} \\ \vdots \\ c_{m2} \end{bmatrix} + \cdots + d_{ri}\begin{bmatrix} c_{1r} \\ c_{2r} \\ \vdots \\ c_{mr} \end{bmatrix},$$

which implies that each column vector of **A** lies in a space spanned by a single set of r vectors.

Sincer is the dimension of the row space of **A**, we get

$$\dim(\text{Column Space of } A) \leq \dim(\text{Row Space of } A).$$

By similar reasoning, we can show that

$$\dim(\text{Row Space of } A) \leq \dim(\text{Column Space of } A).$$

Combining these two results, we see that

$$\dim(\text{Row Space of } A) = \dim(\text{Column Space of } A).$$

Definition 3.9:

The dimension of the *row space* and the *column space* of a matrix **A** is called the Rank of **A**. The Rank of **A** is denoted as Rank$(\mathbf{A})$ or $\rho(\mathbf{A})$.

Example 3.24:

Find the Rank of the matrix

$$\mathbf{A} = \begin{bmatrix} -1 & 2 & 3 \\ 1 & 1 & -2 \\ 2 & -5 & 8 \end{bmatrix}.$$

Solution:

$$\text{Rank}(\mathbf{A}) = 3.$$

Note: It is not applicable in a higher-order matrix.

Theorem 3.15:

The non-zero row vectors of a matrix **A** are in *reduced echelon form* (REF), a basis for the row space of the matrix **A**. Therefore, Rank **A** is the number of non-zero row vectors.

Proof:

Let **A** be a $m \times n$ matrix with its non-zero row vectors be $R_1, R_2, \cdots, R_L$.
Let us consider the identity
$k_1 R_1 + k_2 R_2 + \cdots + k_L R_L = \bar{0}$, where $k_1, k_2, \cdots, k_L$ are scalars.

The first non-zero element of R_1 is one, and is the only one, of the row vectors to have a non-zero number.

Thus, upon adding the vectors $k_1 R_1, k_2 R_2, \cdots, k_L R_L$, we get a vector whose first component is k_1.

On equating this vector to zero, we get $k_1 = 0$.

The identity then reduces to

$$k_2R_2 + \cdots + k_LR_L = \bar{0}.$$

The first non-zero element of R_2 is 1, and it is the only one of these remaining row vectors with a non-zero number in this component. Thus, $k_2 = 0$. Similarly, $k_3, k_4, \cdots, k_L$ are all zero.

The vectors $R_1, R_2, \cdots, R_L$ are therefore linearly independent. These vectors span the row space of $\mathbf{A}$. Thus, $R_1, R_2, \cdots, R_L$ form a basis for the row space of the matrix $\mathbf{A}$.

The dimension of the row space is L. Thus, the Rank of $\mathbf{A}$ is L, i.e., the number of non-zero row vectors in $\mathbf{A}$.

Example 3.25:

Find the Rank of the matrix

$$\mathbf{A} = \begin{bmatrix} 1 & -2 & 0 & 0 \\ 0 & 0 & 1 & 0 \\ 0 & 0 & 0 & 1 \\ 0 & 0 & 0 & 0 \end{bmatrix}.$$

Solution:

The matrix $\mathbf{A}$ is in reduced echelon form. Here the three non-zero row vectors, namely $(1, -2, 0, 0)$, $(0, 0, 1, 0)$, and $(0, 0, 0, 1)$, form a basis for the row space of $\mathbf{A}$. Therefore, by *Theorem 3.15*, the row vectors $(1, -2, 0, 0)$, $(0, 0, 1, 0)$ and $(0, 0, 0, 1)$ form a basis for the row space of the matrix $\mathbf{A}$.

Hence, $\text{Rank}(\mathbf{A}) = 3$.

Theorem 3.16:

Let us consider two row equivalent matrices $\mathbf{A}$ and $\mathbf{B}$. Then the row equivalent matrices $\mathbf{A}$ and $\mathbf{B}$ have the same row space, i.e., $\text{Rank}(\mathbf{A}) = \text{Rank}(\mathbf{B})$.

Proof

Since $\mathbf{A}$ and $\mathbf{B}$ are row equivalent matrices, the rows of the matrix $\mathbf{B}$ can be obtained from the rows of $\mathbf{A}$ through a sequence of elementary row operations. This implies that each row of the matrix $\mathbf{B}$ is a linear combination of the rows of the matrix $\mathbf{A}$. Thus, the row space of the matrix $\mathbf{B}$ is contained in the row space of the matrix $\mathbf{A}$.

Similarly, the rows of the matrix $\mathbf{A}$ can be obtained from the rows of the matrix $\mathbf{B}$ through a sequence of elementary row operations, which results in the row space of $\mathbf{A}$ being equal to the row space of $\mathbf{B}$, which shows that the row spaces of $\mathbf{A}$ and $\mathbf{B}$ are similar. Since the row spaces of matrices are equal to their Ranks, the Rank of both matrices $\mathbf{B}$ must be equal, which implies $\text{Rank}(\mathbf{A}) = \text{Rank}\,(\mathbf{B})$.

Theorem 3.17:

Let E be the *reduced echelon form* (REF) of a matrix $\mathbf{A}$. If the non-zero row vectors of E form a basis for the row space of the matrix $\mathbf{A}$, then the Rank of $\mathbf{A}$ is the number of non-zero row vectors in E.

Example 3.26:

Determine a basis for the row space of the following matrix

$$\mathbf{A} = \begin{bmatrix} 1 & 2 & 3 \\ 2 & 5 & 4 \\ 1 & 1 & 5 \end{bmatrix}$$

and hence find its Rank.

Solution:

Upon using the elementary row operations on the matrix $\mathbf{A}$ to get the reduced echelon form (REF) of the matrix $\mathbf{A}$, we get

$$\begin{bmatrix} 1 & 2 & 3 \\ 2 & 5 & 4 \\ 1 & 1 & 5 \end{bmatrix} \approx \begin{bmatrix} 1 & 2 & 3 \\ 0 & 1 & -2 \\ 0 & -1 & 2 \end{bmatrix} \approx \begin{bmatrix} 1 & 0 & 7 \\ 0 & 1 & -2 \\ 0 & 0 & 0 \end{bmatrix}.$$

As the two vectors $(1, 0, 7)$ and $(0, 1, -2)$ form a basis for the row space of $\mathbf{A}$, $\text{Rank}(\mathbf{A}) = 2$.

Example 3.27:

Determine a basis for the subspace $\mathbf{U}$ of $\mathbf{V} = \mathbf{R}^4$ spanned by the vectors $(1, 2, 3, 4)$, $(-1, -1, -4, -2)$, $(3, 4, 11, 8)$.

Solution:

Here we construct a matrix $\mathbf{A}$ by considering these vectors as their row vectors, i.e.,

$$\mathbf{A} = \begin{bmatrix} 1 & 2 & 3 & 4 \\ -1 & -1 & -4 & -2 \\ 3 & 4 & 11 & 8 \end{bmatrix}.$$

Upon using the elementary row operation on the matrix $\mathbf{A}$, the reduced echelon form of the matrix can be obtained as follows:

$$\begin{bmatrix} 1 & 2 & 3 & 4 \\ -1 & -1 & -4 & -2 \\ 3 & 4 & 11 & 8 \end{bmatrix} \approx \begin{bmatrix} 1 & 2 & 3 & 4 \\ 0 & 1 & -1 & 2 \\ 0 & -2 & 2 & -4 \end{bmatrix}.$$

$$\approx \begin{bmatrix} 1 & 0 & 5 & 0 \\ 0 & 1 & -1 & 2 \\ 0 & 0 & 0 & 0 \end{bmatrix}$$

It can be seen from the reduced echelon form of the matrix that the non-zero vectors of this reduced echelon form, namely, $(1, 0, 5, 0)$ and $(0, 1, -1, 2)$, form a basis for the subspace $\mathbf{U}$ of R^4.

Hence, the dimension of this subspace $\mathbf{U}$ is 2, i.e., $\dim \mathbf{U} = 2$.

Theorem 3.18:

Consider a non-homogenous linear system of equations $AX = B$ having m-linear equations in n-unknown variables.

(1) If $\rho\,[\mathbf{A} : \mathbf{B}] = \rho\,(\mathbf{A}) = r = n$, then the solution of $\mathbf{AX} = \mathbf{B}$ is unique.
(2) If $\rho\,[\mathbf{A} : \mathbf{B}] = \rho\,(\mathbf{A}) = r \neq n$, then the system $\mathbf{AX} = \mathbf{B}$ has infinitely many solutions.
(3) If $\rho\,[\mathbf{A} : \mathbf{B}] \neq \rho\,(\mathbf{A})$, then the solution of $\mathbf{AX} = \mathbf{B}$ does not exist.

Proof:

Let us consider the system of equation $\mathbf{AX} = \mathbf{B}$.

The system can be written as

$$x_1 \begin{bmatrix} a_{11} \\ a_{21} \\ \vdots \\ a_{m1} \end{bmatrix} + x_2 \begin{bmatrix} a_{12} \\ a_{22} \\ \vdots \\ a_{m2} \end{bmatrix} + \ldots\ldots + x_n \begin{bmatrix} a_{1n} \\ a_{2n} \\ \vdots \\ a_{mn} \end{bmatrix} = \begin{bmatrix} b_1 \\ b_2 \\ \vdots \\ b_m \end{bmatrix}.$$

That is,

$$\bar{a}_1 x_1 + \bar{a}_2 x_2 + \ldots + \bar{a}_n x_n = \bar{b}. \tag{3.2}$$

Thus, the existence and uniqueness of the solution depend upon whether $\bar{b}$ can be written as a linear combination of $\bar{a}_1, \bar{a}_2, \ldots, \bar{a}_n$ or this combination is unique or not.

Let us now look at three possibilities that can arise.

(1) Since the matrix of coefficient and augmented matrix Ranks are the same, $\bar{b}$ must be linearly dependent on $\bar{a}_1, \bar{a}_2, \cdots, \bar{a}_n$.
Furthermore, since the Rank is n, the vectors are linearly independent and form a basis for the column space of the augmented matrix.
Therefore, eqn (3.2) has a unique solution. Thus, the solution to the system is unique.

(2) Since the Ranks of the matrix of coefficients and the augmented matrix are the same, $\bar{b}$ must be linearly dependent on $\bar{a}_1, \bar{a}_2, \cdots, \bar{a}_n$.
However, since the vectors are linearly dependent, $\bar{b}$ can therefore be expressed in more than one way as a linear combination of $\bar{a}_1, \bar{a}_2, \cdots, \bar{a}_n$.
Thus, eqn (3.2) has many solutions, which implies that the solutions to the systems exist but are not unique.

(3) Since the Rank of the augmented matrix is not equal to the Rank of the coefficient, it implies that $\bar{b}$ is linearly dependent of $\bar{a}_1, \bar{a}_2, \cdots, \bar{a}_n$.
Thus, eqn (3.2) has no solution, which implies that a solution to the system does not exist.

Geometrical interpretation:

(1) If $\bar{a}_1, \bar{a}_2, \cdots, \bar{a}_n$ are a basis for $\mathbf{V}$ and $\bar{b}$ lies in $\mathbf{V}$, then the solution is unique.

(2) If $\bar{a}_1, \bar{a}_2, \cdots, \bar{a}_n$ are linearly dependent and $\bar{b}$ lies in $\mathbf{V}$, then there are many solutions.

(3) If $\bar{b}$ does not lie in $\mathbf{V}$, then there is no solution.

(4) Space spanned by the column vectors of $A, \bar{a}_1, \bar{a}_2, \cdots, \bar{a}_n$, as shown in Figure 3.4.

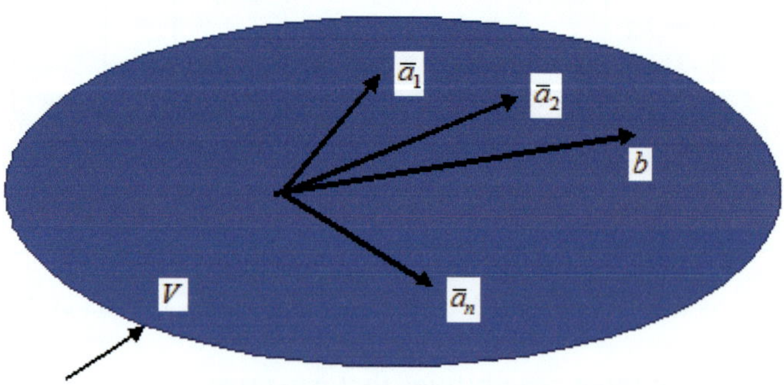

Figure 3.4 Geometrical interpretation of solutions to $\mathbf{AX} = \mathbf{B}$.

Summary of results:

Theorem 3.19:

Let $\mathbf{A}$ be an $n \times n$ matrix. Then the following statements are equivalent.

(1) The matrix $\mathbf{A}$ is non-singular, i.e., $|\mathbf{A}| \neq \mathbf{0}$.
(2) The matrix $\mathbf{A}$ is invertible.
(3) The matrix $\mathbf{A}$ is a row equivalent to an identity matrix $\mathbf{I_n}$.
(4) The non-homogenous system of linear equations $\mathbf{AX} = \mathbf{B}$ has a unique solution.
(5) The Rank of a matrix $\mathbf{A}$ is n, i.e., $\text{Rank}(\mathbf{A}) = n$.
(6) The column vectors of the matrix $\mathbf{A}$ form a basis for $\mathbf{V} = \mathbf{R^n}$.

Example 3.28:

Verify the summary of results as in Theorem 3.19 for the matrix $\mathbf{A}$, i.e.,

$$A = \begin{bmatrix} 1 & -1 & -2 \\ 2 & -3 & -5 \\ -1 & 3 & 5 \end{bmatrix}.$$

3.13 Sum and Intersection of Subspaces

Let $\mathbf{U_1}$ and $\mathbf{U_2}$ be the subspaces of a vector space $\mathbf{V}$. Then these two subspaces $\mathbf{U_1}$ and $\mathbf{U_2}$ can be combined in a natural way to form a new subspace of $\mathbf{V}$. The first one is the intersection of both these subspaces, i.e., $\mathbf{U_1} \bigcap \mathbf{U_2}$, the set of all vectors belonging to both $\mathbf{U_1}$ and $\mathbf{U_2}$.

The second one is the union of both these subspaces, i.e., $\mathbf{U_1} \bigcup \mathbf{U_2}$ that can be formed from $\mathbf{U_1}$ and $\mathbf{U_2}$ is not generally closed under addition. So, it may not be a subspace.

Finally, the subspace we are looking for is the sum $\mathbf{U_1} + \mathbf{U_2}$ defined as the set of all form vectors $\mathbf{U_1} + \mathbf{U_2} = \{u + v : u \in \mathbf{U_1}, v \in \mathbf{U_2}\}$.

Theorem 3.20:

Let $\mathbf{U_1}$ and $\mathbf{U_2}$ be the subspaces of a vector space $\mathbf{V}$. Then their intersection $\mathbf{U_1} \bigcap \mathbf{U_2}$ and the sum $\mathbf{U_1} + \mathbf{U_2}$ are also the subspaces of $\mathbf{V}$.

Proof:

Both $\mathbf{U_1}$ and $\mathbf{U_2}$ are closed under the operations of vector addition and scalar multiplication and contain the zero vector. Hence, their intersection $\mathbf{U_1} \bigcap \mathbf{U_2}$ also has the zero vector and is closed under vector addition and scalar multiplication operations.

Thus, $\mathbf{U_1} \bigcap \mathbf{U_2}$ is a subspace of a vector space $\mathbf{V}$.

Similarly, it can be checked for $\mathbf{U_1} + \mathbf{U_2}$.

Since both $\mathbf{U_1}$ and $\mathbf{U_2}$ contains the zero vector, clearly $\mathbf{U_1} + \mathbf{U_2}$ contains the zero vector, since $0_v + 0_v = 0_v$.

Now, if u_1, u_2 and w_1, w_2 are vectors in $\mathbf{U_1}$ and $\mathbf{U_2}$ respectively and k is any scalar, then

$$(u_1 + w_1) + (u_2 + w_2) = (u_1 + u_2) + (w_1 + w_2) \in \mathbf{U_1} + \mathbf{U_2}$$

and $c\,(u_1 + w_1) = cu_1 + cw_1 \in \mathbf{U_1} + \mathbf{U_2}$.

This implies that $\mathbf{U_1} + \mathbf{U_2}$ is closed under the operations of vector addition and scalar multiplication; so $\mathbf{U_1} + \mathbf{U_2}$ is a subspace of the vector space $\mathbf{V}$.

Example 3.29:

Let us consider the subspaces $\mathbf{U_1} = \begin{bmatrix} a \\ b \\ c \\ 0 \end{bmatrix}$ and $\mathbf{U_2} = \begin{bmatrix} 0 \\ d \\ e \\ f \end{bmatrix}$ of R^4 respectively. Here, a, b, c, d, e and f are arbitrary scalars.

Then their intersection $\mathbf{U_1} \bigcap \mathbf{U_2}$ consists of all vectors of the form $\begin{bmatrix} 0 \\ b \\ c \\ 0 \end{bmatrix}$.

While $\mathbf{U_1} + \mathbf{U_2}$ equals $\mathbf{V} = R^4$, Since every vector in R^4 can be expressed as the sum of a vector in $\mathbf{U_1}$ and a vector in $\mathbf{U_2}$.

Theorem 3.21:

Let $\mathbf{U_1}$ and $\mathbf{U_2}$ be the subspaces of a finitely generated vector space $\mathbf{V}$. Then $\dim(\mathbf{U_1} + \mathbf{U_2}) = \dim(\mathbf{U_1}) + \dim(\mathbf{U_2}) - \dim(\mathbf{U_1} \bigcap \mathbf{U_2})$.

Proof:

If $\mathbf{U_1} = \mathbf{0_V}$, then obviously

$\mathbf{U_1} + \mathbf{U_2} = \mathbf{U_2}$ and $\mathbf{U_1} \bigcap \mathbf{U_2} = \mathbf{0_V}$.

Here, the formula is undoubtedly true, when $\mathbf{U_2} = \mathbf{0_V}$.

Let us assume that $\mathbf{U_1} \neq \mathbf{0}$, $\mathbf{U_2} \neq \mathbf{0}$

and put $m = \dim(\mathbf{U_1})$ and $n = \dim(\mathbf{U_2})$.

Consider the first case, where $\mathbf{U_1} \bigcap \mathbf{U_2} = \mathbf{0_V}$.

Let $\{u_1, u_2, \cdots, u_m\}$ and $\{w_1, w_2, \cdots, w_n\}$ be bases of $\mathbf{U_1}$ and $\mathbf{U_2}$ respectively.

Then the vectors $u_1, u_2, \cdots, u_m, w_1, w_2, \cdots, w_n$ indeed generate $\mathbf{U_1} + \mathbf{U_2}$.

These vectors are also *linearly independent*; for if there is a linear relation between them, say

$$k_1u_1 + k_2u_2 + \cdots + k_mu_m + l_1w_1 + l_2w_2 + \cdots + l_nw_n = 0,$$

then

$$k_1u_1 + k_2u_2 + \cdots + k_mu_m = (-l_1)\, w_1 + (-l_2)\, w_2 + \cdots + (-l_n)\, w_n.$$

A vector that belongs to both $\mathbf{U_1}$, $\mathbf{U_2}$ and to $\mathbf{U_1} \bigcap \mathbf{U_2}$, is the zero subspace. Therefore, this vector must be the zero vector $\mathbf{0_V}$.

Consequently, all the k_i and l_j must be zero since u_i are linearly independent, as is w_j. Thus, the vectors $u_1, u_2, \cdots, u_m, w_1, w_2, \cdots, w_n$ form a basis for $\mathbf{U_1} + \mathbf{U_2}$.

So

$$\dim(\mathbf{U_1} + \mathbf{U_2}) = m + n = \dim(\mathbf{U_1}) + \dim(\mathbf{U_2})$$

(since $\mathbf{U_1} \bigcap \mathbf{U_2} = \mathbf{0_V}$).

Next, let us consider the case where $\mathbf{U_1} \bigcap \mathbf{U_2} \neq \mathbf{0_V}$.

Let us choose a basis for $\mathbf{U_1} \bigcap \mathbf{U_2}$, say $(z_1, z_2, \cdots, z_r)$.

Since **A** linearly independent subset of finitely generated vector space **V**, is contained in some basis of **V**.

So, the basis for $\mathbf{U_1} \bigcap \mathbf{U_2}$ can be extended to bases of $\mathbf{U_1}$ and of $\mathbf{U_2}$, say

$\{z_1, z_2, \cdots, z_r, u_{r+1}, u_{r+2}, \cdots, u_m\}$ and $\{z_1, z_2, \cdots, z_r, w_{r+1}, w_{r+2}, \cdots, w_n\}$ respectively.

Now the vectors

$$z_1, z_2, \cdots, z_r, u_{r+1}, u_{r+2}, \cdots, u_m, w_{r+1}, w_{r+2}, \cdots, w_n$$

generate $\mathbf{U_1} + \mathbf{U_2}$; for which we can express any vector of $\mathbf{U_1}$ or $\mathbf{U_2}$ in terms of them.

Next, to prove that they are linearly independent, let us consider the linear relation, i.e.,

$$\sum_{i=1}^{r} k_i z_i + \sum_{j=r+1}^{m} k_j u_j + \sum_{k=r+1}^{n} l_k w_k = 0\,,$$

where k_i, k_j, and l_k are scalars.

Then,

$$\sum_{k=r+1}^{n} l_k w_k = \sum_{i=1}^{r} (-k_i)\, z_i + \sum_{j=r+1}^{m} (-k_j)\, u_j,$$

belongs to both $\mathbf{U_1}$,$\mathbf{U_2}$ and so to $\mathbf{U_1} \bigcap \mathbf{U_2}$.

Thus, the vector $\sum l_k w_k$ can be expressible as a linear combination of the z_i and u_j since these vectors are known to form a basis of $\mathbf{U_1} \bigcap \mathbf{U_2}$.

However, $z_1, z_2, \cdots, z_r, w_{r+1}, w_{r+2}, \cdots, w_n$ are *linearly independent.* Therefore, all the l_j are zero, and our linear relation becomes

$$\sum_{i=1}^{r} k_i z_i + \sum_{j=r+1}^{m} k_j u_j = 0.$$

But since the vectors $z_1, z_2, \cdots, z_r, u_{r+1}, u_{r+2}, \cdots, u_m$ are linearly independent, it implies that k_j and k_i are also zero, which establishes the linear dependence.

We thus conclude that the set of vectors $z_1, z_2, \cdots, z_r, u_{r+1}, u_{r+2}, \cdots, u_m, w_{r+1}, w_{r+2}, \cdots, w_n$ forms a basis of $\mathbf{U_1} + \mathbf{U_2}$.

Hence,

$$\begin{aligned} \dim(\mathbf{U_1} + \mathbf{U_2}) &= r + (m - r) + (n - r) = m + n - r \\ &= \dim(\mathbf{U_1}) + \dim(\mathbf{U_2}) - \dim(\mathbf{U_1} \bigcap \mathbf{U_2}) \end{aligned}.$$

Example 3.30:

Consider that $\mathbf{U_1}$ and $\mathbf{U_2}$ subspaces of the vector space $\mathbf{V} = R^{10}$ have dimensions 6 and 8 respectively. Then, find the smallest possible dimension of $\mathbf{U_1} \bigcap \mathbf{U_2}$.

Solution:

Since $\dim\left(R^{10}\right) = 10$ and $\mathbf{U_1} + \mathbf{U_2}$ is a subspace of R^{10}, its dimension cannot exceed 10.

Therefore, by the previous theorem

$$\begin{aligned} \dim(\mathbf{U_1} \bigcap \mathbf{U_2}) &= \dim(\mathbf{U_1}) + \dim(\mathbf{U_2}) - \dim(\mathbf{U_1} + \mathbf{U_2}) \\ &\geq 6 + 8 - 10 = 4 \end{aligned}.$$

Thus, the dimension of $\mathbf{U_1} \bigcap \mathbf{U_2}$ is at least 4.

3.14 Direct Sums of Subspaces

Let $\mathbf{U_1}$ and $\mathbf{U_2}$ be two subspaces of a vector space $\mathbf{V}$. Then the vector space $\mathbf{V}$ is said to be the *direct sum* of $\mathbf{U_1}$ and $\mathbf{U_2}$, i.e., $\mathbf{V} = \mathbf{U_1} \oplus \mathbf{U_2}$, if $\mathbf{V} = \mathbf{U_1} + \mathbf{U_2}$ or $\mathbf{U_1} \bigcap \mathbf{U_2} = \mathbf{0}$.

A consequence of the definition:

Let $\mathbf{V}$ be a vector space. Then each vector $v \in \mathbf{V}$ has a unique expression of the form $v = u + w$, where $u \in \mathbf{U_1}$ and $w \in \mathbf{U_2}$.

Indeed, if there are two such expressions, i.e.,

$v = u_1 + w_1 = u_2 + w_2$ with $u_i \in \mathbf{U_1}$ and $w_i \in \mathbf{U_2}$.

Then $u_1 - u_2 = w_2 - w_1$ which belongs to $\mathbf{U_1} \bigcap \mathbf{U_2} = \mathbf{0}$; hence, $u_1 = u_2$ and $w_1 = w_2$.

Example 3.31:

Let $\mathbf{U_1}$ denote the subset of the vector space $\mathbf{V} = R^3$ consisting of all vectors of the form $\begin{bmatrix} a \\ b \\ 0 \end{bmatrix}$ and let $\mathbf{U_2}$ be the subset of all vectors of the form $\begin{bmatrix} 0 \\ 0 \\ c \end{bmatrix}$, where a, b, c are arbitrary scalars.

Then $\mathbf{U_1}$ and $\mathbf{U_2}$ are the subspaces of the vector space $\mathbf{V} = R^3$.

Besides,

$\mathbf{U_1} + \mathbf{U_2} = \mathbf{V} = R^3$ and $\mathbf{U_1} \bigcap \mathbf{U_2} = \mathbf{0}$.

Hence, $R^3 = \mathbf{U_1} \oplus \mathbf{U_2}$.

Theorem 3.22:

Let $\mathbf{V}$ be a finitely generated vector space and $\mathbf{U_1}$ and $\mathbf{U_2}$ be subspaces of the vector space $\mathbf{V}$ such that $\mathbf{V} = \mathbf{U_1} \oplus \mathbf{U_2}$.

Then

$$\dim(\mathbf{V}) = \dim(\mathbf{U_1}) + \dim(\mathbf{U_2}).$$

This follows from the theorem that $\dim(\mathbf{U_1} \bigcap \mathbf{U_2}) = 0$.

3.15 Direct Sums of More Than Two Subspaces

The direct sum of subspaces can be extended to an arbitrary number of subspaces.

Let $\mathbf{U_1}, \mathbf{U_2}, \cdots, \mathbf{U_k}$ be subspaces of a vector space $\mathbf{V}$.

Let us define the sum of these subspaces $\mathbf{U_1} + \mathbf{U_2} + \cdots + \mathbf{U_k}$ as the set of all vectors of the form $u_1 + u_2 + \cdots + u_k$, $u_i \in \mathbf{U_i}$.

This implies that it forms a subspace of $\mathbf{V}$.

The vector space $\mathbf{V}$ is said to be *the direct sum* of the subspaces $\mathbf{U_1}, \mathbf{U_2}, \cdots, \mathbf{U_k}$, i.e., $\mathbf{V} = \mathbf{U_1} \oplus \mathbf{U_2} \oplus ... \oplus \mathbf{U_k}$.

If the following holds

(1) $\mathbf{V} = \mathbf{U_1} + \mathbf{U_2} + \cdots + \mathbf{U_k}$

(2) For each $i = 1, 2, \cdots, k$, the intersection of $\mathbf{U_i}$ with the sum of all other subspaces $\mathbf{U_j}, j \neq i$ equals zero.

These are equivalent to desiring that each element of $\mathbf{V}$ can be expressible in an exclusive way as a sum of the form $u_1 + u_2 + \cdots + u_k$, where each $u_i \in \mathbf{U_i}$.

The practical approach of a direct sum is that it generally allows us to express a vector space as a direct sum of subspaces that are, in some sense, simpler.

3.16 Generating a Basis for a Direct Sum of Two Subspaces

Let us consider an n-dimensional vector space $\mathbf{V} = R^n$ defined over a field $\mathbf{F}$ and suppose a specifically ordered basis.

Assume that the vectors $u_1, u_2, \cdots, u_r$ and $w_1, w_2, \cdots, w_s$ are generating the subspaces of $\mathbf{U_1}$ and $\mathbf{U_2}$ respectively.

Claim: To determine bases for the subspaces $\mathbf{U_1} + \mathbf{U_2}$ and $\mathbf{U_1} \bigcap \mathbf{U_2}$ and hence to compute their dimension.

Initially the problem is to be translated to the vector space F^n.

Associate with each u_i and w_j, its coordinate column vectors C_i and D_j with respect to the given ordered basis of $\mathbf{V}$. Then $C_1, C_2, \cdots, C_r$ and $D_1, D_2, \cdots, D_s$ generates respective subspaces $\mathbf{U_1}$ and $\mathbf{U_2}$ of F^n.

From these bases $\mathbf{U_1}$ and $\mathbf{U_2}$, the bases for $\mathbf{U_1} + \mathbf{U_2}$ and $\mathbf{U_1} \bigcap \mathbf{U_2}$ can be find out.

So here let us assume that $\mathbf{V} = \mathbf{F^n}$ and let us take the case of $\mathbf{U_1} \bigcap \mathbf{U_2}$ first.

Let $\mathbf{A}$ be the matrix whose columns are $u_1, u_2, \cdots, u_r$. Also let B be the matrix whose columns are $w_1, w_2, \cdots, w_s$.

Then $\mathbf{U_1} + \mathbf{U_2}$ is defined the column space of the matrix $\mathbf{M} = [\mathbf{A} \vdots \mathbf{B}]$.

A basis for $\mathbf{U_1} + \mathbf{U_2}$ can therefore be found by putting $\mathbf{M}$ in reduced column echelon form and deleting the zero columns.

Next, consider the case of $\mathbf{U_1} \bigcap \mathbf{U_2}$.

Here for scalars k_i and l_j, every element of $\mathbf{U_1} \bigcap \mathbf{U_2}$, can be expressed as

$$k_1 u_1 + k_2 u_2 + \cdots + k_r u_r = l_1 w_1 + l_2 w_2 + \cdots + l_s w_s.$$

Equivalently, $k_1 u_1 + k_2 u_2 + \cdots + k_r u_r + (-l_1)\, w_1 + (-l_2)\, w_2 + \cdots + (-l_s)\, w_s = 0$.

Now, this equation asserts that the vector

$$\begin{bmatrix} k_1 \\ \vdots \\ k_r \\ -l_1 \\ \vdots \\ -l_s \end{bmatrix} \text{ belongs to the null space of } \left[\mathbf{A} \vdots \mathbf{B}\right].$$

A method for finding a basis for the null space of a matrix is described in an earlier section. To complete the process, read off the first r entries of each vector based on the null space of $\left[\mathbf{A} \vdots \mathbf{B}\right]$ and take these entries to be $k_1, k_2, \cdots, k_r$. The resulting vectors form a basis $\mathbf{U_1} \bigcap \mathbf{U_2}$.

Example 3.32:

Let $\mathbf{M} = \begin{bmatrix} 1 & 0 & 2 & 1 \\ 2 & 1 & 5 & 2 \\ 2 & -2 & -1 & -1 \\ 1 & 1 & 5 & 3 \end{bmatrix}$ and $\mathbf{U_1}$ and $\mathbf{U_2}$ be the subspaces of $\mathbf{V} = R^4$ generated by the columns C_1, C_2 and C_3, C_4 of $\mathbf{M}$ respectively. Find a basis for $\mathbf{U_1} + \mathbf{U_2}$.

Solution:

Upon applying the procedure for finding a basis of the column space of the matrix $\mathbf{M}$ by using the method of reduced echelon form, we obtain

$$\begin{bmatrix} 1 & 0 & 0 & 0 \\ 0 & 1 & 0 & 0 \\ 0 & 0 & 1 & 0 \\ 3 & -\frac{1}{3} & -\frac{2}{3} & 0 \end{bmatrix}.$$

The first three columns of this matrix form a basis of $\mathbf{U_1} + \mathbf{U_2}$.

Hence, $\dim\left(\mathbf{U_1} + \mathbf{U_2}\right) = 3$.

Example 3.33:

Determine a basis of $\mathbf{U_1} \bigcap \mathbf{U_2}$, where $\mathbf{U_1}$ and $\mathbf{U_2}$ are the subspaces of $\mathbf{V} = R^4$ generated by columns C_1, C_2 and by C_3, C_4 of $\mathbf{M}$ respectively. Here the matrix $\mathbf{M}$ is defined as

$$\mathbf{M} = \begin{bmatrix} 1 & 0 & 2 & 1 \\ 2 & 1 & 5 & 2 \\ 2 & -2 & -1 & -1 \\ 1 & 1 & 5 & 3 \end{bmatrix}.$$

Solution:

Upon following the above procedure, here the matrix $\mathbf{M}$ is in reduced row echelon form, i.e.,

$$\begin{bmatrix} 1 & 0 & 0 & -1 \\ 0 & 1 & 0 & -1 \\ 0 & 0 & 1 & 1 \\ 0 & 0 & 0 & 0 \end{bmatrix}.$$

From this, a basis for the null space $\mathbf{M}$ can be read off as described in the preceding paragraph. In our case, the basis has a single element.

$$\begin{bmatrix} 1 \\ 1 \\ -1 \\ 1 \end{bmatrix}.$$

Therefore, a basis for $\mathbf{U_1} \bigcap \mathbf{U_2}$ is obtained by taking the linear combination of the generating vectors of $\mathbf{U_1}$ corresponding to the scalars in the first two rows of this vector, that is to say

$$1. \begin{bmatrix} 1 \\ 2 \\ 2 \\ 1 \end{bmatrix} + 1. \begin{bmatrix} 0 \\ 1 \\ -2 \\ 1 \end{bmatrix} = \begin{bmatrix} 1 \\ 3 \\ 0 \\ 2 \end{bmatrix}.$$

Thus, $\dim(\mathbf{U_1} \bigcap \mathbf{U_2}) = 1$.

Example 3.34:

Determine the bases for the sum and intersection of the subspaces $\mathbf{U_1}$ and $\mathbf{U_2}$ of $\mathbf{P_4}(R)$ generated by the respective sets of polynomials $\{1 + 2x + x^3, 1 - x - x^2\}$ and $\{x + x^2 - 3x^3, 2 + 2x - 2x^3\}$.

Here the first step is to translate the problem to R^4, by writing down the coordinate columns of the given polynomials concerning the standard ordered basis $1, x, x^2, x^3$ of $\mathbf{P_4}(R)$.

Arrange as the columns of a matrix. These are

$$\mathbf{A} = \begin{bmatrix} 1 & 1 & 0 & 2 \\ 2 & -1 & 1 & 2 \\ 0 & -1 & 1 & 0 \\ 1 & 0 & -3 & -2 \end{bmatrix}.$$

Let $\mathbf{U^*}_\mathbf{1}$ and $\mathbf{U^*}_\mathbf{2}$ be the subspaces of R^4 generated by the coordinate columns of the polynomials that generate $\mathbf{U_1}$ and $\mathbf{U_2}$, that is, by columns C_1, C_2 and by C_3, C_4 of the matrix $\mathbf{A}$, respectively.

Now find bases for $\mathbf{U_1}^* + \mathbf{U_2}^*$ and $\mathbf{U_1}^* \bigcap \mathbf{U_2}^*$ just as in the previous example. It emerges that $\mathbf{U_1}^* + \mathbf{U_2}^*$, which is just the column space of $\mathbf{A}$, has a basis

$$\begin{bmatrix} 1 \\ 0 \\ 0 \\ -3 \end{bmatrix}, \begin{bmatrix} 0 \\ 1 \\ 0 \\ 2 \end{bmatrix}, \begin{bmatrix} 0 \\ 0 \\ 1 \\ -5 \end{bmatrix}.$$

On writing down the polynomials with these coordinate vectors, we obtain the basis

$1 - 3x^3, x + 2x^3,\ x^2 - 5x^3$ for $\mathbf{U_1}^* + \mathbf{U_2}^*$.

In the case of $\mathbf{U_1} \bigcap \mathbf{U_2}$, the procedure is to find a basis for $\mathbf{U_1}^* \bigcap \mathbf{U_2}^*$.

This turns out to consist of the single vector

$$\begin{bmatrix} 1 \\ 1 \\ 1 \\ -1 \end{bmatrix}.$$

Finally, read off that polynomial as

$$1.\left(1 + 2x + x^3\right) + 1.\left(1 - x - x^2\right) = 2 + x - x^2 + x^3$$

which forms a basis of $\mathbf{U_1} \bigcap \mathbf{U_2}$.

Exercises

1. Determine whether the vector $(8, 0, 5)$ is a linear combination of the vectors $(1, 2, 3)$, $(0, 1, 4)$, and $(2, -1, 1)$.

2. Determine whether the vector $(4,5,5)$ is a linear combination of the vectors $(1,2,3)$, $(-1,1,4)$, and $(3,3,2)$.
3. Consider the vectors $(-1,5,3)$ and $(2,-3,4)$ in R^3. Let $U = Span\left[(-1,5,3),(2,-3,4)\right]$. Then U will be a subspace of R^3 consisting of all vectors of the form $c_1(-1,5,3)+c_2(2,-3,4)$.
4. Determine whether the matrix $\begin{bmatrix} -1 & 7 \\ 8 & -1 \end{bmatrix}$ is a linear combination of $\begin{bmatrix} 1 & 0 \\ 2 & 1 \end{bmatrix}$, $\begin{bmatrix} 2 & -3 \\ 0 & 2 \end{bmatrix}$, and $\begin{bmatrix} 0 & 1 \\ 2 & 0 \end{bmatrix}$ in the vector space M_{22} of 2×2 matrices.
5. Show that the function $f(x) = 4x^2+3x-7$ lies in the space Span $\{g,h\}$ generated by $g(x)=2x^2-5$ and $h(x)=x+1$.
6. Determine whether the set $\{(1,2,0),(0,1,-1),(1,1,2)\}$ is linearly independent in R^3.
7. Show that the set $\left\{x^2+1, 3x-1, -4x+1\right\}$ is linearly independent in P_2.
8. Show that the set $\{x+1, x-1, -x+5\}$ is linearly dependent in P_1.
9. Let the set $\{v_1,v_2\}$ be linearly independent. Prove that $\{v_1+v_2, v_1-v_2\}$ is also linearly independent.
10. Prove that the set $\{(1,3,-1),(2,1,0),(4,2,1)\}$ is a basis for R^3.
11. Determine a basis for the column space of the following matrix $A = \begin{bmatrix} 1 & 1 & 0 \\ 2 & 3 & -2 \\ -1 & -4 & 6 \end{bmatrix}$.
12. Let U_1, U_2, U_3 be subspaces of the vector space $V = R^5$, which consists of all vectors of the form

$$\begin{bmatrix} 0 \\ 0 \\ a \\ 0 \\ 0 \end{bmatrix}, \begin{bmatrix} 0 \\ b \\ 0 \\ c \\ 0 \end{bmatrix}, \begin{bmatrix} d \\ 0 \\ 0 \\ 0 \\ e \end{bmatrix}$$ respectively, where a,b,c,d,e are arbitrary scalars.

Then $V = R^5 = U_1 \oplus U_2 \oplus U_3$. Find the bases for the sum and intersection of subspaces U_1, U_2, U_3.

4

Eigenvalues and Eigenvectors

This chapter discusses the properties of eigenvalues and eigenvectors along with some properties of inner product spaces. It also includes the Gram–Schmidt orthogonalization process and QR-factorization. Eigenvalues and eigenvectors are particular scalars and vectors associated with matrices. They can be expressed in terms of determinants. Many numerical methods make use of eigenvalues and eigenvectors. We shall use eigenvalues and eigenvectors to make long-term predictions of populations in analyzing oscillating systems and in studying the connectives of road networks.

4.1 Introduction

Eigenvalues and eigenvectors are particular scalars and vectors associated with matrices. These are used in many branches of the natural and social sciences and engineering.

General applications of eigenvectors are:

(1) To rank pages in the search engine Google.
(2) To use in demography to predict long-term trends.
(3) To use meteorology with an example of weather prediction for Tel Aviv.
(4) To study the oscillating system.

We commence our discussion with the definition of an eigenvalue and eigenvector.

Definition 4.1:

Let us consider an $n \times n$ square matrix $\boldsymbol{A}$. Then a scalar λ is said to be an *eigenvalue* of the matrix $\boldsymbol{A}$, if there exists a non-zero vector $\boldsymbol{X} \in \mathbb{R}^n$ such $\boldsymbol{AX} = \lambda\boldsymbol{X}$. The vector $\boldsymbol{X}$ is called an *eigenvector* corresponding to the *eigenvalue* λ.

The eigenvectors and eigenvalues have an actual application in the study of linear differential equations, the system of linear recurrence relation, the Markov process, and many more fields of study.

Let us look at the geometrical significance of an eigenvector that corresponds to a non-zero eigenvalue.

Geometrical significance:

The vector $\boldsymbol{AX}$ is in the same or opposite direction as $\boldsymbol{X}$ depending on the sign of λ.

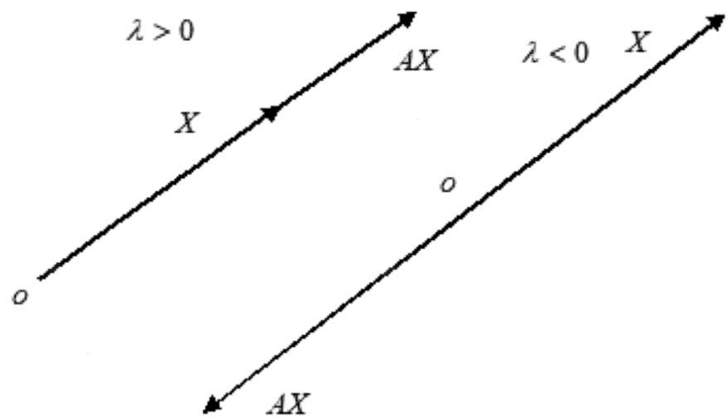

Figure 4.1 Direction of eigen vector $\boldsymbol{X}$ with vector $\boldsymbol{AX}$.
$\boldsymbol{X}$ is an eigenvector of $\boldsymbol{A}$. $\boldsymbol{AX}$ is in the same or opposite direction as $\boldsymbol{X}$.

4.2 Computation of Eigenvalues and Eigenvectors of a Matrix

Let λ be an eigenvalue and $\boldsymbol{X}$ be the corresponding eigenvector of an $n \times n$ matrix $\boldsymbol{A}$.

Thus,

$$\boldsymbol{AX} = \lambda \boldsymbol{X}\,, \text{where } \boldsymbol{X} \in R^n. \tag{4.1}$$

Eqn (4.1) may be rewritten as

$$\begin{aligned} &\boldsymbol{AX} - \lambda \boldsymbol{X} = \boldsymbol{0}_v \\ &\Rightarrow (\boldsymbol{A} - \lambda I_n)\boldsymbol{X} = \boldsymbol{0}_v \end{aligned}. \tag{4.2}$$

This matrix equation (4.2) represents a system of homogenous linear equations having the matrix of coefficient $(\boldsymbol{A} - \lambda I_n)$.

Eqn (4.2) implies either
$|\boldsymbol{A} - \lambda I_n| = \boldsymbol{0}$ or $\boldsymbol{X} = \boldsymbol{0}_v$.

Since the eigenvectors are defined to be the non-zero vectors, the nontrivial solutions to this system (5.2.2) can only exist if the matrix of the coefficient is singular, i.e.,

$$|\boldsymbol{A} - \lambda I_n| = 0. \tag{4.3}$$

Eqn (4.3) is called the *characteristic equation* of the matrix $\boldsymbol{A}$.

Upon solving eqn (4.3),

i.e., $|\boldsymbol{A} - \lambda I_n| = 0$ for λ.

It leads to all the eigenvalues of $\boldsymbol{A}$.

Next, on expanding the determinant $|\boldsymbol{A} - \lambda I_n|$, we get a polynomial in λ called the *characteristic polynomial* of $\boldsymbol{A}$. Then by substituting back the eigenvalues into the equation $(\boldsymbol{A} - \lambda I_n)\boldsymbol{X} = \boldsymbol{0}_v$, we can find the corresponding eigenvectors.

4.3 Properties of the Characteristic Polynomials, Eigenvalues, and Eigenvectors

Let us consider an $n \times n$-matrix $\boldsymbol{A} = [a_{ij}]$. The matrix $M = \boldsymbol{A} - \lambda I_n$ can be obtained by substituting λ down the diagonal of the matrix $\boldsymbol{A}$.

Here, I_n is the identity matrix of order n and λ is an intermediate to be determined.

The determinant of $\boldsymbol{A} - \lambda I_n$,

i.e., $\Delta(\lambda) = \det(\boldsymbol{A} - \lambda I_n)$

is a polynomial in λ of degree n called the *characteristic polynomial* of the matrix $\boldsymbol{A}$.

The characteristic polynomial of degrees 2 and 3:

There are elementary formulas for the characteristic polynomial of matrices of order 2 and 3.

(1) Let us consider $\boldsymbol{A} = \begin{bmatrix} a_{11} & a_{12} \\ a_{21} & a_{22} \end{bmatrix}$. Then

$$\begin{aligned}\Delta(\lambda) &= \lambda^2 - (a_{11} + a_{22}).\lambda + \det(\boldsymbol{A}) \\ &= \lambda^2 - tr(\mathbf{A}).\lambda + \det(\mathbf{A})\end{aligned},$$

where $\operatorname{tr}(\boldsymbol{A})$ denotes the *trace of* $\boldsymbol{A}$, which is the *sum of the diagonal elements* of the matrix $\boldsymbol{A}$.

(2) Suppose

$$\boldsymbol{A} = \begin{bmatrix} a_{11} & a_{12} & a_{13} \\ a_{21} & a_{22} & a_{23} \\ a_{31} & a_{32} & a_{33} \end{bmatrix}.$$

Then

$$\Delta(\lambda) = \lambda^3 - \text{tr}(\boldsymbol{A}).\lambda^2 + (\boldsymbol{A}_{11} + \boldsymbol{A}_{22} + \boldsymbol{A}_{33})\lambda - \det(\boldsymbol{A}),$$

where $\boldsymbol{A}_{11}$,$\boldsymbol{A}_{22}$ and $\boldsymbol{A}_{33}$ denote, respectively, the cofactors of the diagonal elements a_{11} , a_{22} and a_{33}.

It can be written in the following form:

$$\Delta(\lambda) = \lambda^3 - S_1\lambda^2 + S_2\lambda - S_3,$$

where

$S_1 = \text{tr}(\boldsymbol{A})$, $S_2 = (\boldsymbol{A}_{11} + \boldsymbol{A}_{22} + \boldsymbol{A}_{33})$ and $S_3 = \det(\boldsymbol{A})$.

Note: Each S_k is the sum of all principal minors of the matrix $\boldsymbol{A}$ of order k.

The characteristic polynomial of an $n \times n$ matrix:

Next, let us discuss the characteristic polynomial of an $n \times n$ matrix $\boldsymbol{A}$.

Let $\Delta(\lambda)$ denote the characteristic polynomial of $\boldsymbol{A}$ and it can be defined as

$$\Delta(\lambda) = \begin{bmatrix} a_{11} - \lambda & a_{12} & \cdots & a_{1n} \\ a_{21} & a_{22} - \lambda & \cdots & a_{2n} \\ \vdots & \vdots & \vdots & \vdots \\ a_{n1} & a_{n2} & \cdots & a_{nn} - \lambda \end{bmatrix}.$$

Here, upon recalling the definition of a determinant, the term $\Delta(\lambda)$ with the highest degree in λ can be expressed as.

The term $\Delta(\lambda)$ with the highest degree in λ arises from the product $(a_{11} - \lambda)(a_{22} - \lambda)\cdots(a_{nn} - \lambda)$and is clearly $(-\lambda)^n$.

The terms of degree $(n-1)$ can also be obtained from the same product. Thus, the coefficient of λ^{n-1} is

$$(-1)^{n-1}(a_{11} + a_{22} + \cdots + a_{nn}),$$

where the sum of the diagonal entries of the matrix $\boldsymbol{A}$ can be termed as the trace of $\boldsymbol{A}$:

i.e., $\text{tr}(\boldsymbol{A}) = a_{11} + a_{22} + \cdots + a_{nn}$
implies that the coefficient of λ^{n-1}can be expressed of the form $\text{tr}(\boldsymbol{A})(-\lambda)^{n-1}$.

The constant term $\Delta(\lambda)$ can be found by simply putting $\lambda = 0$ in $\Delta(\lambda) = \det(\boldsymbol{A} - \lambda I_n)$, i.e., by merely evaluating $\det(\boldsymbol{A})$.

Hence, the characteristic polynomial $\Delta(\lambda)$ of the matrix $\boldsymbol{A}$ can be summarized by the formula

$$\Delta(\lambda) = (-\lambda)^n + \text{tr}(\boldsymbol{A})(-\lambda)^{n-1} + \cdots + \det(\boldsymbol{A}).$$

The other coefficient in the characteristic polynomials is not so easy to express, but they are signified as sub-determinants $\det(\boldsymbol{A})$.

For example:

If we take the case of λ^{n-2}, then the terms in λ^{n-2} can be summarized in two ways:

Either from the product
$(a_{11}-\lambda)(a_{22}-\lambda)\cdots(a_{nn}-\lambda)$ or from products like

$$-a_{12}\,a_{21}\,(a_{33}-\lambda)\,,\cdots,(a_{nn}-\lambda)\,.$$

Such a typical expression to the coefficient of λ^{n-2} is

$$(-1)^{n-2}\,(a_{11}a_{22}-a_{12}a_{21})=(-1)^{n-2}\begin{vmatrix} a_{11} & a_{12} \\ a_{21} & a_{22} \end{vmatrix}.$$

From this, it is apparent that the term of degree $(n-2)$ in $\Delta(\lambda)$ is just $(-\lambda)^{n-2}$ times the sum of all the 2×2 determinants of the form
$\begin{vmatrix} a_{ii} & a_{ij} \\ a_{ji} & a_{jj} \end{vmatrix}$, where $i<j$.

Example 4.1:

Determine the eigenvalues and eigenvectors of the matrix

$$\boldsymbol{A}=\begin{bmatrix} 2 & -1 \\ 2 & 4 \end{bmatrix}.$$

Solution:

Consider a 2×2 matrix

$$\boldsymbol{A}=\begin{bmatrix} 2 & -1 \\ 2 & 4 \end{bmatrix}. \tag{4.4}$$

Let $\boldsymbol{X}=\begin{bmatrix} x_1 \\ x_2 \end{bmatrix}$ be a column vector.

The case for the vector $\boldsymbol{X}=\begin{bmatrix} x_1 \\ x_2 \end{bmatrix}$ to be an eigenvector of $\boldsymbol{A}$ is that

$$\boldsymbol{AX}=\lambda\boldsymbol{X},\ \text{for some scalar}\ \lambda\in\mathbb{R}. \tag{4.5}$$

Eqn (4.5) is equivalent to

$$(\boldsymbol{A}-\lambda I_2)\,\boldsymbol{X}=\mathbf{0}_v, \tag{4.6}$$

which asserts that $\boldsymbol{X}$ is a solution of the linear system,
i.e.,

$$\begin{bmatrix} 2-\lambda & -1 \\ 2 & 4-\lambda \end{bmatrix} \begin{bmatrix} x_1 \\ x_2 \end{bmatrix} = \begin{bmatrix} 0 \\ 0 \end{bmatrix}. \tag{4.7}$$

For non-trivial solutions x_1 and x_2 of eqn (4.7), the determinant of the coefficient matrix $\boldsymbol{A} - \lambda I_2$ must vanish,
i.e.,

$$\begin{vmatrix} 2-\lambda & -1 \\ 2 & 4-\lambda \end{vmatrix} = 0, \tag{4.8}$$

which implies

$$\lambda^2 - 6\lambda + 10 = 0. \tag{4.9}$$

Upon solving eqn (4.9), the roots of this quadratic equation (4.9) can be obtained as $\lambda_1 = 3 + \sqrt{-1}$ and $\lambda_2 = 3 - \sqrt{-1}$, which are the eigenvalues of the matrix $\boldsymbol{A}$. The corresponding eigenvectors of eqn (4.6) can be found by using these values of λ in the equations
$(\boldsymbol{A} - \lambda \mathrm{I}_2)\boldsymbol{X} = \boldsymbol{0}_v$, i.e., $(\boldsymbol{A} - \lambda_1 I_2)\boldsymbol{X} = \boldsymbol{0}$ and $(\boldsymbol{A} - \lambda_2 I_2)\boldsymbol{X} = \boldsymbol{0}$.
Now for the case of $\lambda = \lambda_1$, we get

$$(\boldsymbol{A} - \lambda_1 I_2)\boldsymbol{X} = \boldsymbol{0}$$

$$\Rightarrow \begin{array}{c} \left(-1-\sqrt{-1}\right)x_1 - x_2 = 0 \\ 2x_1 + \left(1-\sqrt{-1}\right)x_2 = 0 \end{array}. \tag{4.10}$$

Thus, the general solution of system (4.10) is
$x_1 = \frac{k}{2}\left(-1+\sqrt{-1}\right)$ and $x_2 = k$, where k is an arbitrary scalar.
Therefore, the eigenvectors of $\boldsymbol{A}$ associated with the eigenvalue λ_1 are the non-zero vectors of the form

$$k \begin{bmatrix} \frac{\left(-1+\sqrt{-1}\right)}{2} \\ 1 \end{bmatrix}.$$

Similarly, the eigenvectors for the eigenvalue $\lambda = \lambda_2 = 3 - \sqrt{-1}$ can also be found in vector form as

$$l \begin{bmatrix} \frac{\left(-1-\sqrt{-1}\right)}{2} \\ 1 \end{bmatrix},$$

where $l \neq 0$.

These two vectors, jointly with the zero vector, form a one-dimensional subspace of R^2.

Theorem 4.1:

Let λ be an eigenvalue of an $n \times n$ matrix $\boldsymbol{A}$. The set of all eigenvectors analogous to the eigenvalue λ, together with the zero vector $\mathbf{0}_v$, is a subspace of R^n. This subspace is called the *eigenspace* of the eigenvalue λ.

Proof:

Let $\boldsymbol{X}$ be the set of all eigenvectors of the matrix $\boldsymbol{A}$ corresponding to the eigenvalues λ, together with the zero vector $\mathbf{0_v}$.

To show that the set $\boldsymbol{X}$ is a subspace, we have to show that $\boldsymbol{X}$ is closed under the operation of vector addition and scalar multiplication.

Let $\boldsymbol{X}_1$ and $\boldsymbol{X}_2$ be two eigenvectors in $\boldsymbol{X}$ and let λ be a scalar.

Then $\boldsymbol{AX}_1 = \lambda\boldsymbol{X}_1$ and $\boldsymbol{AX}_2 = \lambda\boldsymbol{X}_2$.

This gives

$$\begin{gathered}\boldsymbol{AX}_1 + \boldsymbol{AX}_2 = \lambda\boldsymbol{X}_1 + \lambda\boldsymbol{X}_2 \\ \Rightarrow \boldsymbol{A}\left(\boldsymbol{X}_1 + \boldsymbol{X}_2\right) = \lambda\left(\boldsymbol{X}_1 + \boldsymbol{X}_2\right) .\end{gathered}$$

Thus, $\boldsymbol{X}_1 + \boldsymbol{X}_2$ is an eigenvector corresponding to $\boldsymbol{X}$.

Hence, $\boldsymbol{X}$ is closed under the operation of vector addition.

Further, since

$$\boldsymbol{AX}_1 = \lambda\boldsymbol{X}_1,$$

then for any scalar c,

$$\begin{aligned}&\Rightarrow c\boldsymbol{AX}_1 = c\lambda\boldsymbol{X}_1 \\ &\Rightarrow \boldsymbol{A}\left(cX_1\right) = \lambda\left(cX_1\right)\end{aligned}.$$

Therefore, cX_1 is also an eigenvector corresponding to the eigenvalue λ.

Thus, $\boldsymbol{X}$ is closed under scalar multiplication.

Hence, $\boldsymbol{X}$ is a subspace of R^n called the *eigenspace* of the eigenvalue λ.

Remark: An eigenvalue λ is said to be of *multiplicity* k if the eigenvalue occurs as k times repeated roots of the characteristic equation.

Example 4.2:

Let $\lambda_1, \lambda_2, \cdots, \lambda_n$ be the eigenvalues and $\boldsymbol{X}_1, \boldsymbol{X}_2, \cdots, \boldsymbol{X}_n$ be the corresponding eigenvectors of an $n \times n$ matrix $\boldsymbol{A}$. Prove that if $c \neq 0$, then the eigenvalues of $c\boldsymbol{A}$ are $c\lambda_1, c\lambda_2, \cdots, c\lambda_n$ with corresponding eigenvectors $\boldsymbol{X}_1, \boldsymbol{X}_2, \cdots, \boldsymbol{X}_n$.

Note: Every characteristic polynomial of a real matrix may not have real eigenvalues in R, while the complex matrices always have all their eigenvalues and eigenvectors in C.

Example 4.3:

Let us consider a matrix $\boldsymbol{A} = \begin{bmatrix} 0 & 1 \\ -1 & 0 \end{bmatrix}$ whose characteristic polynomial is $\lambda^2 + 1$, which implies that it has no real roots. Thus, the matrix has no real eigenvalues in $\mathbb{R}$.

However, if the matrix $\boldsymbol{A}$ is complex, its characteristic equation will have n complex roots, some of which may be equal.

Let us define a theorem for the eigenvalue and eigenvectors of a matrix defined on a complex field $\mathbb{C}$.

Theorem 4.2:

Let $\boldsymbol{A}$ be an $n \times n$ complex matrix. Then

(1) The eigenvalues of the matrix $\boldsymbol{A}$ are precisely the n roots of the characteristic polynomial $\det(\boldsymbol{A} - \lambda I_n)$.

(2) The eigenvectors of the matrix $\boldsymbol{A}$ associated with an eigenvalue λ are the non-zero vectors in the *null space* of the matrix $\boldsymbol{A} - \lambda I_n$.

Example 4.4:

Find the eigenvalues and eigenvectors of the matrix

$$\boldsymbol{A} = \begin{bmatrix} 2 & -1 \\ -2 & 4 \end{bmatrix}.$$

Solution:

The characteristic polynomial of the matrix

$\boldsymbol{A} = \begin{bmatrix} 2 & -1 \\ -2 & 4 \end{bmatrix}$ is

$$\begin{vmatrix} 2-\lambda & -1 \\ -2 & 4-\lambda \end{vmatrix} = \lambda^2 - 6\lambda + 10.$$

The eigenvalues of the matrix $\boldsymbol{A}$ are the roots of the characteristic equation, i.e.,

$$\lambda^2 - 6\lambda + 10 = 0,$$

which gives

$\lambda_1 = 3 + \sqrt{-1}$ and $\lambda_2 = 3 - \sqrt{-1}$.

The eigenspaces of λ_1 and λ_2 are generated by the vectors

$r \begin{bmatrix} \frac{(-1+\sqrt{-1})}{2} \\ 1 \end{bmatrix}$ and $s \begin{bmatrix} \frac{-(1+\sqrt{-1})}{2} \\ 1 \end{bmatrix}$, respectively.

Example 4.5:

Determine the *eigenvalues* of the upper triangular matrix

$$A = \begin{bmatrix} a_{11} & a_{12} & \cdots & a_{1n} \\ 0 & a_{22} & \cdots & a_{2n} \\ \vdots & \vdots & \vdots & \vdots \\ 0 & 0 & \cdots & a_{nn} \end{bmatrix}.$$

Solution:

The characteristic polynomial of the upper triangular matrix A is

$$\begin{bmatrix} a_{11}-\lambda & a_{12} & \cdots & a_{1n} \\ 0 & a_{22}-\lambda & \cdots & a_{2n} \\ \vdots & \vdots & \vdots & \vdots \\ 0 & 0 & \cdots & a_{nn}-\lambda \end{bmatrix},$$

which, by definition, equals

$$(a_{11}-\lambda)(a_{22}-\lambda)\cdots(a_{nn}-\lambda).$$

Thus, the eigenvalues of the matrix A are, therefore, just the diagonal entries of the upper triangular matrix, i.e., $a_{11}, a_{22}, \cdots, a_{nn}$.

Example 4.6:

Find the eigenspace of the matrix

$$A = \begin{bmatrix} 2 & -1 & -1 \\ -1 & 2 & -1 \\ -1 & -1 & 0 \end{bmatrix}.$$

Solution:

Consider a 3×3 matrix

$$A = \begin{bmatrix} 2 & -1 & -1 \\ -1 & 2 & -1 \\ -1 & -1 & 0 \end{bmatrix}.$$

The characteristic polynomial of this matrix A is

$$\begin{vmatrix} 2-\lambda & -1 & -1 \\ -1 & 2-\lambda & -1 \\ -1 & -1 & -\lambda \end{vmatrix} = -\lambda^3 + 4\lambda^2 - \lambda - 6.$$

It can be easily guessed that a root of this cubic polynomial is $\lambda = -1$.

Divide the polynomial by $\lambda + 1$ using long division so that we get the quotient

$$-\lambda^2 + 5\lambda - 6 = -(\lambda - 2)(\lambda - 3).$$

Thus, the characteristic polynomial of the matrix $\boldsymbol{A}$ can be entirely factorized as $-(\lambda + 1)(\lambda - 2)(\lambda - 3)$, and the eigenvalues of the matrix $\boldsymbol{A}$ are $-1, -2$ and 3.

The corresponding eigenvectors can be found by solving the equation
$(\boldsymbol{A} + I_3)\boldsymbol{X} = 0$, $(\boldsymbol{A} - 2I_3)\boldsymbol{X} = 0$, and $(\boldsymbol{A} - 3I_3)\boldsymbol{X} = 0$
for different eigenvalues λ.

Upon solving the above equations, we can obtain that the respective eigenvectors $\boldsymbol{X}$, which are the non-zero scalar multiples of the vectors, as

$$k_1\begin{bmatrix}1\\1\\2\end{bmatrix}, k_2\begin{bmatrix}1\\1\\-1\end{bmatrix} \text{ and } k_3\begin{bmatrix}1\\-1\\0\end{bmatrix}.$$

Hence, the eigenspaces of λ are generated by these three vectors

$$k_1\begin{bmatrix}1\\1\\2\end{bmatrix}, k_2\begin{bmatrix}1\\1\\-1\end{bmatrix}, \text{ and } k_3\begin{bmatrix}1\\-1\\0\end{bmatrix}, \text{ where each has dimension 1.}$$

In general, one can prove by considering the properties of characteristic polynomials that the following statement is true.

Corollary 4.1:

Let us consider an $n \times n$ square matrix $\boldsymbol{A}$. Then the eigenvalue product is equal to the determinant of the matrix $\boldsymbol{A}$, and the sum of the eigenvalues equals the *trace of* $\boldsymbol{A}$.

Theorem 4.3:

Let us consider an $n \times n$ square matrix $\boldsymbol{A}$ over the complex field C. Then the matrix $\boldsymbol{A}$ has at least one eigenvalue.

Theorem 4.4:

Suppose $\boldsymbol{X}_1, \boldsymbol{X}_2, \cdots, \boldsymbol{X}_n$ are the non-zero eigenvectors of a matrix $\boldsymbol{A}$ belonging to distinct eigenvalues $\lambda_1, \lambda_2, \cdots, \lambda_n$. Then the eigenvectors $\boldsymbol{X}_1, \boldsymbol{X}_2, \cdots, \boldsymbol{X}_n$ are linearly independent.

Theorem 4.5:

If $\Delta(\lambda)$ is the characteristic *polynomial* of an $n \times n$ matrix $\boldsymbol{A}$, which is a product of n-distinct factors, say

$$\Delta(\lambda) = (\lambda - a_{11})(\lambda - a_{22}) \cdots (\lambda - a_{nn}).$$

Then the matrix $\boldsymbol{A}$ is similar to the diagonal matrix $D = \text{diag}\,(a_{11}, a_{22}, \cdots a_{nn})$.

Definition 4.2:

Let λ be an eigenvalue of a matrix $\boldsymbol{A}$. Then the algebraic multiplicity of the eigenvalue is defined as the multiplicity of the characteristic polynomial of the matrix $\boldsymbol{A}$.

The *geometric multiplicity* of the eigenvalue λ is defined as the *dimension of its eigenspace*, i.e., $\dim E_\lambda$.

Theorem 4.6:

The *geometric multiplicity* of an eigenvalue λ of a matrix $\boldsymbol{A}$ does not exceed its algebraic *multiplicity*, i.e., $G.M \leq \boldsymbol{A}.M$.

Theorem 4.7:

The eigenvalues λ of a real symmetric matrix $\boldsymbol{A}$ are all real.

4.4 Cayley-Hamilton Theorem

Theorem 4.8 (Cayley-Hamilton theorem):

A matrix $\boldsymbol{A}$ satisfies its own characteristic equation.

Proof:

Let $\boldsymbol{P} = \boldsymbol{A} - \lambda I_n$ and

$$\begin{aligned}\Delta_{\mathbf{P}}(\lambda) &= \det \boldsymbol{P} \\ &= p_0 + p_1\lambda + p_2\lambda^2 + \cdots + p_{n-1}\lambda^{n-1}\ .\end{aligned}$$

Consider the adjoint of the matrix $\boldsymbol{P}$ as $adj\,\boldsymbol{P}$. By the definition of determinant, there is an $n \times n$ matrix whose entries are polynomials in λ of degree at most $n-1$; so we have

$$\text{adj}\,\boldsymbol{P} = \boldsymbol{P}_0 + \boldsymbol{P}_1\lambda + \boldsymbol{P}_2\lambda^2 + \cdots + \boldsymbol{P}_{n-1}\lambda^{n-1},$$

for some $n \times n$ matrices $\boldsymbol{P}_0, \boldsymbol{P}_1, \boldsymbol{P}_2, \cdots, \boldsymbol{P}_{n-1}$.

Since, by definition, for every $n \times n$ matrix $\boldsymbol{A}$, we have

$$\boldsymbol{A}.\,\text{adj}\boldsymbol{A} = (\det \boldsymbol{A})\, I_n = \text{adj}\boldsymbol{A}.\boldsymbol{A}.$$

So here we have

$$\boldsymbol{P}.\text{adj}\,\boldsymbol{P} = (\det \boldsymbol{P})\, I_n,$$

which implies

$$\begin{aligned}(\det \boldsymbol{P})\, I_n &= \boldsymbol{P}.\,\text{adj}\,\boldsymbol{P} \\ &= (\boldsymbol{A} - \lambda I_n)\,.\text{adj}\,\boldsymbol{P}\end{aligned}$$

$$= \boldsymbol{A}.\,\text{adj}\,\boldsymbol{P} - \lambda I_n.\,\text{adj}\,\boldsymbol{P}.$$

We have the polynomial identity

$$p_0\mathrm{I_n} + p_1\mathrm{I_n}\lambda + p_2\mathrm{I_n}\lambda^2 + \cdots + p_n\mathrm{I_n}\lambda^n = \boldsymbol{AP}_0 + \boldsymbol{AP}_1\lambda + \boldsymbol{AP}_2\lambda^2 + \cdots + \boldsymbol{AP}_{n-1}\lambda^{n-1} - \boldsymbol{P}_0\lambda - \boldsymbol{P}_1\lambda^2 - \boldsymbol{P}_2\lambda^3 - \cdots - \boldsymbol{P}_{n-1}\lambda^n.$$

On equating the coefficient of like powers, we obtain

$$\begin{aligned}
&\mathrm{p}_0 I_n = \boldsymbol{AP}_0\\
&\mathrm{p}_1 I_n = \boldsymbol{AP}_1 - \boldsymbol{P}_0\\
&\vdots\\
&\mathrm{p}_{n-1} I_n = \boldsymbol{AP}_{n-1} - \boldsymbol{P}_{n-2}\\
&\mathrm{p}_n I_n = -\boldsymbol{P}_{n-1}
\end{aligned}$$

On multiplying the first equation on the left by $\boldsymbol{A}^0 = I_n$ and similarly the second equation by $\boldsymbol{A}$, the third equation by $\boldsymbol{A}^2$ and so on, we obtain

$$\begin{aligned}
&\mathrm{p}_0 I_n = \boldsymbol{AP}_0\\
&\mathrm{p}_1 \boldsymbol{A} = \boldsymbol{A}^2\boldsymbol{P}_1 - \boldsymbol{AP}_0\\
&\vdots\\
&\mathrm{p}_{n-1}\boldsymbol{A}^{n-1} = \boldsymbol{A}^n\boldsymbol{P}_{n-1} - \boldsymbol{A}^{n-1}\boldsymbol{P}_{n-2}\\
&\mathrm{p}_n\boldsymbol{A}^n = -\boldsymbol{A}^n\boldsymbol{P}_{n-1}
\end{aligned}$$

Adding these equations together, we obtain $\Delta_A(\boldsymbol{A}) = 0$.

Note: The Cayley-Hamilton theorem is quite significant, stating that an $n \times n$ matrix should satisfy a polynomial equation of degree n.

4.5 Google, Demography, and Weather Prediction

We now look at applications of eigenvectors. We discuss the role of eigenvalues and eigenvectors play in Google.

4.5.1 The Google search engine

Google is a web search engine that was developed by Lary Page and Sergey Brin, when they were graduate students at Stanford University. There are usually a huge number of webpages that correspond to a specific query. Google finds pages that match that query and lists them in the order of their page rank. Page rank in Google is a primary way of deciding a page's importance to a given query. We now see that page rank is calculated from the eigenvector of a very large matrix ***A*** (2.7×2.7billion in 2002).

Let n be the number of pages that Google examines in a search; number these pages from 1 to n. In searching the web page to page, Google finds that some pages have outgoing links whereas other pages do not (are dead ends). Let p be the fraction of the total number of pages searched that have outgoing links. An $n \times n$ matrix *G* is defined as follows:

$$g_{ij} = \begin{cases} 1 \text{if there is a link from page i to page } j \\ 0 \text{otherwise} \end{cases}$$

(*G* is, in fact, the adjacency matrix of the digraph having pages as vertices and links as arcs.) Let c_j be the sum of the elements in the jth column of G. Then ***A*** is an $n \times n$ matrix defined by

$$a_{ij} = p\left(\frac{g_{ij}}{c_j}\right) + \left(\frac{1-p}{n}\right).$$

Google usually takes $p = 0.85$. A theorem in linear algebra called the Perron–Frobiniuous theorem guarantees a matrix having the largest eigenvalue 1 with the corresponding eigenspace of dimension 1. Let x be an eigenvector corresponding to $\lambda = 1$. Normalize x such that $\sum x_i = 1$. The element *s* of this normalized vector is the Google page rank.

The determinant method introduced in this section can be used for finding eigenvectors of small matrices but is not practical for large matrices and most certainly not for a matrix of the size encountered in a Google search. The Google company has not revealed how it calculates the eigenvector x of this very large matrix ***A***, but it is generally believed that it is based on the power method.

More in-depth information on page rank can be found in

1. The World's Largest Matrix Computations by Cleve Moler, MATLAB News and Notes, October 2002, Pages 12–13.
2. Google Page Rank Explained by Phil Craven, www.webworkshop.net/agerank.html.

Long-term prediction:

We now discuss how eigenvalues and eigenvectors can be used to predict the long-term behavior of certain Markov chains. Applications in demography and weather prediction are given.

Let us return to the population movement model where we found that a sequence of vectors could describe annual population distributions

$x_0, x_1 (= \boldsymbol{P}x_0), x_2 (= \boldsymbol{P}x_1), x_3 (= \boldsymbol{P}x_2), \cdots$, $\boldsymbol{P}$ is a matrix of transition probabilities that takes us from one vector in the sequence to the following vector. Such a sequence (or chain) of vectors called a *Markov chain* of particular interest are Markov chains called *regular Markov chains* where the sequence $x_0, x_1, x_2, \cdots$ converges to some fixed vector x where $\boldsymbol{P}x = x$. The population movement would then be in a "steady state," with the total city population and total suburban population remaining constant after that. We then write

$$x_0, x_1, x_2, \cdots \to x.$$

Since such a vector x satisfies $\boldsymbol{P}x = x$, it would be an eigenvector corresponding to eigenvalue 1. Knowledge of the existence and value of such a vector would give us information about the long-term behavior of the population distribution.

A particular class of Markov chains has these desired properties. We now define this class, discuss their properties, and apply the results to our population movement model to make long-term predictions.

Definition 4.3:

The transition matrix $\boldsymbol{P}$ of a Markov chain is said to be *regular* if all the components' power is positive. The chain is then called a *regular Markov chain*.

For example:

$\boldsymbol{A} = \begin{bmatrix} 0.3 & 0.6 \\ 0.7 & 0.4 \end{bmatrix}$ is regular because all the elements are positive.

$B = \begin{bmatrix} 0.7 & 1 \\ 0.3 & 0 \end{bmatrix}$ is regular, because $B^2 = \begin{bmatrix} 0.79 & 0.7 \\ 0.21 & 0.3 \end{bmatrix}$.

$C = \begin{bmatrix} 0.4 & 0 \\ 0.6 & 1 \end{bmatrix}$ is not regular, because

$$C^2 = \begin{bmatrix} 0.16 & 0 \\ 0.84 & 1 \end{bmatrix} C^3 = \begin{bmatrix} 0.064 & 0 \\ 0.936 & 1 \end{bmatrix} \cdots\cdots$$

The element in row 1, column 2, will always be zero.

The following theorem, which we do not prove, leads to information about the long-term behavior of Markov chains.

Theorem 4.9:

Consider a regular Markov chain having an initial vector x_0 and transition matrix $\boldsymbol{P}$. Then

1. $x_0, x_1, x_2, \cdots \to x$ where x satisfies $\boldsymbol{P}x = x$. Thus, x is an eigenvector of $\boldsymbol{P}$ corresponding to $\lambda = 1$.
2. $\boldsymbol{P}, \boldsymbol{P}^2, \boldsymbol{P}^3, \cdots \to Q$, where Q is a stochastic matrix. The columns of Q are all identical, each being an eigenvector of $\boldsymbol{P}$ corresponding to $\lambda = 1$.

Let us now apply this theorem to population movement.

4.5.2 Population prediction

Example 4.7:

Determine the long-term trends in population movements between US cities and suburbs.

Solution:

We remind the reader of the model that was developed earlier. The population of US cities and suburbs in 2007 were described by the following vector x_0 (in units of one million), and Markov chains give the populations in the following years with the transition matrix $\boldsymbol{P}$.

$$\begin{array}{l}\text{InitialFrom} \\ \text{Populations city suburb to}\end{array}$$

$$x_0 = \begin{bmatrix} 82 \\ 163 \end{bmatrix} \begin{matrix} \text{city} \\ \text{suburb} \end{matrix}, \boldsymbol{P} = \begin{bmatrix} 0.96 & 0.01 \\ 0.04 & 0.99 \end{bmatrix} \begin{matrix} \text{city} \\ \text{suburb} \end{matrix}.$$

Observe that all the elements of $\boldsymbol{P}$ are positive. Therefore, the chain is regular, and the preceding theorem results can be applied to give the long-term trends. The theorem tells us that $\boldsymbol{P}$ will have an eigenvalue of 1 and that the steady-state vector x is a corresponding eigenvector. Thus,

$$\begin{aligned} &\boldsymbol{P}x = x \\ &\Rightarrow (\boldsymbol{P} - I_2)x = 0 \\ &\Rightarrow \begin{bmatrix} 0.96 - 1 & 0.01 \\ 0.04 & 0.99 - 1 \end{bmatrix} \begin{bmatrix} x_1 \\ x_2 \end{bmatrix} = 0 \end{aligned}.$$

This leads to the system of equations

$$\begin{aligned} -0.04x_1 + 0.01x_2 = 0 \\ 0.04x_1 - 0.01x_2 = 0 \end{aligned}.$$

Giving $x_2 = 4x_1$. The solution to this system of equations is $x_1 = r, x_2 = 4r$ where r is a scalar. Thus, the eigenvectors of $\boldsymbol{P}$ that correspond to $\lambda = 1$ are non-zero vectors of the form

$$r \begin{bmatrix} 1 \\ 4 \end{bmatrix}.$$

The steady-state vector x will be a vector of this form. Let us assume that there is no total annual population change over the years. Therefore, the sums of the elements of x and x_0 are equal.

$$\begin{aligned} r + 4r &= 82 + 163 \\ r &= 49 \end{aligned}.$$

The steady-state vector is thus

$$\mathrm{x} = \begin{bmatrix} 49 \\ 196 \end{bmatrix}.$$

This implies the following long-term prediction:
US cities populations $\rightarrow$ 49 million.
US suburban populations $\rightarrow$196 million.

The above theorem gives further information about long-term population trends. Each column of the matrix Q is an eigenvector corresponding to the eigenvalue 1. Let

$$Q = \begin{bmatrix} s & s \\ 4s & 4s \end{bmatrix}.$$

Since Q is a stochastic matrix, the sum of the elements in each column is 1. Thus,

$$\begin{aligned} s + 4s &= 1 \\ s &= 0.2 \end{aligned}.$$

We get (exhibiting the elements to two decimal places for ease of reading)

$$\underset{\boldsymbol{P}}{\begin{bmatrix} 0.96 & 0.01 \\ 0.04 & 0.99 \end{bmatrix}}, \underset{\boldsymbol{P}^2}{\begin{bmatrix} 0.92 & 0.02 \\ 0.08 & 0.98 \end{bmatrix}}, \underset{\boldsymbol{P}^3}{\begin{bmatrix} 0.89 & 0.03 \\ 0.11 & 0.97 \end{bmatrix}} \cdots \rightarrow \underset{Q}{\begin{bmatrix} 0.2 & 0.2 \\ 0.8 & 0.8 \end{bmatrix}}.$$

Let us interpret these results. We focus on the $(2, 1)$ element in each matrix, and a similar interpretation will apply to the other elements. We get the sequence

$$0.04, 0.08, 0.11, \cdots \rightarrow 0.8.$$

These are the probabilities of moving from city to suburbia in one year, two years, three years, etc. The probability gradually increases approaching 0.8. Q is thus the *long-term transition matrix* of the model. It gives the long-term probabilities of living in a city or suburbia.

$$\begin{array}{l} \quad\;\;\text{(from)} \\ \quad\;\;\text{city suburb (to)} \\ Q = \begin{bmatrix} 0.2 & 0.2 \\ 0.8 & 0.8 \end{bmatrix} \begin{array}{c} \text{city} \\ \text{suburb} \end{array} \end{array}.$$

Observe that the long-term probability of living in the city is 0.2, while the long-term probability of living in suburbia is 0.8. These probabilities are independent of initial locations. The long-term probabilities being independent of the initial state are a characteristic of regular Markov chains.

We now discuss an interesting application of Markov chains in a model that describes rainfall in Tel Aviv.

4.5.3 Weather in Tel Aviv

Example 4.8:[1]

The probabilities used were based on daily rainfall data in Tel Aviv (Nahami Street) for the 27 years from 1923 to 1950. Days were classified as wet or dry according to whether or not there had been recorded at least 0.1 mm of precipitations in the 24 hours from 8 a.m. to 8 a.m. the following day. A Markov chain was constructed for each of the months from November through April, these months constituting the rainy season. We discuss the chain developed for November. The model assumes that the probability of rainfall on any day depends only on whether the previous day was wet or dry. The statistics accumulated over the years for November were

A given day	**Following day**
Wet	117 out of 195
Dry	80 out of 615

Thus, the probability of a wet day following a wet day is $\frac{117}{195} = 0.6$. The probability of a wet day following a dry day is $\frac{80}{615} = 0.13$. These probabilities lead to the following transition matrix for the weather pattern in November.

$$\begin{array}{c} \text{(A given day)} \\ \begin{array}{cc} \text{wet} & \text{dry} \end{array} \end{array}$$
$$\boldsymbol{P} = \begin{bmatrix} 0.6 & 0.13 \\ 0.4 & 0.87 \end{bmatrix} \begin{array}{c} \text{wet} \\ \text{dry} \end{array} \text{(following day)}.$$

On any given day in November, one can use $\boldsymbol{P}$ to predict the weather on a future day in November. For example, if today a Wednesday is wet, let us compute the probability that next Saturday will be dry. Saturday is three days; hence, the various probabilities for the weather on Saturday will be

[1] K. R. Gabriel and J. Nemann have developed a Markov chain model for daily rainfall occurrence at Tel Aviv, Quart J. R. Met. Soc., 88(1962), 90–95.

given by the elements of $\boldsymbol{P}^3$. It can be shown that (exhibiting the elements to two decimal places for ease of reading)

$$\begin{array}{c} \text{(Today – Wednesday)} \\ \begin{array}{cc} \text{wet} & \text{dry} \end{array} \end{array}$$

$$\boldsymbol{P}^3 = \begin{bmatrix} 0.32 & 0.22 \\ 0.68 & 0.78 \end{bmatrix} \begin{matrix} \text{wet} \\ \text{dry} \end{matrix} \quad (Saturday)$$

If today is wet, the probability of Saturday being dry is 0.68.

Observe that ***P***, the matrix of transition probabilities, is regular. Eigenvectors can thus be used to obtain long-term weather predictions. The eigenvectors of ***P*** corresponding to the eigenvalue 1 are found to be non-zero vectors of the form

$$r \begin{bmatrix} 0.325 \\ 1 \end{bmatrix}.$$

The column vectors of the long-term transition matrix Q will be eigenvectors of this type, whose components add up to 1, since Q is stochastic. Therefore, $0.325r + r = 1$ giving $r = 0.75$(to 2 decimal places). Thus,

$$Q = \begin{bmatrix} 0.25 & 0.25 \\ 0.75 & 0.75 \end{bmatrix}.$$

We can interpret this matrix as follows:

$$\begin{array}{c} \text{(Today)} \\ \begin{array}{cc} \text{wet} & \text{dry} \end{array} \end{array}$$

$$\begin{bmatrix} 0.32 & 0.22 \\ 0.68 & 0.78 \end{bmatrix} \begin{matrix} \text{wet} \\ \text{dry} \end{matrix} \quad \text{(day in the distant future)}.$$

This implies the following weather forecast for Tel Aviv in November.

Long-range forecast for Tel Aviv in November:

$$\begin{matrix} 0.25 \text{ probility wet} \\ 0.75 \text{ probability dry} \end{matrix}.$$

4.5.4 Weather in Belfast

Example 4.9:

A matrix model for weather in Belfast, Northern Ireland, is given by William J. Stuart in *Introduction to Numerical Solution of Markov Chains*

(Princeton University Press, 1994, 6). Three weather conditions are considered rainy (R), cloudy (C), and sunny (S).

The matrix describes daily changes.

$$\begin{array}{c} (\boldsymbol{A} \text{ given day}) \\ \begin{array}{cccc} & \mathrm{R} & \mathrm{C} & \mathrm{S} \end{array} \\ \boldsymbol{P} = \begin{bmatrix} 0.8 & 0.15 & 0.05 \\ 0.7 & 0.2 & 0.1 \\ 0.5 & 0.3 & 0.2 \end{bmatrix} \begin{array}{l} \mathrm{R} \\ \mathrm{C} \ \ (\text{following day}) \\ \mathrm{S} \end{array} \end{array}.$$

Thus, for example, if today is cloudy, the probability that tomorrow is sunny is 0.3. The eigenvectors corresponding to the eigenvalue 1 are found to be of the form

$$r\begin{bmatrix} 1 \\ 1 \\ 1 \end{bmatrix}.$$

Using $r = \frac{1}{3}$ leads to the stochastic matrix that describes the long-term forecast,

$$\begin{array}{c} (\text{today}) \\ \begin{array}{cccc} & \mathrm{R} & \mathrm{C} & \mathrm{S} \end{array} \\ Q = \begin{bmatrix} \frac{1}{3} & \frac{1}{3} & \frac{1}{3} \\ \frac{1}{3} & \frac{1}{3} & \frac{1}{3} \\ \frac{1}{3} & \frac{1}{3} & \frac{1}{3} \end{bmatrix} \begin{array}{l} \mathrm{R} \\ \mathrm{C} \ \ (\text{day in the distant future}) \\ \mathrm{S} \end{array} \end{array}.$$

There is the same probability for any future day in Belfast to be rainy, cloudy, or sunny.

Exercises

1. Determine the eigenvalues and the corresponding eigenspaces for the following matrices:

(a) $\begin{bmatrix} 3 & 1 \\ 6 & 2 \end{bmatrix}$ (b) $\begin{bmatrix} 0 & -1 \\ 1 & 0 \end{bmatrix}$ (c) $\begin{bmatrix} 0 & 3 \\ 2 & -1 \end{bmatrix}$

(d) $\begin{bmatrix} 3 & 2 & 4 \\ 2 & 0 & 2 \\ 4 & 2 & 3 \end{bmatrix}$ (e) $\begin{bmatrix} 1 & 0 & 0 \\ 2 & 1 & 0 \\ 3 & 2 & 0 \end{bmatrix}$ (f) $\begin{bmatrix} 0 & i & i \\ i & 0 & i \\ i & i & 0 \end{bmatrix}$

(g) $\begin{bmatrix} 3 & 2 & 1 & 0 \\ 0 & 1 & 0 & 1 \\ 0 & 2 & -1 & 0 \\ 0 & 0 & 0 & \frac{1}{2} \end{bmatrix}$.

2. Diagonalize the matrices of (b), (c), and (g) in problem 1.

3. Find the characteristic polynomial of the matrix
$\begin{bmatrix} 1 & 0 & 0 & 0 \\ -1 & i & 0 & 0 \\ 2 & \frac{1}{2} & -i & 0 \\ \frac{1}{3} & -i & \pi & -1 \end{bmatrix}$, diagonalize the matrix, if possible.

4. If λ is an eigenvalue of the matrix $\boldsymbol{A}$, prove that
(a) λ^2 is an eigenvalue of $\boldsymbol{A}^2$.
(b) λ^n is an eigenvalue of $\boldsymbol{A}^n$.
(c) $\alpha\lambda$ is an eigenvalue of $\alpha\boldsymbol{A}$, where α is a scalar.
(d) $g(\lambda)$ is an eigenvalue of $g(\boldsymbol{A})$, where g is a polynomial.

5. If $\boldsymbol{X}$ is an eigenvector of $\boldsymbol{A}$ corresponding to the eigenvalue λ, prove that
(a) $\boldsymbol{X}$ is an eigenvector of $\boldsymbol{A}^n$ corresponding to the eigenvalue λ^n.
(b) $\boldsymbol{X}$ is an eigenvector of $g(\boldsymbol{A})$ corresponding to the eigenvalue $g(\lambda)$.

6. Prove that $\lambda = 0$ is an eigenvalue of the matrix $\boldsymbol{A}$ iff $\boldsymbol{A}$ is singular.

7. If λ is an eigenvalue of the matrix $\boldsymbol{A}$, prove that
(a) λ is also an eigenvalue of $\boldsymbol{A}^T$.
(b) $\frac{1}{\lambda}$ is an eigenvalue of $\boldsymbol{A}^{-1}$, if $\boldsymbol{A}$ is non-singular.

8. If $\boldsymbol{X}$ is an eigenvector of $\boldsymbol{A}$ corresponding to the eigenvalue λ, prove that
(a) $\boldsymbol{X}$ need not be an eigenvector of $\boldsymbol{A}^T$ corresponding to the eigenvalue λ.
(b) $\boldsymbol{X}$ is an eigenvector of $\boldsymbol{A}^{-1}$ (if $\boldsymbol{A}$ is non-singular) corresponding to the eigenvalue $\frac{1}{\lambda}$.

9. Prove that the eigenvalues of a triangular matrix are its diagonal elements.

5

Linear Transformation

This chapter discusses an integral part of this subject, i.e., linear transformations, including the range and Kernel of transformation, one-to-one and invertible transformation, coordinate representation, change of basis, isomorphism, matrix of linear transformations, visualization of the matrix representation, and relation between matrix representation. The significance of vector spaces arises from the fact that we can pass from one vector space to another by means of functions that possess a particular property called linearity. These functions are called linear transformations.

This chapter aims to familiarize the students with the fundamental properties of linear transformations. A topic such as a Kernel, range, and rank nullity theorem are presented here. Linear transformations, range and Kernel, are used to give the reader a geometrical picture of the sets of solutions to systems of linear operations, both homogenous and non-homogenous. The rank nullity theorem and its consequences are presented in detail. The theory developed so far is applied to operator equations and, in particular, to differential equations. This application discloses that the solution space of the nth-order homogenous linear ordinary differential equations is an n-dimensional subspace of n-times continuously differentiable functions.

5.1 The Idea of a Linear Transformation

Definition 5.1:

Let $\mathbf{V}$ and $\mathbf{W}$ be vector spaces defined over the same field of scalar F. Let v_1 and v_2 be vectors in $\mathbf{V}$ and let k be a scalar.

A linear transformation (or linear mapping) from $\mathbf{V}$ to $\mathbf{W}$ is a function $\boldsymbol{T} : \mathbf{V} \to \mathbf{W}$ that is said to be linear. If

$$\boldsymbol{T}(v_1 + v_2) = \boldsymbol{T}(v_1) + \boldsymbol{T}(v_2).$$
$$\boldsymbol{T}(kv) = k\,\boldsymbol{T}(v).$$

For all $v_1, v_2 \in \mathbf{V}$ and $k \in \mathbf{F}$.

Shortly, the transformation $\boldsymbol{T}$ is necessary to act linearly on the sum and scalar multiples of the vectors in $\mathbf{V}$. If $\boldsymbol{T} : \mathbf{V} \to \mathbf{V}$ is a linear transformation, then we shall say that $\boldsymbol{T}$ defines a linear identity operator on $\mathbf{V}$.

The operation of addition and scalar multiplication preserves under a linear transformation.

Example 5.1:

(1) Let $\mathbf{V}$ and $\mathbf{W}$ be two vector spaces defined over the same field F. Then the transformation that sends every vector $\mathbf{V}$ into the zero vector in $\mathbf{W}$ is called a *zero linear transformation.* It is denoted by 0_w or 0.
(2) The transformation $\boldsymbol{I}_v : \mathbf{V} \to \mathbf{V}$ terms as the linear identity operator on $\mathbf{V}$.

Example 5.2:

Prove that the transformation $\boldsymbol{T} : R^2 \to R^2$ defined by $\boldsymbol{T}(x_1, x_2) = (2x_1, x_1 + x_2)$ is linear.

Example 5.3:

Let $\boldsymbol{P}_n(R)$ be the vector space of the real polynomial function deg $\leq n$. Show that the following transformation $\boldsymbol{T} : \boldsymbol{P}_2 \to \boldsymbol{P}_1$ defined by $\boldsymbol{T}\left(ax^2 + bx + c\right) = (a + b)\, x + c$ is linear.

Theorem 5.1:

Let $\boldsymbol{T} : \mathbf{V} \to \mathbf{W}$ be a linear transformation. Let $\mathbf{0_v}$ and $\mathbf{0_w}$ be the zero vectors of $\mathbf{V}$ and $\mathbf{W}$. Then $\boldsymbol{T}(0_v) = 0_w$, i.e., a linear transformation $\boldsymbol{T}$ maps a zero vector into a zero vector.

Proof:

Let $\boldsymbol{T} : \mathbf{V} \to \mathbf{W}$ be a linear transformation.
Let $v \in \mathbf{V}, w \in \mathbf{W}$, and $\boldsymbol{T}(v) = w$.
Let 0 be the zero scalars.
Since $0.0_u =$ and $0.v = 0_v$ and $\boldsymbol{T}$ is linear, so we get

$$\boldsymbol{T}(\mathbf{0_v}) = \boldsymbol{T}(0.v) = 0.\boldsymbol{T}(v) = 0.w = \mathbf{0_w}.$$

5.2 The Range and Kernel of Transformation

Definition 5.2:

The vector sets in $\mathbf{V}$ that mapped into the zero vector in $\mathbf{W}$ under the transformation $\boldsymbol{T} : \mathbf{V} \to \mathbf{W}$ are called the Kernel of $\boldsymbol{T}$, denoted by $Ker(\boldsymbol{T})$.

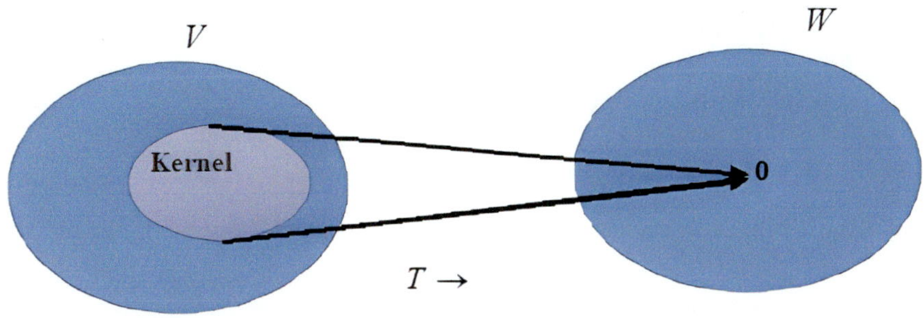

Kernel
All vectors in **V** that are mapped into 0.

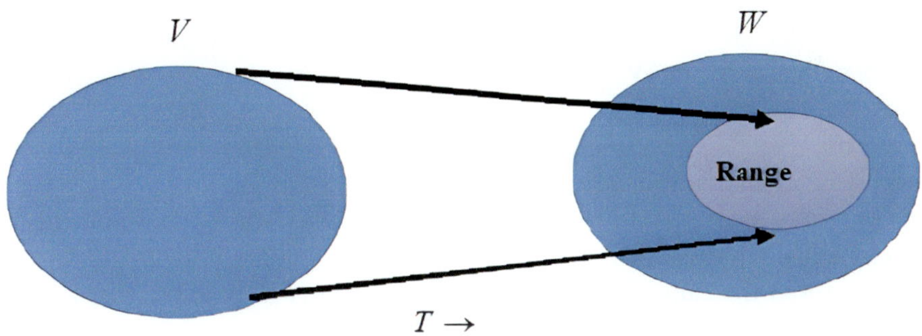

Range
All vectors in **W** that are images of vectors in **V**.

Figure 5.1 The Range and Kernel of Transformation.

Similarly, the vector sets in **W** that are the images of vectors in **V** are called the range of $\boldsymbol{T}$. The range of $\boldsymbol{T}$ is denoted by $\text{Range}\,(\boldsymbol{T})$, as shown Figure 5.1

Theorem 5.2:

Let $\boldsymbol{T} : \mathbf{V} \rightarrow \mathbf{W}$ be a linear transformation. Then

(1) The $\text{Ker}\,(\boldsymbol{T})$ is a subspace of $\mathbf{V}$.
(2) The $\text{Range}\,(\boldsymbol{T})$ is a subspace of $\mathbf{W}$.

Proof:

We know that the $\text{Ker}\,(\boldsymbol{T})$ is non-empty since it contains the zero vector in $\mathbf{V}$.

To prove that the Kernel is a subspace of **V**, it remains to show that it is closed under addition and scalar multiplications.

Closure under addition:

Let $v_1, v_2 \in \text{Ker}\,(\boldsymbol{T})$.

Thus, $\boldsymbol{T}\,(v_1) = 0_w$ and $\boldsymbol{T}\,(v_2) = 0_w$.

Now by using the linearity of $\boldsymbol{T}$, we get

$$\boldsymbol{T}\,(v_1 + v_2) = \boldsymbol{T}\,(v_1) + \boldsymbol{T}\,(v_2) = 0_w + 0_w \; = 0_w.$$

This implies that the vector $v_1 + v_2$ maps into the zero vector $\mathbf{0_w}$ in $\mathbf{W}$. Thus,

$$v_1 + v_2 \in \text{Ker}\,(\boldsymbol{T})\,.$$

Let us now show that the $Ker\,(\boldsymbol{T})$ is closed under scalar multiplication.

Let $k \in \mathbf{F}$ be a scalar.

Again, using the linearity of $\boldsymbol{T}$, we get

$\boldsymbol{T}\,(kv_1) = kT\,(v_1) = k.0_w = 0_w$, which implies that the vector kv_1 maps into the zero vector 0_w in $\mathbf{W}$.

Thus, $kv_1 \in \text{Ker}\,(\boldsymbol{T})$.

Since the $\text{Ker}\,(\boldsymbol{T})$ is closed under addition and scalar multiplication operations, $\text{Ker}\,(\boldsymbol{T})$ is a subspace of $\mathbf{V}$.

The previous theorem tells us the range is non-empty since it contains the zero vector of $\mathbf{V}$.

To prove that the range is a subspace of $\mathbf{V}$, it remains to show that it is closed under addition and scalar multiplications.

Let w_1 and w_2 be elements of $\text{Range}\,(\boldsymbol{T})$.

Thus, there exist vectors v_1 and v_2 in the domain $\mathbf{V}$ such that

$$\boldsymbol{T}\,(v_1) = w_1 \;\text{ and }\; \boldsymbol{T}\,(\mathrm{v}_2) = \mathrm{w}_2$$

.

Using the linearity of T,

$$\boldsymbol{T}\,(v_1 + v_2) = \boldsymbol{T}\,(v_1) + \boldsymbol{T}\,(v_2) \;= w_1 + w_2.$$

The vector $w_1 + w_2$ is the image of $v_1 + v_2$.

Thus, $w_1 + w_2$ is in the range.

Let k be a scalar. By the linearity of T,

$$\boldsymbol{T}\,(kv_1) = kT\,(v_1) \;= kw_1.$$

The vector kw_1 is the image of kv_1. Thus, kw_1 is in the range.

Since the range is closed under addition and scalar multiplication, $\text{Range}\,(\boldsymbol{T})$ is a subspace of $\mathbf{W}$, as shown in Figure 5.2.

Example 5.4:

Determine the $\text{Ker}\,(\boldsymbol{T})$ and $\text{Range}\,(\boldsymbol{T})$ of the linear operator $\boldsymbol{T}\,(x_1, x_2, x_3) = (x_1, x_2, 0)$.

Solution:

Here, $\boldsymbol{T}\,:\,R^3 \rightarrow R^3$. So the $\text{Ker}\,(T)$ and $\text{Range}\,(\boldsymbol{T})$ will both be subspaces of R^3.

Kernel of *T*:

Since the $\text{Ker}\,(\boldsymbol{T})$ is the subset that maps into the zero vector $(0, 0, 0)$ of R^3, we see that

$$T\,(x_1, x_2, x_3) = (x_1, x_2, 0) = (0, 0, 0)\,.$$

If $x_1 = 0 \;\; and \;\; x_2 = 0$, then $Ker\,(\boldsymbol{T})$ is the set of all vectors of the form $(0, 0, x_3)$.

Geometrically, $\text{Ker}\,(\boldsymbol{T})$ is the set of all vectors that lie on the z axis, as shown in Figure 5.2.

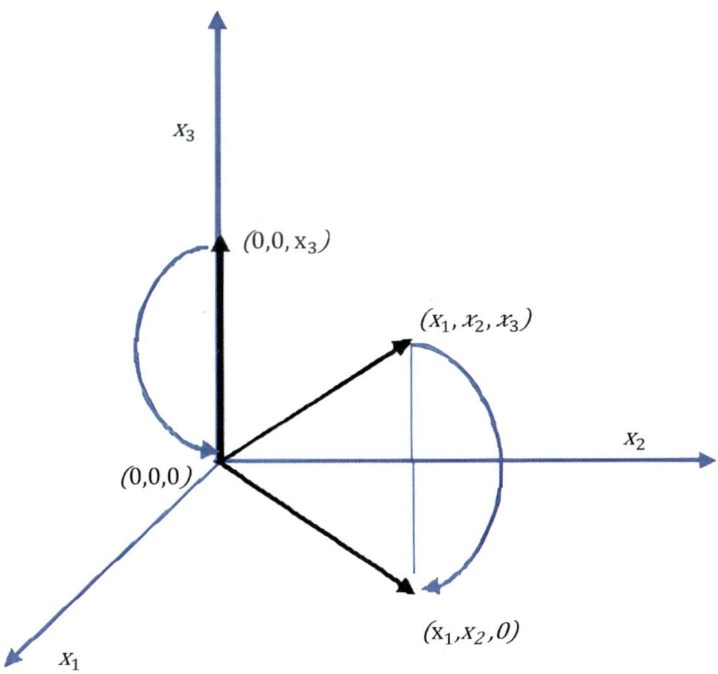

Projection: $f\,(x_1, x_2, x_3) = (x_1, x_2, 0)$

Figure 5.2 Range of ***T*** and Kernel of ***T***.

Range:

The Range $(\boldsymbol{T})$ is the set of all vectors of the form $(x_1, x_2, 0)$. Thus, Range $(\boldsymbol{T}) = \{(x_1, x_2, 0)\}$.

It implies Range $(\boldsymbol{T})$ is the set of all vectors that lie in the xy-plane.

5.3 One-to-One Transformation and Inverse Transformation

An element in the range of a transformation may be the image of a single component or multiple elements in the domain.

Definition 5.3:

A linear transformation $\boldsymbol{T} : \mathbf{V} \rightarrow \mathbf{W}$ is said to be one to one if each element in the range $\mathbf{W}$ is the image of just one component of the domain $\mathbf{V}$.

The transformation $\boldsymbol{T}$ is said to be *one to one* if $\boldsymbol{T}(v_1) = \boldsymbol{T}(v_2)$ implies $v_1 = v_2$, as shown in Figure 5.3.

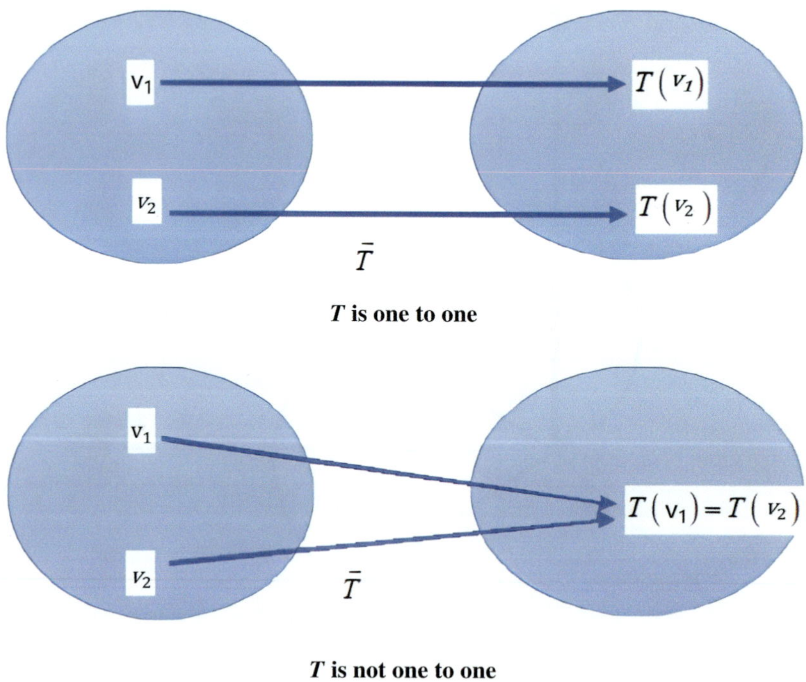

Figure 5.3 One to one transformation.

Example 5.5:

A linear transformation $\boldsymbol{T}: \mathbf{V} \to \mathbf{W}$ defined by

(1) $\boldsymbol{T}(x) = 2x$ is one to one.
(2) $\boldsymbol{T}(x) = x^2$ is not one to one.

Theorem 5.3:

A linear transformation $\boldsymbol{T}: \mathbf{V} \to \mathbf{W}$ is said to be *one to one* if and only if the $\text{Ker}(\boldsymbol{T})$ is a zero vector.

Proof:

Let us consider that the linear transformation $\boldsymbol{T}: \mathbf{V} \to \mathbf{W}$ is *one to one*.

Claim:

The $\text{Ker}(\boldsymbol{T})$ consists of all the vectors in $\mathbf{V}$ that map into zero vector of $\mathbf{W}$.

Since the linear transformation $\boldsymbol{T}: \mathbf{V} \to \mathbf{W}$ is one to one, the $Ker(\boldsymbol{T})$ must consist of a single vector.

Still, we identify that the zero vector must be in the $\text{Ker}(\boldsymbol{T})$.

Thus, the $Ker(\boldsymbol{T})$ is the zero vector.

Conversely, let us suppose that the $\text{Ker}(\boldsymbol{T})$ is the zero vector.

Let v_1 and v_2 be vectors in $\mathbf{V}$ such that

$$\boldsymbol{T}(v_1) = \boldsymbol{T}(v_2) .$$

Using the linear property of $\boldsymbol{T}$, we get

$$\begin{aligned} \boldsymbol{T}(v_1) - T(v_2) &= 0_w \\ \Rightarrow \boldsymbol{T}(v_1 - v_2) &= 0_w. \end{aligned}$$

Thus, $v_1 - v_2$ is in the $\text{Ker}(\boldsymbol{T})$.

But the $\text{Ker}(\boldsymbol{T})$ is a zero vector. Therefore,

$$\begin{aligned} v_1 - v_2 &= 0_w \\ \Rightarrow v_1 &= v_2. \end{aligned}$$

Thus, the linear transformation $\boldsymbol{T}: \mathbf{V} \to \mathbf{W}$ is *one to one*.

Theorem 5.4:

Let $\boldsymbol{T}: \mathbf{V} \to \mathbf{W}$ be a linear transformation. Then

$$\dim(\ker(\boldsymbol{T})) + \dim(\text{range}(\boldsymbol{T})) = \dim \text{domain}(\boldsymbol{T}) .$$

Proof:

Let us assume that the Kernel consists of more than the zero vector and that it is not the whole of **V**.

Let $v_1, v_2, \ldots, v_m$ be a basis for $\ker(\boldsymbol{T})$.

Add vectors $v_{m+1}, v_{m+2}, \ldots, v_n$ to this set to give a basis $v_1, v_2, \ldots, v_n$ for **V**.

We shall show that $\boldsymbol{T}(v_{m+1}), \boldsymbol{T}(v_{m+2}), \ldots, \boldsymbol{T}(v_n)$ form a basis for the range.

Thus, the theorem is proved.

Let $v \in V$ be a vector in **V**.

Then the vector v can be expressed as a linear combination of the basis vectors $v_1, v_2, \ldots, v_n$ as follows:

$$v = a_1v_1 + a_2v_2 + \cdots + a_mv_m + a_{m+1}v_{m+1} + a_{m+2}v_{m+2} + \cdots + a_nv_n.$$

Thus,

$$\boldsymbol{T}(v) = \boldsymbol{T}(a_1v_1 + a_2v_2 + \cdots + a_mv_m + a_{m+1}v_{m+1} + a_{m+2}v_{m+2} + \cdots + a_nv_n).$$

The linearity of $\boldsymbol{T}$ gives

$$\boldsymbol{T}(v) = a_1\boldsymbol{T}(v_1) + a_2\boldsymbol{T}(v_2) + \cdots + a_m\boldsymbol{T}(v_m) + a_{m+1}\boldsymbol{T}(v_{m+1}) + a_{m+2}\boldsymbol{T}(v_{m+2}) + \cdots + a_n\boldsymbol{T}(v_n).$$

Since $v_1, v_2, \ldots, v_m$ are in the Kernel, this reduces to

$$\boldsymbol{T}(v) = a_{m+1}\boldsymbol{T}(v_{m+1}) + a_{m+2}\boldsymbol{T}(v_{m+2}) + \cdots + a_n\boldsymbol{T}(v_n),$$

which implies that $\boldsymbol{T}(v)$ represents an arbitrary vector in the range of $\boldsymbol{T}$.

Thus, the vectors $\boldsymbol{T}(v_{m+1}), \boldsymbol{T}(v_{m+2}), \ldots, \boldsymbol{T}(v_n)$ span the range of $\boldsymbol{T}$.

It remains to prove that these vectors $\boldsymbol{T}(v_{m+1}), \boldsymbol{T}(v_{m+2}), \ldots, \boldsymbol{T}(v_n)$ are also linearly independent.

Consider the identity:

$$b_{m+1}\boldsymbol{T}(v_{m+1}) + b_{m+2}\boldsymbol{T}(v_{m+2}) + \cdots + b_n\boldsymbol{T}(v_n) = 0_w,$$

where $b_i's$ are scalars. $(i = m+1, m+2, ..., n)$.

The linearity of $\boldsymbol{T}$ implies that

$$\boldsymbol{T}(b_{m+1}v_{m+1} + b_{m+2}v_{m+2} + \cdots + b_nv_n) = 0_w.$$

This means that the vectors $b_{m+1}v_{m+1} + b_{m+2}v_{m+2} + \cdots + b_n v_n$ are in the Kernel. Thus, it can be expressed as a linear combination of the basis of the Kernel.

Let

$$b_{m+1}v_{m+1} + b_{m+2}v_{m+2} + \cdots + b_n v_n = c_1 v_1 + c_2 v_2 + \cdots + c_m v_m.$$

$\Rightarrow b_{m+1}, b_{m+2}, \ldots, b_n$ are all zero.

$\Rightarrow \boldsymbol{T}(v_{m+1}), \boldsymbol{T}(v_{m+2}), \ldots, \boldsymbol{T}(v_n)$ are linearly dependent.

$\Rightarrow$ The set of vectors $\boldsymbol{T}(v_{m+1}), \boldsymbol{T}(v_{m+2}), \ldots, \boldsymbol{T}(v_n)$ is a basis for the range.

Hence proved.

Theorem 5.5:

The linear transformation $\boldsymbol{T} : R^n \rightarrow R^n$ defined by $\boldsymbol{T}(\boldsymbol{X}) = \boldsymbol{AX}$ is *one to one* if and only if the matrix $\boldsymbol{A}$ is non-singular.

Proof:

Let us consider the linear transformation $\boldsymbol{T} : R^n \rightarrow R^n$ to be one to one.

Thus, $Ker(\boldsymbol{T}) = 0_w$.

Hence, the rank nullity theorem implies $\dim(\text{Range}\,\boldsymbol{T}) = n$.

It implies that the n columns of the matrix $\boldsymbol{A}$ are linearly independent.

Thus, the Rank is n, and hence the matrix $\boldsymbol{A}$ is non-singular, i.e., $|\boldsymbol{A}| \neq 0$.

Conversely:

Assume that the matrix $\boldsymbol{A}$ is non-singular, i.e., $|\boldsymbol{A}| \neq 0$.

This implies that the Rank of $\boldsymbol{A}$ is n.

Since the Rank of $\boldsymbol{A}$ is n, it implies that the n column vectors of the matrix $\boldsymbol{A}$ are linearly independent.

Thus, $\dim(\text{Range}\,\boldsymbol{T}) = n$.

Hence, the rank nullity theorem implies that $\text{Ker}(\boldsymbol{T}) = 0_w$.

This implies that the linear transformation $\boldsymbol{T} : R^n \rightarrow R^n$ is one to one.

Remark:

A linear transformation $\boldsymbol{T} : R^n \rightarrow R^n$ defined by $\boldsymbol{T}(\boldsymbol{X}) = \boldsymbol{AX}$ with $|\boldsymbol{A}| \neq 0$ is said to be a *non-singular transformation*.

Example 5.6:

Determine whether the linear transformations $\boldsymbol{T_A}$ and $\boldsymbol{T_B}$ defined by the following matrices are one to one or not.

$$(1)\ \boldsymbol{A} = \begin{bmatrix} 1 & -2 & 5 & 7 \\ 0 & 1 & 9 & 8 \\ 0 & 0 & 1 & 3 \end{bmatrix}$$

(2) $\boldsymbol{B} = \begin{bmatrix} 2 & 0 & -1 \\ 3 & 4 & 2 \\ 0 & 7 & 5 \end{bmatrix}$.

Solution:

Since the rows of the matrix $\boldsymbol{A}$ are linearly independent.
It implies that Rank $(\boldsymbol{A}) = 3$.
$\Rightarrow$dim $(\text{Range}(\boldsymbol{T_A})) = 3$.
Now the domain of $\boldsymbol{T_A}$ is R^4.
So, it implies that dim domain $(\boldsymbol{T_A}) = 4$.
Thus, by the rank nullity theorem, dim $(\ker(\boldsymbol{T_A})) = 1$.
And since dim $(\ker(\boldsymbol{T_A})) \neq 0$,
it implies that the Kernel is not the zero vector.
Thus, the mapping is not one to one.
Since $|\boldsymbol{B}| = -9 \neq 0$,
$\Rightarrow$ the matrix $\boldsymbol{B}$ is non-singular;
$\Rightarrow$ the transformation $\boldsymbol{T_B}$ is one to one.

Theorem 5.6:

Let $\boldsymbol{T} : \mathbf{V} \to \mathbf{W}$ be a one-to-one linear transformation. If the set $\{v_1, v_2, \ldots, v_n\}$ is linearly independent in $\mathbf{V}$, then $\{\boldsymbol{T}(v_1), \boldsymbol{T}(v_2), \ldots, \boldsymbol{T}(v_n)\}$ is linearly independent in $\mathbf{W}$, that is, one-to-one linear transformation preserves linear independence.

Proof:

Let us consider the identity

$$a_1\boldsymbol{T}(v_1) + a_2\boldsymbol{T}(v_2) + \cdots + a_nT(v_n) = 0_w,$$

having scalars $a_1, a_2, \ldots, a_n \in \mathbf{F}$.

Since the transformation $\boldsymbol{T} : \mathbf{V} \to \mathbf{W}$ is linear, it may be written as

$$\boldsymbol{T}(a_1v_1 + a_2v_2 + \cdots + a_nv_n) = 0_w.$$

Since the transformation $\boldsymbol{T}$ is one to one, the Kernel is the zero vector, i.e., 0_w. Therefore,

$$a_1v_1 + a_2v_2 + \cdots + a_nv_n = 0_w.$$

But the set $\{v_1, v_2, \ldots, v_n\}$ is linearly independent.
Thus, $a_1 = 0, a_2 = 0, \ldots, a_n = 0$.
$\Rightarrow$The set $\{\boldsymbol{T}(v_1), \boldsymbol{T}(v_2), \ldots, \boldsymbol{T}(v_n)\}$ is linearly independent.

5.4 Invertible Linear Transformation

Definition 5.4:

Let $\boldsymbol{T} : R^n \to R^n$ be a linear transformation. Then the linear transformation T is said to be invertible, if there is a linear transformation $S : R^n \to R^n$, such that $S(\boldsymbol{T}(v)) = v$ and $\boldsymbol{T}(S(v)) = v$ for every vector v in R^n.

Remark:

If such a transformation S exists, then it is unique and linear.

Theorem 5.7:

Let $\boldsymbol{T} : R^n \to R^n$ be a linear transformation. Then the transformation T is *invertible* if and only if it is *non-singular.*

Proof:

Assume that $\boldsymbol{T}$ is invertible, So, by definition, there is a transformation S such that $\boldsymbol{T}(S(u)) = u$ for every vector u in $\mathbb{R}^n$.

Let $S(u) = v$ imply that there is a vector v such that $\boldsymbol{T}(v) = u$ for every vector u in R^n.

$\Rightarrow$ The range of T is thus R^n.

$\Rightarrow \mathrm{Ker}(\boldsymbol{T})$ is the zero vector.

$\Rightarrow$ The transformation $\boldsymbol{T}$ is one to one.

$\Rightarrow$ The transformation $\boldsymbol{T}$ is non-singular.

Conversely:

Assume that the transformation $\boldsymbol{T}$ is non-singular.

Let the standard matrix of $\boldsymbol{T}$ be $\boldsymbol{A}$.

Since $\boldsymbol{T}$ is non-singular, $\boldsymbol{A}$ is invertible.

Thus, $\boldsymbol{A}^{-1}(\boldsymbol{A}u) = u$ and $\boldsymbol{A}\left(\boldsymbol{A}^{-1}u\right) = u$, for every vector u in R^n.

Let S be the linear transformation with the standard matrix $\boldsymbol{A}^{-1}$.

Then $S(\boldsymbol{T}(u)) = u$ and $T(S(u)) = u$.

$\Rightarrow$ The transformation S is invertible of $\boldsymbol{T}$.

$\Rightarrow$ The transformation $\boldsymbol{T}$ is invertible.

Properties of invertible linear transformation:

Let $\boldsymbol{T} : \boldsymbol{X} \to \boldsymbol{AX}$ be a linear transformation defined by $\boldsymbol{T}(u) = Au$. Then the following statements are equivalent.

(1) The transformation $\boldsymbol{T}$ is invertible.
(2) The transformation $\boldsymbol{T}$ is non-singular $(|\boldsymbol{A}| \neq 0)$.
(3) The transformation $\boldsymbol{T}$ is one to one.
(4) The Kernel of the transformation $\boldsymbol{T}$ is the zero vector.

(5) The range of the transformation $\boldsymbol{T}$ is R^n.
(6) The inverse transformation $\boldsymbol{T}^{-1}$ is linear.
(7) The matrix defines the inverse transformation $\boldsymbol{T}^{-1}$ is $\boldsymbol{A}^{-1}$.

Example 5.7:

Let us consider the linear transformation $\boldsymbol{T} : R^2 \to R^2$ defined by

$$\boldsymbol{T}(x, y) = (3x + 4y, 5x + 7y).$$

(1) Prove that the linear transformation $\boldsymbol{T}$ is invertible and find the inverse of the linear transformation $\boldsymbol{T}$.
(2) Find the pre-image of the vector (x, y) where $(x, y) = (1, 2)$.

Solution:

(1) Let the linear transformation $\boldsymbol{T}$ be defined by

$$\boldsymbol{T}\left(\begin{bmatrix} x \\ y \end{bmatrix}\right) = \begin{bmatrix} 3x + 4y \\ 5x + 7y \end{bmatrix}.$$

We find the images of the vectors on the standard basis

$$\boldsymbol{T}\left(\begin{bmatrix} 1 \\ 0 \end{bmatrix}\right) = \begin{bmatrix} 3 \\ 5 \end{bmatrix}, \boldsymbol{T}\left(\begin{bmatrix} 0 \\ 1 \end{bmatrix}\right) = \begin{bmatrix} 4 \\ 7 \end{bmatrix}.$$

These are the columns of the standard matrix $\boldsymbol{A}$ of $\boldsymbol{T}$.
We find that $\boldsymbol{A}$ is invertible, proving that $\boldsymbol{T}$ has an inverse.

$$\boldsymbol{A} = \begin{bmatrix} 3 & 4 \\ 5 & 7 \end{bmatrix}, \boldsymbol{A}^{-1} = \begin{bmatrix} 7 & -4 \\ -5 & 3 \end{bmatrix}.$$

$\boldsymbol{A}^{-1}$ is the standard matrix of $\boldsymbol{T}^{-1}$, and we get

$$\boldsymbol{T}^{-1}\left(\begin{bmatrix} x \\ y \end{bmatrix}\right) = \begin{bmatrix} 7 & -4 \\ -5 & 3 \end{bmatrix}\begin{bmatrix} x \\ y \end{bmatrix} = \begin{bmatrix} 7x - 4y \\ -5x + 3y \end{bmatrix}.$$

Writing in row form for convenience

$$\boldsymbol{T}^{-1}(x, y) = (7x - 4y, -5x + 3y).$$

(2) The pre-image of $(1, 2)$will be $\boldsymbol{T}^{-1}(1, 2)$, i.e., $\boldsymbol{T}^{-1}(1, 2) = (-1, 1)$, as shown in Figure 5.4.

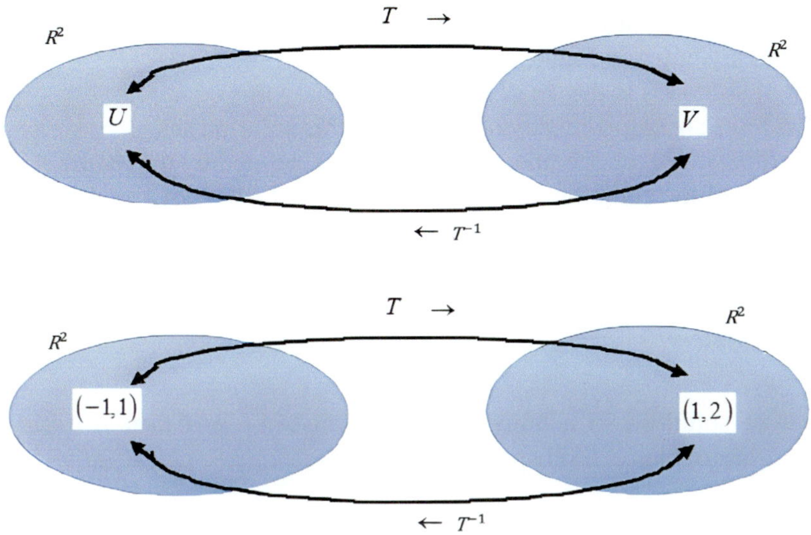

Figure 5.4 Image and Preimage of transformation.

5.5 Transformation and Systems of Linear Equations

Linear transformations and the concepts of Kernel and the range of a linear transformation play an essential role in analyzing linear equations.

We find that they enable us to visualize the set of solutions.

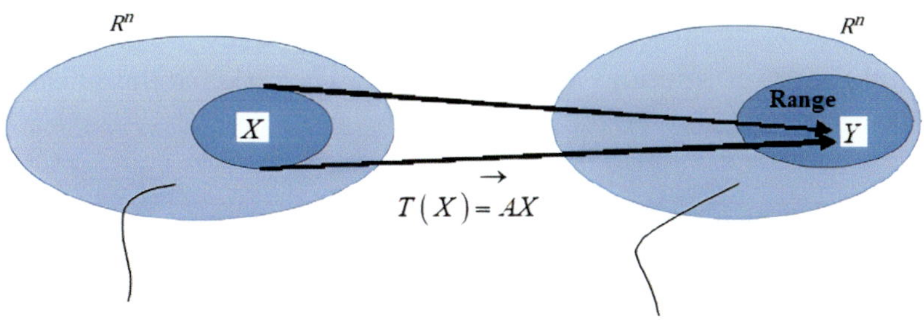

Set of all solutions to $\boldsymbol{AX} = \boldsymbol{Y}$. If $\boldsymbol{Y}$ is not in the range (all vectors mapped into $\boldsymbol{Y}$), then there is no solution.

Figure 5.5 Transformation of Systems of Linear Equations.

A system of m-linear equations in n-variables can be written in the matrix form $\boldsymbol{AX} = \boldsymbol{Y}$, where $\boldsymbol{A}$ is an $m \times n$ matrix, which is the matrix of coefficient of the system, as shown in Figure 5.5.

The set of solutions is the set of all $\boldsymbol{X}$'s that satisfy this equation.

We now have an exquisite way of looking at this solution set.

Let $\boldsymbol{T} : R^n \rightarrow R^m$ be the linear transformation defined by the matrix $\boldsymbol{A}$.

The system of the equation can now be written as $\boldsymbol{T}(\boldsymbol{X}) = \boldsymbol{AX} = \boldsymbol{Y}$.

It implies that the set of solutions is thus the set of vectors in R^n that are mapped by $\boldsymbol{T}$ into the vector $\boldsymbol{Y}$. If $\boldsymbol{Y}$ is not in the range of $\boldsymbol{T}$, then the system has no solution.

Homogenous system:

Theorem 5.8:

The set of solutions to a homogenous system $\boldsymbol{AX} = \mathbf{0_v}$ of n-linear equations in n-variables is a subspace of R^n.

Proof:

Let $\boldsymbol{T} : R^n \rightarrow R^n$ be a linear transformation defined by $\boldsymbol{T}(\boldsymbol{X}) = \boldsymbol{AX}$.

The set of solution vectors $\boldsymbol{X}$ is the set of vectors in R^n mapped by the linear transformation $\boldsymbol{T}$ into the zero vector $\mathbf{0_V}$.

The set of solutions $\boldsymbol{X}$ is the Kernel of the transformation $\boldsymbol{T}$ and thus forms a subspace.

Example 5.8:

Solve the following homogenous system of linear equations.

$$\begin{aligned} x_1 + 2x_2 + 3x_3 &= 0 \\ -x_2 + x_3 &= 0 \\ x_1 + x_2 + 4x_3 &= 0 \end{aligned} \quad .$$

Interpret the set of solutions $\boldsymbol{X}$ as a subspace. Then, sketch the subspace of solutions.

Solution:

Using the Gauss elimination method, we get

$$\begin{bmatrix} 1 & 2 & 3 & 0 \\ 0 & -1 & 1 & 0 \\ 1 & 1 & 4 & 0 \end{bmatrix} \approx \begin{bmatrix} 1 & 2 & 3 & 0 \\ 0 & -1 & 1 & 0 \\ 0 & -1 & 1 & 0 \end{bmatrix}$$

$$\approx \begin{bmatrix} 1 & 2 & 3 & 0 \\ 0 & 1 & -1 & 0 \\ 0 & -1 & 1 & 0 \end{bmatrix} \approx \begin{bmatrix} 1 & 0 & 5 & 0 \\ 0 & 1 & -1 & 0 \\ 0 & 0 & 0 & 0 \end{bmatrix}.$$

Thus,

$$x_1 + 5x_3 = 0$$
$$x_2 - x_3 = 0$$
$$\Rightarrow x_1 = -5x_3\ , x_2 = x_3.$$

On assigning the value r to x_3, an arbitrary solution is thus

$$x_1 = -5r\ , x_2 = r\ , x_3 = r.$$

The solutions of the system are vectors of the form $(-5r\ , r\ , r)$.

These vectors $(-5r\ , r\ , r)$ form a one-dimensional subspace of R^3 having a basis $(-5, 1, 1)$.

This subspace is the Kernel of the transformation T defined by the matrix of coefficient **A** of the system, as shown in Figure 5.6.

$$\begin{bmatrix} 1 & 2 & 3 \\ 0 & -1 & 1 \\ 1 & 1 & 4 \end{bmatrix}.$$

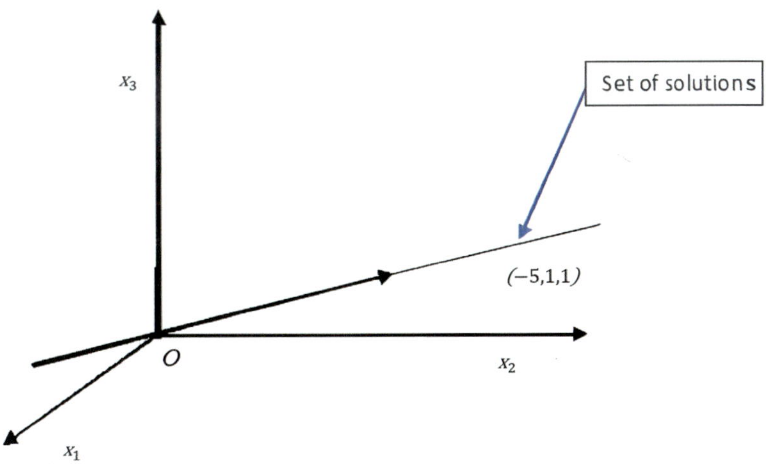

Figure 5.6 Subspace of solutions.

Non-homogenous system:

We now find that the set of solutions to a non-homogenous system of linear equations does not form a subspace.

Let $\boldsymbol{AX} = \boldsymbol{Y}, (\boldsymbol{Y} \neq 0)$ be a non-homogenous system of linear equations.
Let $\boldsymbol{X}_1$ and $\boldsymbol{X}_2$ be solutions.
Thus,

$$\begin{aligned} \boldsymbol{AX}_1 = \boldsymbol{Y} \text{ and } \boldsymbol{AX}_2 = \boldsymbol{Y}. \\ \Rightarrow \boldsymbol{AX}_1 + \boldsymbol{AX}_2 = 2\boldsymbol{Y} \\ \Rightarrow \boldsymbol{A}(\boldsymbol{X}_1 + \boldsymbol{X}_2) = 2\boldsymbol{Y}. \end{aligned}$$

Therefore, $\boldsymbol{X}_1 + \boldsymbol{X}_2$ does not satisfy $\boldsymbol{AX} = \boldsymbol{Y}$.
$\Rightarrow$ It is not a solution.
The set of solutions is not closed under addition.
$\Rightarrow$ It is not a subspace.

Remarks:

The set of solutions to a *non-homogenous* system is not itself a subspace. The collection can be obtained by sliding a specific subspace.

Theorem 5.9:

Let us consider a non-homogenous system $\boldsymbol{AX} = \boldsymbol{Y}$ of m-linear equations in n-variables. Suppose there is a particular solution $\boldsymbol{X}_1$ of the non-homogenous system. Every other solution of $\boldsymbol{AX} = \boldsymbol{Y}$ is expressed in the form $\boldsymbol{X} = z + \boldsymbol{X}_1$, where z is an element of the Kernel of the transformation $\boldsymbol{T}$ and the solution are unique if the Kernel $\boldsymbol{T}$ consists of the zero vector only.

Proof:

Let $\boldsymbol{X}_1$ be a solution to the non-homogenous system $\boldsymbol{AX} = \boldsymbol{Y}$.
Then $\boldsymbol{AX}_1 = \boldsymbol{Y}$ and let $\boldsymbol{X}$ be an arbitrary solution of $\boldsymbol{AX} = \boldsymbol{Y}$.
On equating both the systems $\boldsymbol{AX}_1$ and $\boldsymbol{AX}$, we get

$$\begin{aligned} \boldsymbol{AX}_1 = \boldsymbol{AX} \\ \Rightarrow \boldsymbol{AX} - \boldsymbol{AX}_1 = \mathbf{0_w} \\ \Rightarrow \boldsymbol{A}(\boldsymbol{X} - \boldsymbol{X}_1) = \mathbf{0_w} \\ \Rightarrow \boldsymbol{T}(\boldsymbol{X} - \boldsymbol{X}_1) = \mathbf{0_w} \end{aligned}.$$

Thus, $\boldsymbol{X} - \boldsymbol{X}_1 \in \operatorname{Ker}(\boldsymbol{T})$ and call it z.

$$\begin{aligned} \Rightarrow \boldsymbol{X} - \boldsymbol{X}_1 = z \\ \Rightarrow \boldsymbol{X} = z + \boldsymbol{X}_1 \end{aligned}.$$

Note that this solution is unique if and only if the value of z is $\mathbf{0_v}$, that is, if the $\operatorname{Ker}(\boldsymbol{T})$ is the zero vector.

Remark:

This result implies that the set of solutions to a non-homogenous system of the linear equation $\boldsymbol{AX} = \boldsymbol{Y}$ can be generated from the Kernel of the transformation defined by the matrix of coefficient and a particular solution $\boldsymbol{X}_1$. If we take any vector z in the Kernel and add $\boldsymbol{X}_1$ to it, then we get a solution.

Geometrically:

It means that the set of solutions is obtained by sliding the Kernel in the direction and distance defined by the vector $\boldsymbol{X}_1$, as shown in Figure 5.7.

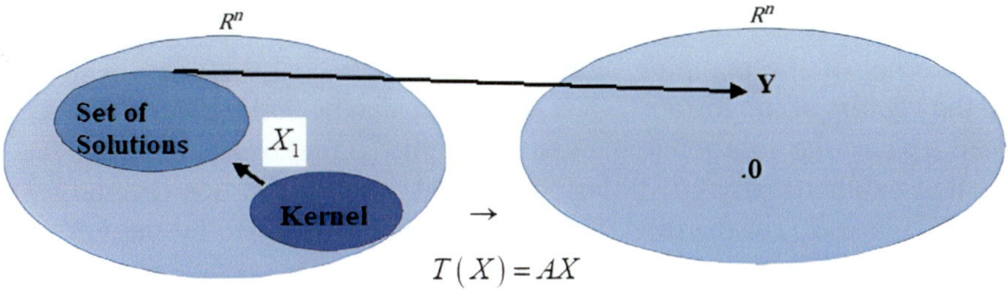

Set of solutions to *AX=Y*.

Figure 5.7 Set of solutions to $\boldsymbol{AX} = \boldsymbol{Y}$.

Example 5.9:

Solve the following system of linear equations and sketch the set of solutions:

$$\begin{aligned} x_1 + 2x_2 + 3x_3 &= 11 \\ -x_2 + x_3 &= -2 \\ x_1 + x_2 + 4x_3 &= 9 \end{aligned}$$

Solution:

Using the Gauss-Jordan elimination method, we get

$$\begin{bmatrix} 1 & 2 & 3 & 11 \\ 0 & -1 & 1 & -2 \\ 1 & 1 & 4 & 9 \end{bmatrix} \approx \begin{bmatrix} 1 & 2 & 3 & 11 \\ 0 & -1 & 1 & -2 \\ 0 & -1 & 1 & -2 \end{bmatrix}$$

$$\approx \begin{bmatrix} 1 & 2 & 3 & 11 \\ 0 & 1 & -1 & 2 \\ 0 & -1 & 1 & -2 \end{bmatrix} \approx \begin{bmatrix} 1 & 0 & 5 & 7 \\ 0 & 1 & -1 & 2 \\ 0 & 0 & 0 & 0 \end{bmatrix}$$

$$\Rightarrow x_1 + 5x_3 = 7$$
$$x_2 - x_3 = 2$$

$$\Rightarrow x_1 = -5x_3 + 7$$
$$x_2 = x_3 + 2$$

Hence, the solution set is $(-5r + 7, r + 2, r)$.

Here, we separate the part involving the parameter r from the constant part. Thus, the part involving r will be in the Kernel of the transformation $\boldsymbol{T}$ defined by $\boldsymbol{T}(\boldsymbol{X}) = \boldsymbol{AX}$, while the continual part will be a particular solution $\boldsymbol{X}_1$ to $\boldsymbol{AX} = \boldsymbol{Y}$.

$$(-5r + 7, r + 2, r) = r(-5, 1, 1) + (7, 2, 0).$$

The Kernel of the mapping defined by the matrix of the coefficient is indeed the vectors of the form $r(-5, 1, 1)$. It can also be verified that $(7, 2, 0)$ is indeed a particular solution to the given system. The set of solutions can be represented geometrically by sliding the Kernel, namely the line defined by the vector $(-5, 1, 1)$, in the direction and distance determined by the vector $(7, 2, 0)$, as shown in Figure 5.8.

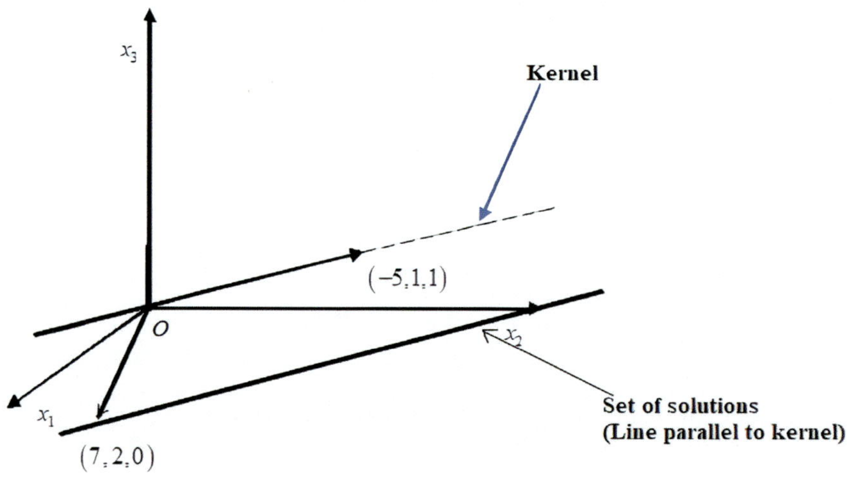

Figure 5.8 Set of solutions to system of linear equations.

Many systems:

Consider several linear systems of homogenous equations

$$\boldsymbol{AX} = \boldsymbol{Y}_1$$

$$
\begin{aligned}
AX &= Y_2 \\
AX &= Y_3 \\
&\ldots\ldots \\
AX &= Y_n
\end{aligned}
$$

All systems are having the same matrix of coefficient A. Let T be the linear transformation defined by the matrix coefficient A and let $X_1, X_2, X_3, \ldots$ be particular solutions to these systems.

Then the sets of the solution to these systems are

$$
\begin{aligned}
&Ker\,(T) + X_1 \\
&Ker\,(T) + X_2 \\
&Ker\,(T) + X_3 \\
&\ldots\ldots
\end{aligned}
$$

These sets are parallel sets, each being the $Ker\,(T)$ that is translated by the amounts $X_1, X_2, X_3, \ldots$

Thus, for example, the solutions to the systems

$$
\begin{aligned}
x_1 + 2x_2 + 3x_3 &= a_1 \\
-x_2 + x_3 &= a_2 \\
x_1 + x_2 + 4x_3 &= a_3
\end{aligned}
$$

will all be straight lines parallel to the line defined by the vector $(-5, 1, 1)$.

5.6 Coordinate Representation

This section discusses the relationship between coordinate system and bases. We have found that a linear transformation can be represented by a matrix relative to a standard basis. We shall see that there is a matrix representation relative to every basis. It will be interesting to find diagonal representation, if possible, and determine the basis (or coordinate system) for which this applies.

Eigenvalues and eigenvectors play an essential role in these discussions. These techniques will enable us to find the most suitable coordinate system for discussing physical situations such as vibrating string.

5.6.1 Coordinate vectors

We have discussed various types of vector spaces, the vector spaces R^n, the space of matrices, and the space of functions.

This section shall find how we can use vectors in R^n called coordinate vectors to describe vectors in any real finite-dimensional vector spaces.

This discussion concludes that all finite-dimensional vector spaces are in some mathematical sense the same as R^n.

Definition 5.5:

Let $\mathbf{V}$ be a vector space with a basis $\boldsymbol{B} = \{u_1, u_2, \ldots, u_n\}$, and let u be a vector in $\mathbf{V}$.

We know that there exist scalars $a_1, a_2, \ldots, a_n$ such that

$$u = a_1u_1 + a_2u_2 + \cdots + a_nu_n.$$

The column vector $\boldsymbol{V_B} = \begin{bmatrix} a_1 \\ a_2 \\ \vdots \\ a_n \end{bmatrix}$ is called the *coordinate vector* of $\mathbf{V}$ relative to this basis. The scalars $a_1, a_2, \ldots, a_n$ are called the coordinates of $\mathbf{V}$ relative to the basis.

Remark:

Here we shall use a column vector form for coordinate vectors rather than row vectors.

Example 5.10:

Find the coordinate vectors of $\bar{u} = (4, 5)$ relative to the following bases $\boldsymbol{B}$ and $\boldsymbol{B}_r$ of R^2.

(1) The standard basis $\boldsymbol{B} = \{(1, 0), (0, 1)\}$.
(2) $\boldsymbol{B}_r = \{(2, 1), (-1, 1)\}$.

Solution:

By observation, we see that

$$(4, 5) = 4(1, 0) + 5(0, 1).$$

Thus, $\boldsymbol{V_B} = \begin{bmatrix} 4 \\ 5 \end{bmatrix}$.

The given representation of $\mathbf{U}$ is, in fact, relative to the standard basis.

Let us now find the coordinate vector of $\mathbf{V}$ relative to $\boldsymbol{B}_r$, which is not the standard basis.

Let $(4, 5) = a_1(2, 1) + a_2(-1, 1)$.

Thus,

$$\begin{aligned}(4,5) &= (2a_1, a_1) + (-a_2, a_2) \\ &= (2a_1 - a_2, a_1 + a_2)\end{aligned}$$

Comparing components leads to the following system of equations:

$$\begin{aligned}2a_1 - a_2 &= 4 \\ a_1 + a_2 &= 5\end{aligned}$$

This system has a unique solution $a_1 = 3$, $a_2 = 2$.

Thus, $\boldsymbol{V}_{\boldsymbol{B}_r} = \begin{bmatrix} 3 \\ 2 \end{bmatrix}$.

These coordinate vectors have geometrical interpretation, as shown in Figure 5.9.

Denote the basis vectors as follows:

$$u_1 = (1,0) \ , u_2 = (1,0) \ \text{ and}$$

$$u_1' = (2,1) \ , u_2' = (-1,1) \, .$$

We can write

$$u = 4u_1 + 5u_2 \ \text{ and } \ \mathrm{u} = 3\mathrm{u}_1' + 2\mathrm{u}_2'$$

Geometrical representation:

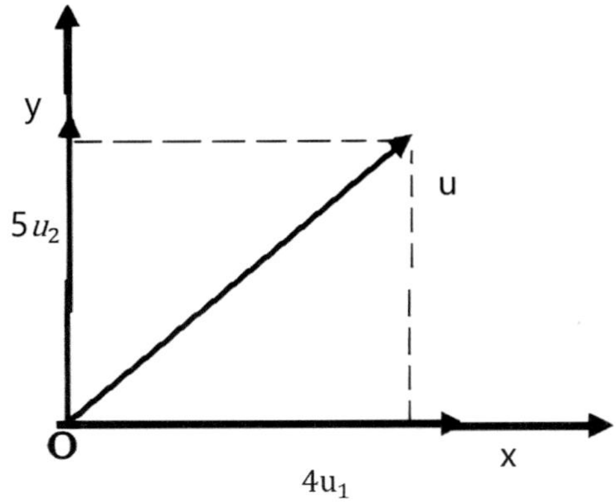

$u = 4u_1 + 5u_2$

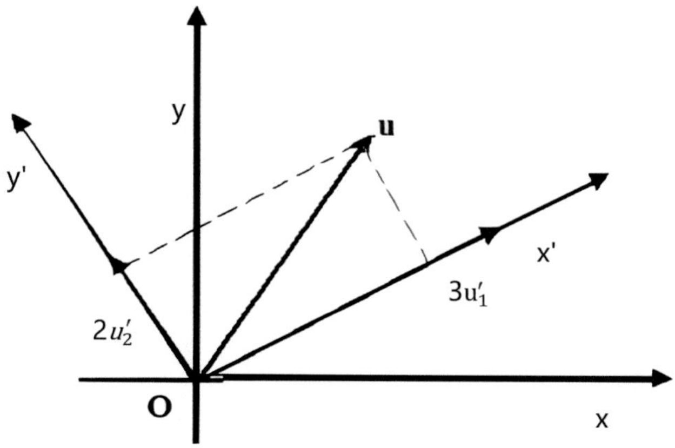

$$u = 3u_1' + 2u_2'$$

Figure 5.9 Coordinate vectors relative to Basis.

In general, a linear equation system has to be solved to find a coordinate vector relative to another basis. However, coordinate vectors can efficiently compute relative to an orthonormal basis.

Example 5.11:

If $\boldsymbol{B} = \{u_1, u_2, \ldots, u_n\}$ is an orthonormal basis for a vector space $\mathbf{V}$, then an arbitrary vector v in $\mathbf{V}$ can be expressed as

$$v = (v.u_1)\, u_1 + (v.u_2)\, u_2 + \cdots + (v.u_n)\, u_n.$$

Thus, the coordinate vector is

$$\boldsymbol{V_B} = \begin{bmatrix} v.u_1 \\ v.u_2 \\ \vdots \\ v.u_n \end{bmatrix}.$$

Example 5.12:

Find the coordinate vector of $(2, -5, 10)$ relative to the orthonormal basis,

$$\boldsymbol{B} = \left\{ (1, 0, 0), \left(0, \frac{3}{5}, \frac{4}{5}\right), \left(0, \frac{4}{5}, -\frac{3}{5}\right) \right\}.$$

Solution:

We get

$$(2,-5,10)\,.\,(1,0,0)=2$$
$$(2,-5,10)\,.\left(0,\frac{3}{5},\frac{4}{5}\right)=5$$
$$(2,-5,10)\,.\left(0,\frac{4}{5},-\frac{3}{5}\right)=-10.$$

Thus, $\boldsymbol{V_B}=\begin{bmatrix}2\\5\\-10\end{bmatrix}.$

Note:

There are occasions when bases other than orthonormal bases better fit the situation. It becomes necessary to know how coordinate vectors relative to different bases are related.

This is the topic of our subsequent discussion.

5.7 Change of Basis

Let $\boldsymbol{B}=\{u_1,u_2,\ldots,u_n\}$ and $\boldsymbol{B}_r=\{u'_1,u'_2,\ldots,u'_n\}$ be bases for a vector space $\mathbf{V}$. A vector u in $\mathbf{V}$ will have coordinate vectors $\boldsymbol{V_B}$ and $\boldsymbol{V_{B_r}}$ relative to these bases. We now discuss the relationship between $\boldsymbol{V_B}$ and $\boldsymbol{V_{B_r}}$.

Let the coordinate vectors of $u_1,u_2,\ldots,u_n$ relative to the basis $\boldsymbol{B}_r=\{u'_1,u'_2,\ldots,u'_n\}$ be $(u_1)_{\boldsymbol{B}_r},(u_2)_{\boldsymbol{B}_r},\ldots,(u_n)_{B_r}$.

The matrix $\boldsymbol{P}$ having these vectors as columns plays a central role in our discussion. It is called the *transition matrix* from the basis $\boldsymbol{B}$ to the basis $\boldsymbol{B}_r$.

Transition matrix:

$$\boldsymbol{P}=\left\{(u_1)_{\boldsymbol{B}_r},(u_2)_{\boldsymbol{B}_r},\ldots,(u_n)_{\boldsymbol{B}_r}\right\}.$$

Theorem 5.10:

Let $\boldsymbol{B}=\{u_1,u_2,\ldots,u_n\}$ and $\boldsymbol{B}_r=\{u'_1,u'_2,\ldots,u'_n\}$ be bases for a vector space $\mathbf{V}$. If u is a vector in $\boldsymbol{V}$ having coordinate vectors $\boldsymbol{V_B}$ and $\boldsymbol{V_{B_r}}$ relative to these bases, then $\boldsymbol{V_{B_r}}=\boldsymbol{PV_B}$. Here $\boldsymbol{P}$ is the transition matrix from B to B_r, that is, $\boldsymbol{P}=\left\{(u_1)_{B_r},(u_2)_{B_r},\ldots,(u_n)_{B_r}\right\}$.

Proof:

Since $\{u'_1,u'_2,\ldots,u'_n\}$ is the basis for $\mathbf{V}$, each of the vectors $u_1,u_2,\ldots,u_n$ of $\mathbf{V}$ can be expressed as a linear combination of these vectors.

Let

$$\begin{array}{l} u_1 = c_{11}u'_1 + c_{21}u'_2 + \cdots + c_{n1}u'_n \\ u_2 = c_{12}u'_1 + c_{22}u'_2 + \cdots + c_{n2}u'_n \\ \cdots\cdots\cdots\cdots\cdots\cdots\cdots\cdots\cdots\cdots \\ u_n = c_{1n}u'_1 + c_{2n}u'_2 + \cdots + c_{nn}u'_n \end{array}.$$

If $u = a_1u_1 + a_2u_2 + \cdots + a_nu_n$, then we get

$$\begin{aligned} u &= a_1u_1 + a_2u_2 + \cdots + a_nu_n \\ &= a_1\left(c_{11}u'_1 + c_{21}u'_2 + \cdots + c_{n1}u'_n\right) + a_2\left(c_{12}u'_1 + c_{22}u'_2 + \cdots + c_{n2}u'_n\right) \\ &\qquad + \cdots + a_n\left(c_{1n}u'_1 + c_{2n}u'_2 + \cdots + c_{nn}u'_n\right) \\ &= \left(a_1c_{11} + a_2c_{12} + \cdots + a_nc_{1n}\right)u'_1 + \left(a_1c_{21} + a_2c_{22} + \cdots + a_nc_{2n}\right)u'_2 \\ &\qquad + \cdots + \left(a_1c_{n1} + a_2c_{n2} + \cdots + a_nc_{nn}\right)u'_n \end{aligned}.$$

The coordinate vector of **V** relative to $\boldsymbol{B}_r$ can therefore be written as

$$\begin{aligned} \boldsymbol{V}_{\boldsymbol{B}_r} &= \begin{bmatrix} a_1c_{11} + a_2c_{12} + \cdots + a_nc_{1n} \\ a_1c_{21} + a_2c_{22} + \cdots + a_nc_{2n} \\ \vdots \\ a_1c_{n1} + a_2c_{n2} + \cdots + a_nc_{nn} \end{bmatrix} \\ &= \begin{bmatrix} c_{11}, c_{12}, \ldots, c_{1n} \\ c_{21}, c_{22}, \ldots, c_{2n} \\ \vdots \\ c_{n1}, c_{n2}, \ldots, c_{nn} \end{bmatrix} \begin{bmatrix} a_1 \\ a_2 \\ \vdots \\ a_n \end{bmatrix} \\ &= \left[(u_1)_{\boldsymbol{B}_r}, (u_2)_{\boldsymbol{B}_r}, \ldots, (u_n)_{\boldsymbol{B}_r}\right] \boldsymbol{V}_{\boldsymbol{B}} \end{aligned}.$$

Example 5.13:

Consider the bases $\boldsymbol{B} = \{(1,2), (3,-1)\}$ and $\boldsymbol{B}_r = \{(1,0), (0,1)\}$of R^2. If u is a vector in $\boldsymbol{B}$ such that $\boldsymbol{V}_{\boldsymbol{B}} = \begin{bmatrix} 3 \\ 4 \end{bmatrix}$, then find $\boldsymbol{V}_{\boldsymbol{B}_r}$.

Solution:

We express the vectors of $\boldsymbol{B}$ in terms of the vector of $\boldsymbol{B}_r$ to get the transition matrix, i.e.,

$$\begin{array}{r} (1,2) = 1\,(1,0) + 2\,(0,1) \\ (3,-1) = 3\,(1,0) - 1\,(0,1) \end{array}.$$

The coordinate vectors of $(1,2)$ and $(3,-1)$ are $\begin{bmatrix} 1 \\ 2 \end{bmatrix}$ and $\begin{bmatrix} 3 \\ -1 \end{bmatrix}$.

The transition matrix $\boldsymbol{P}$ is thus

$$\boldsymbol{P} = \begin{bmatrix} 1 & 3 \\ 2 & -1 \end{bmatrix}.$$

(Observe that the columns of $\boldsymbol{P}$ are the vectors of the basis $\boldsymbol{B}$.)
We get

$$\boldsymbol{V}_{\boldsymbol{B}_r} = \begin{bmatrix} 1 & 3 \\ 2 & -1 \end{bmatrix} \begin{bmatrix} 3 \\ 4 \end{bmatrix} = \begin{bmatrix} 15 \\ 2 \end{bmatrix}.$$

Let $\boldsymbol{B}$ and $\boldsymbol{B}_r$ be bases for a vector space. We now see that the transition matrices from $\boldsymbol{B}$ to $\boldsymbol{B}_r$ and $\boldsymbol{B}_r$ to $\boldsymbol{B}$ are related.

Theorem 5.11:

Let $\boldsymbol{B}$ and $\boldsymbol{B}_r$ be bases for a vector space $\mathbf{V}$ and let $\boldsymbol{P}$ be the transition matrix from the bases $\boldsymbol{B}$ to $\boldsymbol{B}_r$, then the transition matrix $\boldsymbol{P}$ is invertible and the transition matrix from $\boldsymbol{B}_r$ to $\boldsymbol{B}$ is $\boldsymbol{P}^{-1}$.

Proof:

Let u be a vector in $\mathbf{V}$ having column coordinate vectors and relative to the bases $\boldsymbol{B}$ and $\boldsymbol{B}_r$.

Given $\boldsymbol{P}$ is a transition matrix from $\boldsymbol{B}$ to $\boldsymbol{B}_r$, and let $\boldsymbol{Q}$ be the transition matrix from B_r to $\boldsymbol{B}$. Then we know that

$$\mathbf{V}_{\boldsymbol{B}_r} = \boldsymbol{P}\mathbf{V}_{\boldsymbol{B}} \text{ and } \mathbf{V}_{\boldsymbol{B}} = \boldsymbol{Q}\boldsymbol{V}_{\boldsymbol{B}_r}$$

.

We can combine these equations in two ways.

On substituting $\mathbf{V}_{\boldsymbol{B}}$ from the second equation into the first, and similarly substituting for $\mathbf{V}_{\boldsymbol{B}_r}$ from the first into the second, we get

$$\mathbf{V}_{\boldsymbol{B}_r} = \boldsymbol{P}\boldsymbol{Q}\boldsymbol{V}_{\boldsymbol{B}_r} \text{ and } \mathbf{V}_{\boldsymbol{B}} = \boldsymbol{Q}\boldsymbol{P}\boldsymbol{V}_{\boldsymbol{B}}$$

.

Since these results hold for all values of $\mathbf{V}_{\boldsymbol{B}}$ and $\mathbf{V}_{\boldsymbol{B}_r}$, we have that

$$\boldsymbol{P}\boldsymbol{Q} = \boldsymbol{Q}\boldsymbol{P} = I$$

$$\Rightarrow \boldsymbol{Q} = \boldsymbol{P}^{-1}.$$

The following example introduces a beneficial technique for finding the transition matrix from one basis to another if neither basis is the standard basis.

Example 5.14:

Consider the bases $\boldsymbol{B} = \{(1,2),(3,-1)\}$ and $\boldsymbol{B}_r = \{(3,1),(5,2)\}$ of R^2. Then, find the transition matrix from $\boldsymbol{B}$ to $\boldsymbol{B}_r$, if $\mathbf{V}$ is a vector such that $\mathbf{V}_{\boldsymbol{B}} = \begin{bmatrix} 2 \\ 1 \end{bmatrix}$ and find $\mathbf{V}_{\boldsymbol{B}_r}$.

Solution:

Let us use the standard basis $S = \{(1,0),(0,1)\}$ as an intermediate basis.

The transition matrix $\boldsymbol{P}$ from $\boldsymbol{B}$ to S and the transition matrix $\boldsymbol{P}_r$ from $\boldsymbol{B}_r$ to S are

$$\boldsymbol{P} = \begin{bmatrix} 1 & 3 \\ 2 & -1 \end{bmatrix}, \boldsymbol{P}_r = \begin{bmatrix} 3 & 5 \\ 1 & 2 \end{bmatrix}.$$

The transition matrix from S to $\boldsymbol{B}_r$ will be $(\boldsymbol{P}_r)^{-1}$.

The transition matrix from $\boldsymbol{B}$ to $\boldsymbol{B}_r$ is $(\boldsymbol{P}_r)^{-1}\boldsymbol{P}$.

Thus, the transition matrix from $\boldsymbol{B}$ to $\boldsymbol{B}_r$ is

$$\begin{bmatrix} 3 & 5 \\ 1 & 2 \end{bmatrix}^{-1} \begin{bmatrix} 1 & 3 \\ 2 & -1 \end{bmatrix} = \begin{bmatrix} 2 & -5 \\ -1 & 3 \end{bmatrix} \begin{bmatrix} 1 & 3 \\ 2 & -1 \end{bmatrix} = \begin{bmatrix} -8 & 11 \\ 5 & -6 \end{bmatrix}$$

$$\Rightarrow \mathbf{U}_{\boldsymbol{B}_r} = \begin{bmatrix} -8 & 11 \\ 5 & -6 \end{bmatrix} \begin{bmatrix} 2 \\ 1 \end{bmatrix} = \begin{bmatrix} -5 \\ 4 \end{bmatrix}.$$

Remark from the last theorem of isomorphism:

Every real finite-dimensional vector space $\mathbf{V}$ is isomorphic to R^n for some value of n. Thus, every real finite-dimensional vector space is identical from the algebraic viewpoint to R^n.

Example 5.15:

Prove that the linear map $\boldsymbol{T} : V_3 \to V_3$ defined by $\boldsymbol{T}(e_1) = e_1 - e_2$, $\boldsymbol{T}(e_2) = 2e_2 + e_3$, and $\boldsymbol{T}(e_3) = e_1 + e_2 + e_3$ is neither one-one nor onto.

Since $[e_1, e_2, e_3] = V_3$,

$$\begin{aligned} R(\boldsymbol{T}) &= [\boldsymbol{T}(e_1), \boldsymbol{T}(e_2), \boldsymbol{T}(e_3)] \\ &= [e_1 - e_2, 2e_2 + e_3, e_1 + e_2 + e_3] \; . \\ &= [e_1 - e_2, 2e_2 + e_3] \end{aligned}$$

Because $e_1 + e_2 + e_3$ is a linear combination of $e_1 - e_2$ and $2e_2 + e_3$, now we see that $e_1 - e_2$ and $2e_2 + e_3$ are LI. So, dim $R(\boldsymbol{T}) = 2$. Therefore, $R(\boldsymbol{T})$ is a proper subset of V_3. Hence, $\boldsymbol{T}$ is not onto.

To prove that $\boldsymbol{T}$ is not one-one, we check $N(\boldsymbol{T}).N(\boldsymbol{T})$ consists of those vectors (x_1, x_2, x_3) in V_3 for which

$$\boldsymbol{T}(x_1, x_2, x_3) = 0$$

or $$\boldsymbol{T}(x_1e_1 + x_2e_2 + x_3e_3) = 0$$
or $$x_1\boldsymbol{T}(e_1) + x_2\boldsymbol{T}(e_2) + x_3\boldsymbol{T}(e_3) = 0.$$
Because $\boldsymbol{T}$ is linear,

$$(x_1 + x_3, -x_1 + 2x_2 + x_3, x_2 + x_3) = (0, 0, 0),$$

i.e., $x_1 + x_3 = 0$, $x_2 + x_3 = 0$ and $-x_1 + 2x_2 + x_3 = 0$.
Solving these, we get $x_1 = x_2 = -x_3$. Therefore,

$$N(\boldsymbol{T}) = \{(x_1, x_1, -x_1) : x_1 \; is \, an \, arbitrary \, scalar\} = [(1, 1, -1)].$$

Hence, the linear map is not one-one.

Example 5.16:

Let $\boldsymbol{T} : V_4 \to V_3$ be a linear map defined by
$\boldsymbol{T}(e_1) = (1, 1, 1)$, $\boldsymbol{T}(e_2) = (1, -1, 1)$, $\boldsymbol{T}(e_3) = (1, 0, 0)$, $\boldsymbol{T}(e_4) = (1, 0, 1)$, and then verify that

$$r(\boldsymbol{T}) + n(\boldsymbol{T}) = \dim U \, (= V_4) = 4.$$

We know that $R(\boldsymbol{T}) = [(1, 1, 1), (1, -1, 1), (1, 0, 0), (1, 0, 1)]$.

$(1, 1, 1), (1, -1, 1), (1, 0, 0)$ and $(1, 0, 1)$ are LD, because a set of four vectors of $V_3(\dim V_3 = 3)$ is always LD. We find that

$$(1, 0, 1) = \frac{1}{2}(1, 1, 1) + \frac{1}{2}(1, -1, 1) + 0\,(1, 0, 0).$$

Hence, we can discard the vector $(1, 0, 1)$, so that

$$R(\boldsymbol{T}) = [(1, 1, 1), (1, -1, 1), (1, 0, 0)].$$

To check whether $(1, 1, 1), (1, -1, 1)$, and $(1, 0, 0)$ are LI, we suppose

$$\alpha_1(1, 1, 1) + \alpha_2(1, -1, 1) + \alpha_3(1, 0, 0) = 0 = (0, 0, 0)$$

or $$(\alpha_1 + \alpha_2 + \alpha_3, \alpha_1 - \alpha_2, \alpha_1 + \alpha_2) = (0, 0, 0).$$
Solving this, we get $\alpha_1 = 0 = \alpha_2 = \alpha_3$. Hence, $(1, 1, 1), (1, -1, 1)$, and $(1, 0, 0)$ are LI and

$$\dim R(\boldsymbol{T}) = r(\boldsymbol{T}) = 3.$$

Now to find $N(\boldsymbol{T})$, we suppose that $\boldsymbol{T}(u) = 0 = (0, 0, 0)$. If

$$u = (x_1, x_2, x_3, x_4) = x_1e_1 + x_2e_2 + x_3e_3 + x_4e_4.$$

Then $\boldsymbol{T}(x_1, x_2, x_3, x_4) = \boldsymbol{T}(x_1e_1 + x_2e_2 + x_3e_3 + x_4e_4) = (0, 0, 0)$.

Or $(x_1 + x_2 + x_3 + x_4, x_1 - x_2, x_1 + x_2 + x_4) = (0, 0, 0)$.

Solving this, we get $x_1 = x_2 = -\frac{x_4}{2}\ x_3 = 0$. So $N(\boldsymbol{T})$ contains the vectors of the form $(x_1, x_1, 0, -2x_1)$, i.e., $N(\boldsymbol{T}) = [(1, 1, 0, -2)]$. So $n(\boldsymbol{T}) = \dim N(\boldsymbol{T}) = 1$. Hence, $r(\boldsymbol{T}) + n(\boldsymbol{T}) = 3 + 1 = 4$ and the theorem is verified.

The inverse of a linear transformation:

Example 5.17:

Prove that the linear map $\boldsymbol{T} : V_3 \to V_3$ defined by

$\boldsymbol{T}(e_1) = e_1 + e_2$, $\boldsymbol{T}(e_2) = e_2 + e_3$, $\boldsymbol{T}(e_3) = e_1 + e_2 + e_3$ is non-singular, and find its inverse.

First, let us find the value of $\boldsymbol{T}$ at a general element $u = (x_1, x_2, x_3)$:

$$\begin{aligned} \boldsymbol{T}(x_1, x_2, x_3) &= \boldsymbol{T}(x_1 e_1 + x_2 e_2 + x_3 e_3) \\ &= (x_1 + x_3, x_1 + x_2 + x_3, x_2 + x_3) \end{aligned}.$$

If $\boldsymbol{T}(x_1, x_2, x_3) = 0$, then

$$x_1 + x_3 = 0, x_1 + x_2 + x_3 = 0, x_2 + x_3 = 0.$$

Solving these, we get $x_1 = 0 = x_2 = x_3$.

So $N(\boldsymbol{T}) = \{0_{v_3}\}$ and hence $\boldsymbol{T}$ is one to one. It follows that $\boldsymbol{T}$ is also onto. Hence, $\boldsymbol{T}$ is non-singular and $\boldsymbol{T}^{-1}$ exists.

Now we shall give two methods to find the inverse $\boldsymbol{T}^{-1}$, which are also linear, one-one, and onto map from V_3 to V_3.

Method-I:

We have $\boldsymbol{T}(e_1) = e_1 + e_2$, $\boldsymbol{T}(e_2) = e_2 + e_3$, $\boldsymbol{T}(e_3) = e_1 + e_2 + e_3$. Therefore,

$$\begin{aligned} e_1 &= \boldsymbol{T}^{-1}(e_1 + e_2) = \boldsymbol{T}^{-1}(e_1) + \boldsymbol{T}^{-1}(e_2) \\ e_2 &= \boldsymbol{T}^{-1}(e_2 + e_3) = \boldsymbol{T}^{-1}(e_2) + \boldsymbol{T}^{-1}(e_3) \\ e_3 &= \boldsymbol{T}^{-1}(e_1 + e_2 + e_3) = \boldsymbol{T}^{-1}(e_1) + \boldsymbol{T}^{-1}(e_2) + \boldsymbol{T}^{-1}(e_3) \end{aligned}.$$

Because $\boldsymbol{T}^{-1}$ is linear, one-one, and onto, solving these three equations for $\boldsymbol{T}^{-1}(e_1)$,$\boldsymbol{T}^{-1}(e_2)$,$\boldsymbol{T}^{-1}(e_2)$, we get

$$\begin{aligned} \boldsymbol{T}^{-1}(e_1) &= e_3 - e_2 = (0, -1, 1) \\ \boldsymbol{T}^{-1}(e_2) &= e_1 + e_2 - e_3 = (1, 1, -1) \\ \boldsymbol{T}^{-1}(e_3) &= e_3 - e_1 = (-1, 0, 1) \end{aligned}.$$

Now we extend $\boldsymbol{T}^{-1}$ linearly and obtain

$$\begin{aligned}\boldsymbol{T}^{-1}(x_1,x_2,x_3) &= \boldsymbol{T}^{-1}(x_1e_1+x_2e_2+x_3e_3)\\ &= x_1\boldsymbol{T}^{-1}(e_1)+x_2\boldsymbol{T}^{-1}(e_2)+x_3\boldsymbol{T}^{-1}(e_3)\ .\\ &= (x_2-x_3, x_2-x_1, x_1-x_2+x_3)\end{aligned}$$

Method 2:

Let $\boldsymbol{T}^{-1}(x_1,x_2,x_3)=(y_1,y_2,y_3)$. Then

$$\boldsymbol{T}(y_1,y_2,y_3)=(x_1,x_2,x_3)$$

or $$\boldsymbol{T}(y_1e_1+y_2e_2+y_3e_3)=(x_1,x_2,x_3)$$

or $$y_1\boldsymbol{T}(e_1)+y_2\boldsymbol{T}(e_2)+y_3\boldsymbol{T}(e_3)=(x_1,x_2,x_3)$$

or $$(y_1+y_3)e_1+(y_1+y_2+y_3)e_2+(y_2+y_3)e_3=(x_1,x_2,x_3)$$

or $$(y_1+y_3, y_1+y_2+y_3, y_2+y_3)=(x_1,x_2,x_3).$$

This gives $y_1+y_3=x_1$, $y_1+y_2+y_3=x_2$ and $y_2+y_3=x_3$.
Solving these, we get $y_1=x_2-x_3$, $y_2=x_2-x_1$ and $y_3=x_1-x_2+x_3$.
So $\boldsymbol{T}^{-1}(x_1,x_2,x_3)=(x_2-x_3, x_2-x_1, x_1-x_2+x_3)$.

5.8 Isomorphism

Let $\boldsymbol{T}:\mathbf{U}\to\mathbf{V}$ be a linear transformation. The linear transformation $\boldsymbol{T}$ is *one to one* if each element in the range $\boldsymbol{T}$ corresponds to just one domain component $\boldsymbol{T}$.

If every element of **V** is the *image of a component* of **U**, then the transformation $\boldsymbol{T}$ is said to be *onto*.

Let $\boldsymbol{T}$ be a *one-to-one* linear transformation of **U** on to **V**; then $\boldsymbol{T}$ is called an *isomorphism*. **U** and **V** are called isomorphic vector spaces.

Theorem 5.12:

Let **V** be a real vector space with basis $\{u_1,u_2,\ldots,u_n\}$. Let u be an arbitrary element of **V** with coordinate vector $\begin{bmatrix}a_1\\a_2\\\vdots\\a_n\end{bmatrix}$ relative to this basis.

The following transformation $\boldsymbol{T}$ is an isomorphism of **V** on to R^n defined by

$$\boldsymbol{T}(u)=\begin{bmatrix}a_1\\a_2\\\vdots\\a_n\end{bmatrix}.$$

Proof:

First, to show that the transformation $\boldsymbol{T}$ is one to one.

For it, if $\boldsymbol{T}(v_1) = \boldsymbol{T}(v_2)$, then it implies that $v_1 = v_2$.

Let $\boldsymbol{T}(v_1) = \begin{bmatrix} a_1 \\ a_2 \\ \vdots \\ a_n \end{bmatrix}$ and $\boldsymbol{T}(v_2) = \begin{bmatrix} a_1 \\ a_2 \\ \vdots \\ a_n \end{bmatrix}$.

Thus, $v_1 = a_1u_1 + a_2u_2 + \cdots + a_nu_n$

and $v_2 = a_1u_1 + a_2u_2 + \cdots + a_nu_n$.

This implies that $v_1 = v_2$.

Thus, the linear transformation $\boldsymbol{T}$ is one to one.

We now prove that the linear transformation $\boldsymbol{T}$ is onto by showing that every element of R^n is the image of some element $\mathbf{V}$.

Let $\begin{bmatrix} b_1 \\ b_2 \\ \vdots \\ b_n \end{bmatrix}$ be an element of $\mathbb{R}^n$.

Then

$$\boldsymbol{T}(b_1u_1 + b_2u_2 + \cdots + b_nu_n) = \begin{bmatrix} b_1 \\ b_2 \\ \vdots \\ b_n \end{bmatrix}.$$

Thus, $\begin{bmatrix} b_1 \\ b_2 \\ \vdots \\ b_n \end{bmatrix}$ is the image of the vector $b_1u_1 + b_2u_2 + \cdots + b_nu_n$.

Therefore, the linear transformation $\boldsymbol{T}$ is onto.

Since the transformation $\boldsymbol{T}$ is linear, one-one, and onto, the linear transformation implies that the transformation is isomorphism.

5.9 Transformations in Computer Graphics

Computer graphics is the field that analyzes the creation and manipulation of pictures with the help of computers. The effect of computer graphics is going through in many homes through video games; its applications in industry, research, and business are vast and are ever-expanding. Architects employ computer graphics to explore designs; molecular biologists display and manipulate pictures of molecules to gain insight into their structures. Pilots are competent using graphics flight simulators, and transportation engineers use computer-generated transforms in their planning work to refer to a few applications.

The manipulation of pictures in computer graphics is carried out using a sequence of transformations. Matrices define rotations, reflections, dilations, and contractions. A sequence of such transformations can be carried out by a single transformation described by the product of the matrices. Inappropriately, translation as it now stands uses matrix addition, and any sequence of transformation involving translations cannot be combined in this manner into a single matrix. However, if coordinates called *homogenous coordinates* are used to describe points in a plane, translation can also be accomplished through matrix multiplication. Any sequence of these transformations can be described in terms of a single matrix. In homogenous coordinates, the third component of 1 is added to each coordinate and rotation, reflection, dilation/contraction, and translation *R*, *F*, *D*, and ***T*** are defined by the following matrices.

$$\underset{\text{point}}{\overset{x}{\begin{bmatrix} x \\ y \\ 1 \end{bmatrix}}} \qquad \boldsymbol{A} = \underset{\text{rotation}}{\overset{R}{\begin{bmatrix} \cos\theta & -\sin\theta & 0 \\ \sin\theta & \cos\theta & 0 \\ 0 & 0 & 1 \end{bmatrix}}} \qquad \boldsymbol{B} = \underset{\text{reflection}}{\overset{F}{\begin{bmatrix} 1 & 0 & 0 \\ 0 & -1 & 0 \\ 0 & 0 & 1 \end{bmatrix}}}$$

$$\mathrm{C} = \underset{\substack{\text{dilation/contraction} \\ (r>0)}}{\overset{D}{\begin{bmatrix} r & 0 & 0 \\ 0 & r & 0 \\ 0 & 0 & 1 \end{bmatrix}}} \qquad E = \underset{\text{translation}}{\overset{\boldsymbol{T}}{\begin{bmatrix} 1 & 0 & h \\ 0 & 1 & k \\ 0 & 0 & 1 \end{bmatrix}}}.$$

Thus, a dilation D followed by a translation $\boldsymbol{T}$ and then a rotation R would be defined by $R^{\circ}\boldsymbol{T}^{\circ}D\,(x) = AEC\,(x)$. A single matrix AEC would characterize the composite transformation $R^{\circ}\boldsymbol{T}^{\circ}D$.

Some programming languages determine subroutines for rotation, translation, and dilation/contraction that can be used to move pictures on the screen. The subroutines convert screen coordinates into homogenous coordinates and use the matrices that define these transformations in homogenous coordinates to carry out this movement.

We now demonstrate how the transformations are used to rotate a geometrical figure about a point other than the origin.

Example 5.18:

Determine the matrix that illustrates a rotation of points in a plane through an angle θ about a point $\boldsymbol{P}(h, k)$. Use this typical result to find the matrix that describes a rotation of the points through an angle of $\frac{\pi}{2}$ about the point $(5, 4)$. Determine the image of the triangle having the following vertices $\boldsymbol{A}(1, 2)$, $\boldsymbol{B}(2, 8)$, and $\boldsymbol{C}(3, 2)$ under this rotation (Figure 5.10).

Solution:

The rotation about $\boldsymbol{P}$ can be accomplished by a sequence of three of the above transformations:

(1) A translation $\boldsymbol{T}_1$ that takes $\boldsymbol{P}$ to the origin O.
(2) A rotation R about the origin through an angle θ.
(3) A translation $\boldsymbol{T}_2$ that takes back O to $\boldsymbol{P}$.

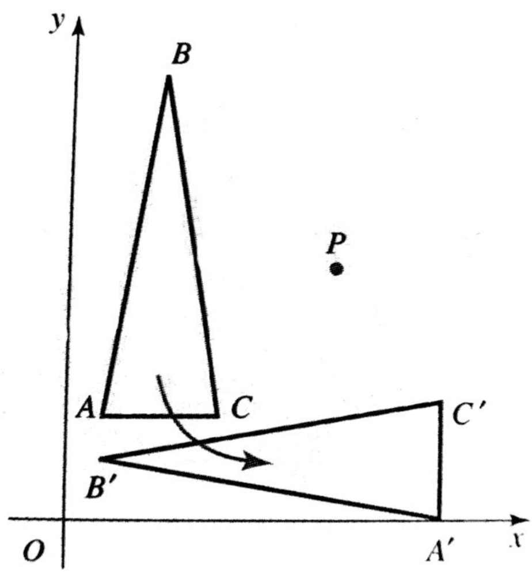

Figure 5.10 Rotation about ***P***.

The matrices that determine these transformations are as follows:

$$\underset{\boldsymbol{T}_1}{\begin{bmatrix} 1 & 0 & -h \\ 0 & 1 & -k \\ 0 & 0 & 1 \end{bmatrix}} \quad \underset{R}{\begin{bmatrix} \cos\theta & -\sin\theta & 0 \\ \sin\theta & \cos\theta & 0 \\ 0 & 0 & 1 \end{bmatrix}} \quad \underset{\boldsymbol{T}_2}{\begin{bmatrix} 1 & 0 & h \\ 0 & 1 & k \\ 0 & 0 & 1 \end{bmatrix}}.$$

The rotation $R_{\boldsymbol{P}}$ about $\boldsymbol{P}$ can be accomplished as follows:

$$R_P\left(\begin{bmatrix} x \\ y \\ 1 \end{bmatrix}\right) = \boldsymbol{T}_2{}^\circ R^\circ \boldsymbol{T}_1\left(\begin{bmatrix} x \\ y \\ 1 \end{bmatrix}\right)$$

$$= \begin{bmatrix} 1 & 0 & h \\ 0 & 1 & k \\ 0 & 0 & 1 \end{bmatrix}\begin{bmatrix} \cos\theta & -\sin\theta & 0 \\ \sin\theta & \cos\theta & 0 \\ 0 & 0 & 1 \end{bmatrix} \times \begin{bmatrix} 1 & 0 & -h \\ 0 & 1 & -k \\ 0 & 0 & 1 \end{bmatrix}\begin{bmatrix} x \\ y \\ 1 \end{bmatrix}$$

$$= \begin{bmatrix} \cos\theta & -\sin\theta & -h\cos\theta + k\sin\theta + h \\ \sin\theta & \cos\theta & -h\sin\theta - k\cos\theta + k \\ 0 & 0 & 1 \end{bmatrix}\begin{bmatrix} x \\ y \\ 1 \end{bmatrix}.$$

Let $h = 5$, $k = 4$, and $\theta = \frac{\pi}{2}$, to get the specific matrix that describes the rotation of the plane through an angle $\frac{\pi}{2}$ about the point $\boldsymbol{P}(5, 4)$, as shown in Figure 5.10.

For example, let $h = 5$, $k = 4$, and $\theta = \frac{\pi}{2}$. The rotation matrix is

$$M = \begin{bmatrix} 0 & -1 & 9 \\ 1 & 0 & -1 \\ 0 & 0 & 1 \end{bmatrix}.$$

To find the images of the vertices of the triangles, write these vertices in columns form as homogenous coordinates and multiply by M. On performing the matrix multiplications, we get

$$\overset{\boldsymbol{A}}{\begin{bmatrix} 1 \\ 2 \\ 1 \end{bmatrix}} \to \overset{A'}{\begin{bmatrix} 7 \\ 0 \\ 1 \end{bmatrix}} \quad \overset{\boldsymbol{B}}{\begin{bmatrix} 2 \\ 8 \\ 1 \end{bmatrix}} \to \overset{\boldsymbol{B}'}{\begin{bmatrix} 1 \\ 1 \\ 1 \end{bmatrix}} \quad \overset{C}{\begin{bmatrix} 3 \\ 2 \\ 1 \end{bmatrix}} \to \overset{\boldsymbol{C}'}{\begin{bmatrix} 7 \\ 2 \\ 1 \end{bmatrix}}.$$

The triangle with vertices $\boldsymbol{A}(1, 2), \boldsymbol{B}(2, 8), \boldsymbol{C}(3, 2)$ is transformed into triangles with vertices $\boldsymbol{A}'(7, 0), \boldsymbol{B}'(1, 1), \boldsymbol{C}'(7, 2)$ (Figure 5.10).

5.10 Fractal Pictures of Nature

Computer graphics systems positioned on traditional Euclidean geometry are suitable for creating pictures of artificial objects such as machinery, building and airplanes. Images of such objects can be formed using circles, lines, and so on. However, these techniques are not suitable for constructing images of natural objects such as trees, animals, and landscapes. In the words of mathematician Benoit B. Mandelbrot, "Clouds are not spheres, mountains are not cones, Coastlines are not circles, and bark is not smooth nor does lightning travelling in straight lines." However, nature does get on its abnormality in an abruptly orderly fashion; it is full of shapes that repeat themselves on different scales within the same object. In 1975, Mandelbrot introduced a new geometry called *fractal geometry* that can be used to characterize natural phenomena. A *fractal* is an appropriate label for irregular and fragmented self-similar shapes. Fractal objects include structures nested within one another. Each smaller structure is a miniature, though not necessarily identical version of the larger form. The story behind the word fractal is appealing. Mandelbrot came across the Latin adjective *fractus* from the verb *frangere* to break in his son's Latin book. The resonance of the foremost English cognates fracture and fraction seemed appropriate, and he coined the word fractal!

We now discuss the methods developed by a research team at the Georgia Institute of Technology to form images of natural objects using fractals. These fractal images of nature are created using affine transformations. The figure shows a fractal image of the fern being gradually generated. Let us see how this is done.

Figure 5.11 A fractal image of the fern.

Consider the following four affine transformations $\boldsymbol{T}_1, \ldots, \boldsymbol{T}_4$ and associated probabilities $p_1, \ldots, p_4$ with these transformations

$$\boldsymbol{T}_1\left(\begin{bmatrix} x \\ y \end{bmatrix}\right) = \begin{bmatrix} 0.86 & 0.03 \\ -0.03 & 0.86 \end{bmatrix}\begin{bmatrix} x \\ y \end{bmatrix} + \begin{bmatrix} 0 \\ 1.5 \end{bmatrix}, p_1 = 0.83$$

$$\boldsymbol{T}_2\left(\begin{bmatrix} x \\ y \end{bmatrix}\right) = \begin{bmatrix} 0.2 & -0.25 \\ 0.21 & 0.23 \end{bmatrix}\begin{bmatrix} x \\ y \end{bmatrix} + \begin{bmatrix} 0 \\ 1.5 \end{bmatrix}, p_2 = 0.08$$

$$\boldsymbol{T}_3\left(\begin{bmatrix} x \\ y \end{bmatrix}\right) = \begin{bmatrix} -0.15 & 0.27 \\ 0.25 & 0.26 \end{bmatrix}\begin{bmatrix} x \\ y \end{bmatrix} + \begin{bmatrix} 0 \\ 0.45 \end{bmatrix}, p_3 = 0.08$$

$$T_4\left(\begin{bmatrix} x \\ y \end{bmatrix}\right) = \begin{bmatrix} 0 & 0 \\ 0 & 0.17 \end{bmatrix}\begin{bmatrix} x \\ y \end{bmatrix} + \begin{bmatrix} 0 \\ 0 \end{bmatrix}, p_4 = 0.01 \ .$$

The following algorithm is used on a computer to produce the images of the form.

1. Let $x = 0, y = 0$.
2. Use a random generator to select one of the affine transformations $\boldsymbol{T}_i$ according to the given probabilities.
3. Let $(x', y') = \boldsymbol{T}_i\,(x, y)$.
4. Plot (x', y').
5. Let $(x, y) = (x', y')$.
6. Repeat steps 2, 3, 4, and 5 twenty thousand times.

As step 4 is executed, each of the twenty thousand times, the image of the fern gradually appears.

Each affine transformation $\boldsymbol{T}_i$ involves six parameters a, b, c, d, e, f and a probability p_i as follows:

$$\boldsymbol{T}_i\left(\begin{bmatrix} x \\ y \end{bmatrix}\right) = \begin{bmatrix} a & b \\ c & d \end{bmatrix}\begin{bmatrix} x \\ y \end{bmatrix} + \begin{bmatrix} e \\ f \end{bmatrix}, p_i.$$

The affine transformation corresponding probabilities that generate a fractal are written as rows of a matrix called an *iterated function system* (IFS). The IFS for the fern is as follows:

IFS for a fern

T	a	b	c	d	e	f	p
1	0.86	0.03	−0.03	0.86	0	1.5	0.83
2	0.2	−0.25	0.21	0.23	0	1.5	0.08
3	−0.15	0.27	0.25	0.26	0	0.45	0.08
4	0	0	0	0.17	0	0	0.01

The appropriate affine transformation that produces a given fractal object is found by determining the transformation that maps the objects (called the *attractor*) into various disjoint images, the union of which is the whole fractal. A theorem called the *collage theorem* guarantees that the transformations could be grouped into an IFS that produces that fractal.

Different probabilities do not generally lead to different images, but they affect the rate at which the image is produced. Appropriate probabilities are

$$p_i = \frac{\text{area of the image under transformation } \boldsymbol{T}_{\text{i}}}{\text{area of image of object}}.$$

These techniques are very valuable because they can be used to produce an image to any desired degree of accuracy using a highly compressed dataset. A fractal image containing infinitely many points whose organization is too complicated to describe directly can be reproduced using mathematical formulas.

Exercises

1. Let U and V be vector spaces over the same field of scalars and $\boldsymbol{T}$ be a map from U to V. Then prove that $\boldsymbol{T}$ is linear iff $\boldsymbol{T}(\alpha u_1 + u_2) = \alpha \boldsymbol{T}(u_1) + \boldsymbol{T}(u_2)$ for all $u_1, u_2 \in U$ and scalar α.
2. Which of the following maps are linear?
   (a) $\boldsymbol{T} : V_1 \to V_3$ defined by $\boldsymbol{T}(x) = (x, 2x, 3x)$.
   (b) $\boldsymbol{T} : V_1 \to V_3$ defined by $\boldsymbol{T}(x) = \left(x, x^2, x^3\right)$.
   (c) $\boldsymbol{T} : {V_2}^C \to {V_3}^C$ defined by $\boldsymbol{T}(x, y) = (x + \alpha, y, 0)\ , \alpha \neq 0$.
   (d) $\boldsymbol{T} : V_2 \to V_2$ defined by $\boldsymbol{T}(x, y) = (2x + 3y, 3x - 4y)$.
   (e) $\boldsymbol{T} : V_3 \to V_3$ defined by $\boldsymbol{T}(x, y, z) = \left(x^2 + xy, xy, yz\right)$.
   (f) $\boldsymbol{T} : V_3 \to V_2$ defined by $\boldsymbol{T}(x, y, z) = (x, y)$.
   (g) $\boldsymbol{T} : {V_2}^C \to {V_2}^C$ defined by $\boldsymbol{T}(x, y) = (y, x)$.
   (h) $\boldsymbol{T} : V_3 \to V_2$ defined by $\boldsymbol{T}(x, y, z) = (x + y + z, 0)$.
   (i) $\boldsymbol{T} : \boldsymbol{P} \to \boldsymbol{P}$ defined by $\boldsymbol{T}(\boldsymbol{P}) = p^2 + p$.
   (j) $\boldsymbol{T} : \boldsymbol{P} \to \boldsymbol{P}$ defined by $\boldsymbol{T}(p)(x) = xp(x) + p(1)$.
   (k) $\boldsymbol{T} : C[0, 1] \to V_2$ defined by $\boldsymbol{T}(f) = (f(0), f(1))$.
   (l) $\boldsymbol{T} : \boldsymbol{P} \to \boldsymbol{P}$ defined by $\boldsymbol{T}(p)(x) = 2 + 3x + 7x^2 p(x)$.
   (m) $\boldsymbol{T} : \boldsymbol{P} \to \boldsymbol{P}$ defined by $\boldsymbol{T}(p)(x) = p(0) + xp'(0) + \frac{x^2}{2!}p''(0)$.
   (n) $\boldsymbol{T} : C^{(n)}(a, b) \to C(a, b)$ defined by $\boldsymbol{T}(f) = a_0 f + a_1 f' + \cdots + a_n f^{(n)}$, a_i's are fixed scalars.
   (o) $\boldsymbol{T} : C^{(2)}(a, b) \to C(a, b)$ defined by $\boldsymbol{T}(f) = \left(3x^2 + 4\right) f'' + (7x + 3) f' + (3x + 5) f$.
   (p) $\boldsymbol{T} : \boldsymbol{P} \to \boldsymbol{P}$ defined by $\boldsymbol{T}(p) = p(0)$.
   (q) $\boldsymbol{T} : \boldsymbol{P} \to \boldsymbol{P}$ defined by $\boldsymbol{T}(p) = p'$.
3. Determine whether there exists a linear map in the following cases and where it does exist give the general formula.
   (a) $\boldsymbol{T} : V_2 \to V_2$ such that $\boldsymbol{T}(1, 2) = (3, 0)$ and $\boldsymbol{T}(2, 1) = (1, 2)$.
   (b) $\boldsymbol{T} : V_2 \to V_2$ such that $\boldsymbol{T}(2, 1) = (2, 1)$ and $\boldsymbol{T}(1, 2) = (4, 2)$.

(c) $\boldsymbol{T} : V_2 \to V_2$ such that $\boldsymbol{T}(0,1) = (3,4)$, $\boldsymbol{T}(3,1) = (2,2)$ and $\boldsymbol{T}(3,2) = (5,7)$.

(d) $\boldsymbol{T} : V_3 \to V_3$ such that $\boldsymbol{T}(0,1,2) = (3,1,2)$ and $\boldsymbol{T}(1,1,1) = (2,2,2)$.

(e) $\boldsymbol{T} : \boldsymbol{P}_3 \to \boldsymbol{P}_3$ such that $\boldsymbol{T}(1+x) = 1+x$, $\boldsymbol{T}(2+x) = x+3x^2$ and $\boldsymbol{T}\left(x^2\right) = 0$.

(f) $\boldsymbol{T} : \boldsymbol{P}_4 \to \boldsymbol{P}_3$ such that $\boldsymbol{T}(1+x) = 1$, $\boldsymbol{T}(x) = 3$ and $\boldsymbol{T}\left(x^2\right) = 4$.

(g) $\boldsymbol{T} : V_2^C \to V_2^C$ such that $\boldsymbol{T}(i,i) = (1+i,1)$.

4. Determine a non-zero linear transformation $\boldsymbol{T} : V_2 \to V_2$, which maps all the vectors on the line $x = y$ on to the origin.
5. Determine a linear transformation, $\boldsymbol{T} : V_2 \to V_2$, which maps all the vectors on the line $x + y = 0$ on to themselves ($\boldsymbol{T} \neq \boldsymbol{I}$).
6. Let $\boldsymbol{T} : {V_2}^C \to {V_2}^C$ be defined by $\boldsymbol{T}(\alpha_1 + i\beta_1, \alpha_2 + i\beta_2) = (\alpha_1, \alpha_2)$, and then prove or disprove that $\boldsymbol{T}$ is linear.
7. Prove that a linear transformation on a one-dimensional vector space is nothing but multiplication by a fixed scalar.

Range and Kernel:

8. Determine the range of the following linear transformations. Also, find the rank of $\boldsymbol{T}$, where it exists.

(a) $\boldsymbol{T} : V_2 \to V_2$ defined by $\boldsymbol{T}(x_1, x_2) = (x_1 + x_2, x_1)$.

(b) $\boldsymbol{T} : V_2 \to V_3$ defined by $\boldsymbol{T}(x_1, x_2) = (x_1, x_1 + x_2, x_2)$.

(c) $\boldsymbol{T} : V_3 \to V_3$ defined by $\boldsymbol{T}(x_1, x_2, x_3) = \left(\frac{1}{2}x_1 + x_2 + x_3, x_1 - \frac{1}{3}x_2, x_3\right)$.

(d) $\boldsymbol{T} : V_3 \to V_3$ defined by $\boldsymbol{T}(x_1, x_2, x_3) = (x_1, x_3, x_2)$.

(e) $\boldsymbol{T} : V_4 \to V_3$ defined by $\boldsymbol{T}(x_1, x_2, x_3, x_4) = (x_1 - x_4, x_2 + x_3, x_3 - x_4)$.

(f) $\boldsymbol{T} : V_3 \to V_4$ defined by $\boldsymbol{T}(x_1, x_2, x_3) = (x_1, x_1 + x_2, x_1 + x_2 + x_3, x_3)$.

(g) $\boldsymbol{T} : V_4 \to V_4$ defined by $\boldsymbol{T}(x_1, x_2, x_3, x_4) = \left(3x_1 + 2x_2, x_1 - x_3, \frac{1}{3}x_1 - x_4, x_2\right)$.

(h) $\boldsymbol{T} : \boldsymbol{P} \to \boldsymbol{P}$ defined by $\boldsymbol{T}(p)(x) = xp(x)$.

(i) $\boldsymbol{T} : \boldsymbol{P} \to \boldsymbol{P}$ defined by $\boldsymbol{T}(p)(x) = xp'(x)$.

(j) $\boldsymbol{T} : \boldsymbol{P} \to \boldsymbol{P}$ defined by $\boldsymbol{T}(p)(x) = p''(x) - 2p(x)$.

(k) $\boldsymbol{T} : C(0,1) \to C(0,1)$ defined by $\boldsymbol{T}(f)(x) = f(x)\sin x$.

(l) $\boldsymbol{T} : C^{(1)}(0,1) \to C(0,1)$ defined by $\boldsymbol{T}(f)(x) = f'(x)e^x$.

9. Determine the Kernel of the linear transformations of problem 1(a) –(i). Also, find the nullity of $\boldsymbol{T}$, where it exists.
10. Let $\boldsymbol{T} : V \to W$ be a linear map and U be a subspace of V. Define $\boldsymbol{T}(U) = \{w \in W : w = \boldsymbol{T}(u) \ for \ some \ u \in U\}$. Then prove that $\boldsymbol{T}(U)$ is a subspace of W.
11. Let $\boldsymbol{T} : V \to W$ be a linear map and W_1 be a subspace of W. Then prove that the set $\{v \in V : \boldsymbol{T}(v) \in W_1\}$ is a subspace of V.
12. Let $\alpha_{11}, \alpha_{12}, \ldots, \alpha_{1n}, \alpha_{21}, \ldots, \alpha_{2n}, \ldots, \alpha_{p1}, \ldots, \alpha_{pn}$ be any pn fixed scalars. Let $\boldsymbol{T} : V_n \to V_p$ be a linear map defined by $\boldsymbol{T}(e_i) = (\alpha_{1i}, \alpha_{2i}, \ldots, \alpha_{pi})$, $i = 1, 2, \ldots, n$. Then prove that
 $\boldsymbol{T}$ is not one-one if $p < n$.
 $\boldsymbol{T}$ is onto
 when $p = n$ and $(\alpha_{11}, \ldots, \alpha_{p1}), (\alpha_{12}, \ldots, \alpha_{p2}), ..., (\alpha_{1p}, \ldots, \alpha_{pp})$ are LI.
13. Find a linear transformation $\boldsymbol{T} : V_3 \to V_3$ such that the set of all vectors (x_1, x_2, x_3) satisfying the equation $4x_1 - 3x_2 + x_3 = 0$ is the Kernel of $\boldsymbol{T}$.
14. Find a linear transformation $\boldsymbol{T} : V_3 \to V_3$ such that the set of all vectors (x_1, x_2, x_3) satisfying the equation $4x_1 - 3x_2 + x_3 = 0$ is the range of $\boldsymbol{T}$.
15. Pick out the maps in problem 1 that are
 (a) one-one (b) onto (c) one-one and onto.
16. Let U be a vector space of dimension n and $\boldsymbol{T} : U \to V$ be linear and onto map. Then prove that $\boldsymbol{T}$ is one-one iff $\dim V = n$.
17. If $\boldsymbol{T} : U \to V$ is a linear map, where U is finite-dimensional, prove that (a) $n(\boldsymbol{T}) \leq \dim U$ (b) $r(\boldsymbol{T}) \leq \min(\dim U, \dim V)$
18. Let Z be a subspace of a finite-dimensional vector space U and $\mathbf{V}$ a finite-dimensional vector space. Then prove that Z will be the Kernel of a linear map $\boldsymbol{T} : U \to V$ iff $\dim Z \geq \dim U - \dim V$.
19. Let R, S, and $\boldsymbol{T}$ be three linear maps from V_3 to V_3 defined by the values in the given table. Determine which of them are non-singular and in each such case find the inverse.

Value at	e_1	e_2	e_3
Linear map			
R	$e_1 + e_2$	$e_1 - e_2 + e_3$	$3e_1 + 4e_3$
S	$e_1 - e_3$	e_2	$e_1 + e_2 - 7e_3$
$\boldsymbol{T}$	$e_1 - e_2 + e_3$	$3e_1 - 5e_3$	$3e_1 - 2e_3$

20. Show each of the following maps is non-singular and find its inverse:
    (a) $\boldsymbol{T} : V_2 \to V_2$ defined by $\boldsymbol{T}(x_1, x_2) = (\alpha_1 x_1, \alpha_2 x_2)$ where α_1 and α_2 are both non-zero.
    (b) $\boldsymbol{T} : V_3 \to V_3$ defined by $\boldsymbol{T}(x_1, x_2, x_3)$
    $= (x_1 + x_2 + x_3,\ x_3 + x_2, x_3)$.
    (c) $\boldsymbol{T} : \boldsymbol{P}_2 \to \boldsymbol{P}_2$ defined by $\boldsymbol{T}\left(\alpha_0 + \alpha_1 x + \alpha_2 x^2\right) = (\alpha_0 + \alpha_1) + (\alpha_1 + 2\alpha_2)\, x + (\alpha_0 + \alpha_1 + 3\alpha_2)\, x^2$.
21. Let U be the subset $\{p \in \boldsymbol{P} : p(0) = 0\}$ of $\boldsymbol{P}$.
    Then prove that the derivative D is a non-singular linear map from U to $\boldsymbol{P}$ and the integral $(\boldsymbol{I}(p)(x)) = \int_0^x p(x)\, dx$ is its inverse.
22. Let $\boldsymbol{T} : U \to V$ be a non-singular linear transformation. Then prove that $\left(\boldsymbol{T}^{-1}\right)^{-1} = \boldsymbol{T}$.

The space $L(U, V)$**:**

23. Let the linear maps $\boldsymbol{T} : V_2 \to V_2$ and $S : V_2 \to V_2$ be defined by

$$\boldsymbol{T}(x_1, x_2) = (x_1 + x_2, 0)$$
$$S(x_1, x_2) = (2x_1, 3x_1 + 4x_2)$$

    Determine the linear maps (a) $2S + 3\boldsymbol{T}$ and (b) $3S - 7\boldsymbol{T}$.
24. Let the linear maps $\boldsymbol{T} : V_3 \to V$ and $S : V_3 \to V_3$ be defined by

$$\boldsymbol{T}(x_1, x_2, x_3) = (2x_1 - 3x_2, 4x_1 + 6x_2, x_3)$$

$$S(e_1) = e_2 - e_3\,, S(e_2) = e_1\,,\ S(e_3) = e_1 + e_2 + e_3.$$

    Determine the linear maps
    (a) $S + \boldsymbol{T}$ (b) $3S - 2\boldsymbol{T}$ (c) αS and find their values at (x_1, x_2, x_3).
25. Prove that the set of all linear maps from V_2 to V_2, which maps the vectors on the line $x + y = 0$ onto the origin, is a subspace of $L(V_2, V_2)$.
26. Let U be a subspace of a vector space V. Then prove that the set of all linear transformations from V to V that vanish on U is a subspace of $L(V, V)$.

6

Inner Product Spaces

This chapter discusses the generalization process of the algebraic properties to R^n, i.e., extending the concepts of dot products of two vectors to the norm, the angle between vectors, and the distance between points in general vector spaces.

This study will enable us to discuss the magnitude of orthogonal functions and approximate functions by polynomials, a technique used to implement functions on calculations and computers.

It will no longer be restricted to Euclidean geometry and create geometries on Rn, such as introducing the geometry of special relativity and discussing the implication of this theory for space travel by looking at a voyage to the star Alpha Centauri.

6.1 Inner Product Spaces

The primary objective of this section is to generalize the scalar product R^n to a general vector space with a mathematical symbol, i.e., an inner product. Later on, it can define the norm, angle, and distance for a general vector space. An inner product space can be defined on a vector space V with the definition of the inner product.

The following example illustrates another inner product on R^2.

Definition 6.1:

An inner product defined on a real vector space $\mathbf{V}$ is a function that associates a number denoted by $\langle u, v\rangle$ each pair of vectors of $\mathbf{V}$.

It satisfies the following conditions for vectors u, v, and w, for any scalar k, i.e.,

Symmetry axiom: $\langle u, v\rangle = \langle v, u\rangle$
Addition axiom: $\langle u + v, w\rangle = \langle u, w\rangle + \langle v, w\rangle$
Homogeneity axiom: $\langle ku, v\rangle = k\langle u, v\rangle$
Positive-definite axiom: $\langle u, u\rangle \geq 0$ and $\langle u, u\rangle = 0$ if and only if $u = 0$.

Example 6.1:

Let $u = (x_1, x_2)$, $v = (y_1, y_2)$, and $w = (z_1, z_2)$ be arbitrary vectors in R^2. Prove that $\langle u, v\rangle$ defined by $\langle u, v\rangle = x_1y_1 + 4x_2y_2$ is an inner product on R^2. Determine the inner product of the vectors $(-2, 5)$ and $(3, 1)$.

Solution:

Upon using the axioms of inner product space, we can check the inner product of the vectors on R^2. Let us consider the following hypotheses:

Axiom 1:

$\langle u, v\rangle = x_1y_1 + 4x_2y_2 = y_1x_1 + 4y_2x_2 = \langle v, u\rangle$

Axiom 2:

$$\begin{aligned}
\langle u + v, w\rangle &= \langle (x_1, x_2) + (y_1, y_2)\,, (z_1, z_2)\rangle \\
&= \langle (x_1 + y_1, x_2 + y_2)\,, (z_1, z_2)\rangle \\
&= (x_1 + y_1)\, z_1 + 4\,(x_2 + y_2)\, z_2 \\
&= x_1z_1 + y_1z_1 + 4x_2z_2 + 4y_2z_2 \\
&= x_1z_1 + 4x_2z_2 + y_1z_1 + 4y_2z_2 \\
&= \langle (x_1, x_2)\,, (z_1, z_2)\,\rangle + \langle (y_1, y_2)\,, (z_1, z_2)\rangle \\
&= \langle u, v\rangle + \langle v, w\rangle\,.
\end{aligned}$$

Axiom 3:

$$\begin{aligned}
\langle cu, v\rangle &= \langle c\,(x_1, x_2)\,, (y_1, y_2)\rangle \\
&= \langle (cx_1, cx_2)\,, (y_1, y_2)\rangle \\
&= cx_1y_1 + 4cx_2y_2 \\
&= c\,(x_1y_1 + 4x_2y_2) = c\,\langle u, v\rangle\,.
\end{aligned}$$

Axiom 4:

$$\begin{aligned}
\langle u, u\rangle &= \langle (x_1, x_2)\,, (x_1, x_2)\rangle \\
&= {x_1}^2 + 4{x_2}^2 \geq 0
\end{aligned}$$

Further, ${x_1}^2 + 4{x_2}^2 = 0$ if and only if $x_1 = 0$ and $x_2 = 0$.

That is $u = \mathbf{0_V}$. Thus, $\langle u, u \rangle \geq 0$ and $\langle u, u \rangle = 0$ if and only if $u = \mathbf{0_V}$.

Hence, the four inner product axioms are satisfied, so $\langle u, v \rangle = x_1y_1 + 4x_2y_2$ forms an inner product on R^2.

The inner product of the vectors $(-2, 5)$ and $(3, 1)$ is

$$\begin{aligned} &\langle (-2,5), (3,1) \rangle \\ &= (-2 \times 3) + 4(5 \times 1) \\ &= 14 \end{aligned}$$

Now we illustrate the inner products on the real vector spaces of matrices and functions.

Example 6.2:

Consider the vector space M_{22} of 2×2 matrices.

Let u and v be defined as follows by the arbitrary matrices defined as follows

$$u = \begin{bmatrix} a_1 & b_1 \\ c_1 & d_1 \end{bmatrix}, \quad v = \begin{bmatrix} a_2 & b_2 \\ c_2 & d_2 \end{bmatrix}.$$

Prove that the following function forms an inner product on M_{22}, where the inner product is defined by $\langle u, v \rangle = a_1a_2 + b_1b_2 + c_1c_2 + d_1d_2$. Determine the inner product of the matrices $\begin{bmatrix} 2 & -3 \\ 0 & 1 \end{bmatrix}$ and $\begin{bmatrix} 5 & 2 \\ 9 & 0 \end{bmatrix}$.

Solution:

We shall verify hypotheses 1 and 3 of an inner product, leaving axioms 2 and 4 for the readers to check.

Axiom 1:

$$\begin{aligned} \langle u, v \rangle &= a_1a_2 + b_1b_2 + c_1c_2 + d_1d_2 \\ &= a_2a_1 + b_2b_1 + c_2c_1 + d_2d_1 \\ &= \langle v, u \rangle \end{aligned}$$

Axiom 2:

Let k be a scalar. Then

$$\begin{aligned} \langle ku, v \rangle &= ka_1a_2 + kb_1b_2 + kc_1c_2 + kd_1d_2 \\ &= k(a_1a_2 + b_1b_2 + c_1c_2 + d_1d_2) = k\langle u, v \rangle \end{aligned}$$

Next, the computation of the inner product on the matrices gives

$$\begin{aligned}&\left\langle \begin{bmatrix} 2 & -3 \\ 0 & 1 \end{bmatrix}, \begin{bmatrix} 5 & 2 \\ 9 & 0 \end{bmatrix} \right\rangle \\ &= (2 \times 5) + (-3 \times 2) + (0 \times 9) + (1 \times 0) \\ &= 4\end{aligned}.$$

Example 6.3:

Consider the vector space $P_n(R)$ of polynomials of deg $\leq n$.

Let f_1 and f_2 be elements of P_n. Prove that the following function defines an inner product on P_n, i.e., $\langle f_1, f_2 \rangle = \int_0^1 f_1(x) f_2(x)\, dx$.

Determine the inner product of the polynomials $f_1(x) = x^2 + 2x - 1$ and $f_2(x) = 4x + 1$.

Solution:

Here, hypotheses 1 and 2 of an inner product have been verified, leaving hypotheses 3 and 4 for the readers to check.

Axiom 1:

$$\begin{aligned}\langle f_1, f_2 \rangle &= \int_0^1 f_1(x) f_2(x)\, dx \\ &= \int_0^1 f_2(x) f_1(x)\, dx \\ &= \langle f_2, f_1 \rangle .\end{aligned}$$

Axiom 2:

$$\begin{aligned}\langle f_1 + f_2, f_3 \rangle &= \int_0^1 [f_1(x) + f_2(x)] f_3(x)\, dx \\ &= \int_0^1 f_1(x) f_3(x)\, dx + \int_0^1 f_2(x) f_3(x)\, dx \\ &= \langle f_1, f_3 \rangle + \langle f_2, f_3 \rangle .\end{aligned}$$

Next, the inner product of the function

$f_1(x) = x^2 + 2x - 1$ and $f_2(x) = 4x + 1$ gives

$$\begin{aligned}\left\langle x^2 + 2x - 1, 4x + 1 \right\rangle &= \int_0^1 \left(x^2 + 2x - 1\right)(4x + 1)\, dx \\ &= \int_0^1 \left(4x^3 + 9x^2 - 2x - 1\right) dx = 2.\end{aligned}$$

6.1.1 Norm of a vector

The norm of a vector $v = (x_1, x_2, \ldots, x_n)$ in R^n can be expressed in terms of the dot product as follows:

$$\begin{aligned}\|(x_1, x_2, \ldots, x_n)\| &= \sqrt{{x_1}^2 + {x_2}^2 + \cdots + {x_n}^2}\\ &= \sqrt{(x_1, x_2, \ldots, x_n)\,(y_1, y_2, \ldots, y_n)}\end{aligned}$$

To obtain the norm of a vector in a general inner product space, these definitions can be generalized using the inner product in place of the dot product. It has sufficient applications in numerical work, although it does not necessarily have geometric interpretations.

Definition 6.2:

Let **V** be an inner product space. The norm of a vector u is defined by $\|u\| = \sqrt{\langle u, u\rangle}$ where the norm is denoted by $\|u\|$.

Example 6.4:

Consider the vector space $P_n(R)$ of polynomials with an inner product is defined by $\langle f_1, f_2\rangle = \int_0^1 f_1(x)\, f_2(x)\, dx$,
where the norm of the function $f(x)$ is defined by

$$\|f\| = \sqrt{\langle f, f\rangle} = \sqrt{\int_0^1 [f(x)]^2\, dx}.$$

Find the norm of the function$f(x) = 5x^2 + 1$.

Solution:

Upon using the definition of norm, we get

$$\begin{aligned}\|5x^2 + 1\| &= \sqrt{\int_0^1 [5x^2 + 1]^2\, dx}\\ &= \sqrt{\int_0^1 (25x^4 + 10x^2 + 1)\, dx}\\ &= \sqrt{\frac{28}{3}}.\end{aligned}$$

Thus, the norm of $f(x)$ is $\sqrt{\frac{28}{3}}$.

6.2 Angle Between Two Vectors

Generally, the angle between vectors can be found using the dot product of vectors in R^n, but here we generalize this to define an angle for a real vector space, using the inner product in place of the dot product.

The angle θ between vectors u and v in R^n is defined by

$$Cos\theta = \frac{u.v}{\|u\|\,\|v\|}.$$

Definition 6.3:

Let $\mathbf{V}$ be a real inner product space. The angle θ between two non-zero vectors u_1 and u_2 in $\mathbf{V}$ can be defined by $Cos\theta = \frac{\langle u_1, u_2\rangle}{\|u_1\|\|u_2\|}$.

Example 6.5:

Consider the inner product space $P_n(R)$ of polynomials with an inner product defined by $\langle f_1, f_2\rangle = \int_0^1 f_1(x) f_2(x)\,dx$ and the angle between f_1 and f_2 is defined by $Cos\theta = \frac{\langle f_1, f_2\rangle}{\|f_1\|\|f_2\|}$, where f_1 and f_2 are non-zero functions. Then determine the angle between the functions $f_1(x) = 5x^2$ and $f_2(x) = 3x$.

Solution:

We get

$$\begin{aligned}
\langle f_1, f_2\rangle &= \int_0^1 \left(5x^2\right)(3x)\,dx \\
&= \int_0^1 \left(15x^3\right)dx = \frac{15}{4}. \\
\|f_1\| &= \left\|5x^2\right\| \\
&= \sqrt{\int_0^1 [5x^2]^2\,dx.} \\
&= \sqrt{5} \\
\|f_2\| &= \|3x\| \\
&= \sqrt{\int_0^1 [3x]^2\,dx.} \\
&= \sqrt{3}
\end{aligned}$$

These give

$$Cos\theta = \frac{15}{4\sqrt{5}\sqrt{3}}$$
$$= \frac{\sqrt{15}}{4} \quad (\theta = 14.48^{\circ}) .$$

6.3 Orthogonal Vectors

Let **V** be an inner product space in R^n. Then the two non-zero vectors u_1 and u_2 in an inner product space **V** are said to be orthogonal if $\langle u_1, u_2 \rangle = 0$.

Example 6.6:

Show that the functions $f_1(x) = 3x - 2$ and $f_2(x) = x$ are orthogonal in the space of polynomials $P_n(R)$ with an inner product defined by $\langle f_1, f_2 \rangle = \int_0^1 f_1(x) f_2(x)\, dx$.

Solution:

The inner product gives

$$\langle 3x - 2, x \rangle = \int_0^1 (3x - 2)(x)\, dx$$
$$= \left[x^3 - x^2\right]_0^1 = 0.$$

Hence, the functions f_1 and f_2 are orthogonal in this inner product space.

6.4 Distance

The next task is to extend the Euclidean concept of distance from a general vector space to an inner product space with the help of a norm. It is also helpful in numerical mathematics to discuss the distance between two functions.

Definition 6.4:

Let **V** be an inner product space defined with the vector norm defined by $\|u\| = \sqrt{\langle u, u \rangle}$.

The distance between two vectors (points) u_1 and u_2 is denoted by $d(u_1, u_2)$ and is defined by

$$d(u_1, u_2) = \|u_1 - u_2\| \left(= \sqrt{\langle u_1 - u_2, u_1 - u_2 \rangle}\right).$$

Example 6.7:

Consider the inner product $P_n(x)$ of polynomials. Determine which of the functions $f_2(x) = x^2 - 3x + 5$ or $f_3(x) = x^2 + 4$ is closest to $f_1(x) = x^2$.

Solution:

We compute the distances between f_1 and f_2 and between f_1 and f_3 using the definition as

$$\begin{aligned}
[d(f_1, f_2)]^2 &= \langle f_1 - f_2, f_1 - f_2 \rangle \\
&= \langle 3x - 5, 3x - 5 \rangle \\
&= \int_0^1 (3x - 5)^2 \, dx = 13 \\
[d(f_1, f_3)]^2 &= \langle f_1 - f_3, f_1 - f_3 \rangle \\
&= \langle -4, -4 \rangle \\
&= \int_0^1 (-4)^2 \, dx = 16
\end{aligned}$$

Thus, $d(f_1, f_2) = \sqrt{13}$ and $d(f_1, f_3) = 4$.

The distance between f_1 and f_3 is 4, and as we might suspect, f_2 is closer than f_3 to f_1.

In practice, relative distances between functions are often more significant than absolute distances.

Example 6.8:

Consider $f_1(t) = 3t - 5$ and $f_2(t) = t^2$ in the polynomial space $\mathbf{P}(t)$ with an inner product defined by $\langle f_1, f_2 \rangle = \int_0^1 f_1(t) f_2(t) \, dt$.

Then find (a) $\langle f_1, f_2 \rangle$ and (b) find $\|f_1\|$ and $\|f_2\|$.

6.5 Cauchy-Schwarz Inequality

Theorem 6.1 (Cauchy-Schwarz Inequality):

Let u_1 and u_2 be any two vectors defined on an inner product space $\mathbf{V}$ i.e., $\langle u_1, u_2 \rangle^2 \leq \langle u_1, u_2 \rangle \langle u_1, u_2 \rangle$ or $|\langle u_1, u_2 \rangle| \leq \|u_1\| \|u_2\|$.

The following examples examine the application of inequality for different cases.

Example 6.9:

Consider any real number $a_1, a_2, a_3, \ldots, b_1, b_2, b_3, \ldots$. Then by the Cauchy–Schwarz inequality, we can write

$$(a_1 b_1 + a_2 b_2 + \cdots + a_n b_n)^2 \\ \leq \left(a_1{}^2 + a_2{}^2 + \cdots + a_n{}^2\right)\left(b_1{}^2 + b_2{}^2 + \cdots + b_n{}^2\right).$$

That is, $\langle u_1, u_2\rangle^2 \leq \|u_1\|^2 \|u_2\|^2$, where $u_1 = (a_i)$, $u_2 = (b_i)$.

Let f_1 and f_2 be continuous functions defined on the unit interval $[0, 1]$. Then by the Cauchy–Schwarz inequality, we find

$$\left[\int_0^1 f_1(x) f_2(x)\,dx\right]^2 \leq \int_0^1 f_1^2(x)\,dx \int_0^1 f_2^2(x)\,dx.$$

That is, $(\langle f_1, f_2\rangle)^2 \leq \|f_1\|^2 \|f_2\|^2$.

Next, we will discuss the basic properties of the norm where for the proof of the third property, it requires the Cauchy–Schwarz inequality.

Theorem 6.2:

Let $\mathbf{V}$ be an inner product space. Then the norm in an inner product space $\mathbf{V}$ satisfies the following properties:

$[N_1]$ $\|u\| \geq 0$ and $\|u\| = 0$ if and only if $u = 0$.

$$[N_2]\ \|ku\| = |k|\ \|u\|$$

$$[N_3]\ \|u_1 + u_2\| \leq \|u_1\| + \|u_2\|.$$

The third property $[N_3]$ is called the *triangle inequality* because if we view $u_1 + u_2$ as the side of the triangle formed with sides u_1 and u_2, then this triangle inequality states that the length of one side of a triangle cannot be greater than the sum of the lengths of the other two sides.

Example 6.10:

Consider vectors $u_1 = (2, 3, 5)$ and $u_2 = (1, -4, 3)$ in $\mathbb{R}^3$.

Then $\langle u_1, u_2\rangle = 2 - 12 + 15 = 5$

$$\|u_1\| = \sqrt{4 + 9 + 25} = \sqrt{38}$$

$$\|u_2\| = \sqrt{1 + 16 + 9} = \sqrt{26}.$$

Then the angle θ between u_1 and u_2 is given by $Cos\theta = \frac{5}{\sqrt{38}\sqrt{26}}$.

Note that θ is an acute angle, since $Cos\theta$ is positive.

Let $f_1(x) = 3x - 5$ and $f_2(x) = x^2$ in the polynomial space $\mathbf{P}_2(R)$ with an inner product

$\langle f_1, f_2\rangle = \int_0^1 f_1(x) f_2(x)\,dx.$

It gives $\langle f_1,\ f_2\rangle = -\frac{11}{12}$, $\|f_1\| = \sqrt{13}$, $\|f_2\| = \frac{1}{5}\sqrt{5}$.
Then the angel θ between f_1 and f_2 is given by

$$Cos\theta = \frac{-\frac{11}{12}}{(\sqrt{13})\left(\frac{1}{5}\sqrt{5}\right)} = -\frac{55}{12\sqrt{13\sqrt{5}}}.$$

Note that the angle θ is obtuse, since $Cos\theta$ is negative.

6.6 Orthogonal Complements

Let $\mathbf{V}$ be an inner product space and $S \subseteq \mathbf{V}$. Then the orthogonal complement of S, denoted by $S^{\perp}$, consists of those vectors in $\mathbf{V}$ that are orthogonal to every vector $u \in S$, i.e.,

$$S^{\perp} = \{v \in \mathbf{V} : \langle v, u\rangle = 0 \ \text{ for every } \ u \in S\}.$$

In particular, for a given vector u in $\mathbf{V}$, we have

$$u^{\perp} = \{v \in V : \langle v, u\rangle = 0\}.$$

That is, $u^{\perp}$ consists of all vectors in $\mathbf{V}$ that are orthogonal to the given vector u.

6.6.1 Subspace

We show that $S^{\perp}$ is a subspace of an inner product space $\mathbf{V}$.

Clearly, $0 \in S^{\perp}$, since 0 is orthogonal to every vector in $\mathbf{V}$.

Now suppose $u_1, u_2 \in S^{\perp}$. Then, for any scalars a and b and any vector $u \in S$, we have

$$\begin{aligned}\langle au_1 + bu_2, u\rangle &= a\langle u_1, u\rangle + b\langle u_2, u\rangle \\ &= a.0 + b.0 = 0\end{aligned}$$

Thus, $au_1 + bu_2 \in S^{\perp}$ and therefore $S^{\perp}$ is a subspace of $\mathbf{V}$.

Proposition 6.1:

Let S be a subset of inner product space V. Then $S^{\perp}$ is a subspace of $\mathbf{V}$.

Example 6.11:

Find a basis for the subspace $u^{\perp}$ of R^3, where $u = (1, 3, -4)$.

Solution:

Note that $u^{\perp}$ consists of all vectors $v = (x_1, x_2, x_3)$ such that $\langle u, v\rangle = 0$ or $x_1 + 3x_2 - 4x_3 = 0$.

The free variables are x_2 and x_3.

Set $x_2 = 1,\ x_3 = 0$ to obtain the solution $v_1 = (-3, 1, 0)$.

Set $x_2 = 0,\ x_3 = 1$ to obtain the solution $v_2 = (4, 0, 1)$.

Thus, the vectors v_1, v_2 form a basis for the solution space of the equation and form a basis for $u^{\perp}$.

Remark:

Suppose $\mathbf{U}$ is a subspace of $\mathbf{V}$. Then both $\mathbf{U}$ and $\mathbf{U}^{\perp}$ are subspaces of $\mathbf{V}$.

Theorem 6.3:

Let $\mathbf{U}$ be a subspace of $\mathbf{V}$. Then $\mathbf{V}$ is the direct sum of $\mathbf{U}$ and $\mathbf{U}^{\perp}$, that is, $\mathbf{V} = \mathbf{U} \oplus \mathbf{U}^{\perp}$.

Definition 6.5:

A basis that forms an orthogonal set is said to be an *orthogonal basis*. Similarly, a basis that includes an orthonormal set is said to be an *orthonormal basis*.

Note:

The standard bases of R^n are orthonormal.

Standard bases:

The following standard bases

$$R^2 : \{(1,0)\,(0,1)\}$$
$$R^3 : \{(1,0,0)\,(0,1,0)\,(0,0,1)\}$$
$$R^n : \{(1,0,0,\ldots 0)\,(0,1,0,\ldots 0) \ldots\ldots (0,0,\ldots 1)\}$$

form an orthonormal base.

6.7 Orthogonal Sets and Bases

Consider a set $S = \{u_1, u_2, \ldots, u_r : u_i \neq 0\}$ defined on an inner product space $\mathbf{V}$. Then the set S is said to be orthogonal if each pair of vectors (u_i, u_j) in S are *orthogonal*, and the set S is said to be *orthonormal* if S is *orthogonal* and each vector in S has a unit length, i.e.,

Orthogonal: $\langle u_i, u_j\rangle = 0$ for $i \neq j$.

Orthonormal: $\langle u_i, u_j\rangle = \begin{cases} 0 & \text{for } i \neq j \\ 1 & \text{for } i = j \end{cases}$.

Theorem 6.4 (Pythagoras):

Suppose $\{u_1, u_2, \ldots, u_r\}$ is an orthogonal set vector. Then $\|u_1 + u_2 + \cdots + u_r\|^2 = \|u_1\|^2 + \|u_2\|^2 + \cdots + \|u_r\|^2$.

Example 6.12:

Let $\mathbf{E} = \{e_1, e_2, e_3\} = \{(1, 0, 0), (0, 1, 0), (0, 0, 1)\}$ be the usual basis of Euclidean space R^3.

Since$\langle e_1, e_2 \rangle = \langle e_2, e_3 \rangle = \langle e_1, e_3 \rangle = 0$ and $\langle e_1, e_1 \rangle = \langle e_2, e_2 \rangle = \langle e_3, e_3 \rangle = 1$.

So it implies that $\mathbf{E}$ forms an orthonormal basis of R^3, which is the usual basis of R^n for every n.

Example 6.13:

Let $\mathbf{V} = C[-\pi, \pi]$ be the vector space of continuous functions defined on the interval $-\pi \leq x \leq \pi$ with inner product defined by

$$\langle f_1, f_2 \rangle = \int_{-\pi}^{\pi} f_1(x) f_2(x)\, dx.$$

Then $\{1, \cos x, \cos 2x, \cos 3x, \ldots, \sin x, \sin 2x, \ldots\}$ is an example of an orthogonal set in the vector space $\mathbf{V}$, and it has essential applications in the Fourier series expansion.

Theorem 6.5:

Let us consider $\mathbf{V}$ to be an inner product space. Let $\{u_1, u_2, \ldots, u_n\}$ be an orthonormal basis for $\mathbf{V}$ and $v \in \mathbf{V}$ be a vector in $\mathbf{V}$. Then the vector $v \in \mathbf{V}$ can be written as a linear combination of these basis vectors as follows, i.e., $v = (v.u_1) u_1 + (v.u_2) u_2 + \cdots + (v.u_n) u_n$.

Proof:

Since $\{u_1, u_2, \ldots, u_n\}$ is a basis, then there exist scalars $k_1, k_2, \ldots, k_n$ such that $v = k_1u_1 + k_2u_2 + \cdots + k_nu_n$.

We shall show that $k_1 = v.u_1,\ k_2 = v.u_2, \ldots, k_n = v.u_n$.

Let u_i be the ith base vector. On taking the dot product on each side of this equation with u_i and using the properties of the dot products, we get

$$\begin{aligned} v.u_i &= (k_1u_1 + k_2u_2 + \ldots + k_nu_n).u_i \\ &= k_1u_1.u_i + k_2u_2.u_i + \ldots + k_nu_n.u_i \end{aligned}$$

The vectors $u_1, u_2, \ldots, u_n$ are mutually orthogonal,
i.e., $u_j.u_i = 0$ unless $j = i$.

Thus, $v.u_i = k_i u_i.u_i$.

Furthermore, since the basis is orthonormal $u_j.u_i = 1$, therefore $v.u_j = k_j$.

Thus, letting $i = 1, 2, \ldots, n$, we get

$$k_1 = v.u_1, k_2 = v.u_2, \ldots, k_n = v.u_n.$$

Hence, we can write $v = (v.u_1)\, u_1 + (v.u_2)\, u_2 + \cdots + (v.u_n)\, u_n$.

Example 6.14:

Show that the following vectors $u_1 = (1, 0, 0)\,,\ u_2 = \left(0, \frac{3}{5}, \frac{4}{5}\right),\ u_3 = \left(0, \frac{4}{5}, -\frac{3}{5}\right)$ form an orthonormal basis for R^3.

Express the vector $v = (7, -5, 10)$ as a linear combination of these vectors $u_1 = (1, 0, 0)\,,\ u_2 = \left(0, \frac{3}{5}, \frac{4}{5}\right),\ u_3 = \left(0, \frac{4}{5}, -\frac{3}{5}\right)$.

Solution:

We get

$$v.u_1 = (7, -5, 10)\,.\,(1, 0, 0) = 7$$

$$v.u_2 = (7, -5, 10)\,.\left(0, \frac{3}{5}, \frac{4}{5}\right) = 5$$

$$v.u_3 = (7, -5, 10)\,.\left(0, \frac{4}{5}, -\frac{3}{5}\right) = -10$$

$$\Rightarrow (7, -5, 10) = 7.\,(1, 0, 0) + 5.\left(0, \frac{3}{5}, \frac{4}{5}\right) - 10\left(0, \frac{4}{5}, -\frac{3}{5}\right).$$

6.8 Projection of One Vector onto Another Vector

Let **V** and **U** be vectors in R^n with angles α between them.

The vector $O\bar{A}$ indicates the vector v' momentum in the direction of **U**, as shown in Figure 6.1(c)(a),(b) and (c).

We call the projection of v onto u. Let us find an expression for $O\bar{A}$; we see that

$$A = OBCos\alpha = \|v\|\, Cos\alpha$$
$$= \|v\|\left(\frac{v}{\|v\|}, \frac{u}{\|u\|}\right) = v.\frac{u}{\|u\|}.$$

The unit vector $\frac{u}{\|u\|}$ defines the direction of the vector $O\bar{A}$.

Thus, $O\bar{A} = \left(v.\frac{u}{\|u\|}\right)\frac{u}{\|u\|} = \frac{v.u}{u.u}u$.

This expression for projection also holds if $\alpha > 90°$.

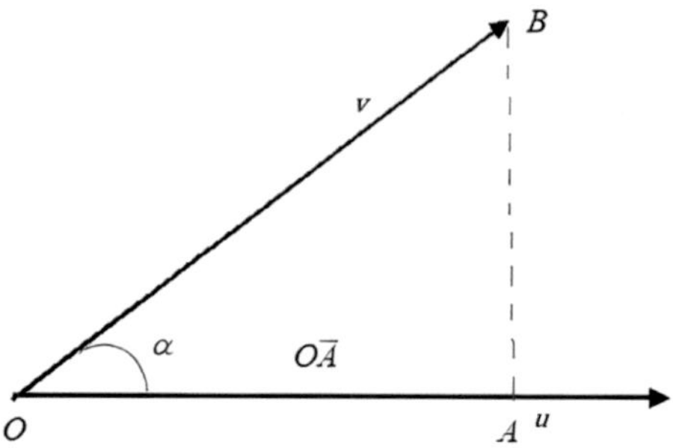

Figure 6.1(a) Projection of One Vector onto Another Vector for $\alpha = 90°$.

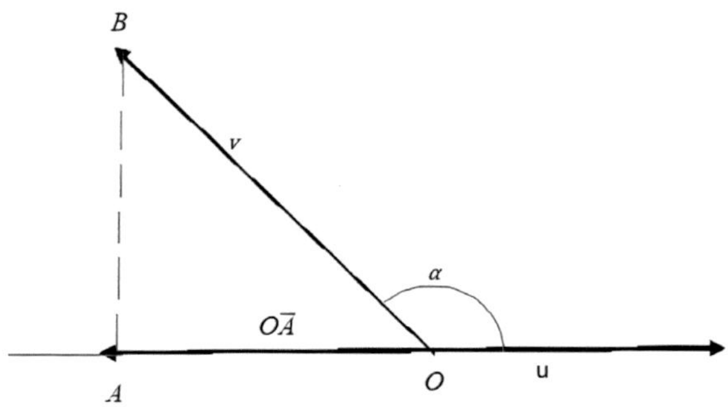

Projection of v onto u $O\bar{A} = \frac{v.u}{u.u}u$.

Figure 6.1(b) Projection of One Vector onto Another Vector for $\alpha > 90°$.

In this case, the projection is opposite to u, and the sign of $\frac{v.u}{u.u}$ is negative.

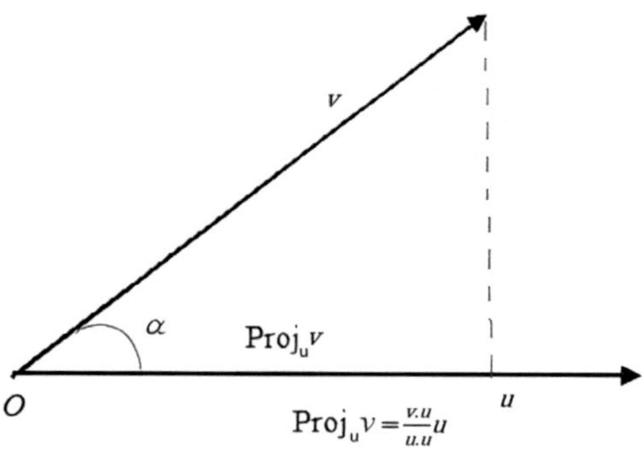

Figure 6.1(c) Projection of One Vector onto Another Vector for $\alpha < 90°$.

Definition 6.6:

The projection of a vector v onto a non-zero vector u in R^n is denoted by $\text{Proj}_u v$ and is defined by $\text{Proj}_u v = \frac{v.u}{u.u}u$.

Example 6.15:

Determine the projection of the vector $v = (6, 7)$ onto the vector $u = (1, 4)$.

Solution:

For these vectors, we get

$$v.u = (6, 7) . (1, 4) = 6 + 28 = 34$$

$$u.u = (1, 4) . (1, 4) = 1 + 16 = 17.$$

Thus, $\text{Proj}_u v = \frac{v.u}{u.u}u = \frac{34}{17}(1, 4) = (2, 8)$.

The projection of v onto u is $(2, 8)$.

Suppose the vector $v = (6, 7)$ represents a force acting on a body located at the origin $(0, 0)$. Then $\text{Proj}_u v = (2, 8)$ is the component of the force in the direction of the vector $u = (1, 4)$.

Physically, the projection vector $(2, 8)$ is the effect of the force in that direction. We now discuss a method for constructing an orthonormal basis from a given basis. The technique uses vector projection.

Theorem 6.6:

Let $\{u_1, u_2, u_3, \ldots, u_n\}$ be an orthogonal basis of an inner product space $\mathbf{V}$. Then for any $v \in \mathbf{V}$, it can be expressed as

$$v = \frac{\langle v, u_1 \rangle}{\langle u_1, u_1 \rangle}u_1 + \frac{\langle v, u_2 \rangle}{\langle u_2, u_2 \rangle}u_2 + \cdots + \frac{\langle v, u_n \rangle}{\langle u_n, u_n \rangle}u_n.$$

Remark:

The scalar $k_i = \frac{\langle v, u_i \rangle}{\langle u_i, u_i \rangle}$ is called the *Fourier c-efficient* of v with respect to u_i as it is analogous to a coefficient in the *Fourier series* of a function.

Theorem 6.7:

Let us consider the vectors $u_1, u_2, u_3, \ldots, u_r$ form an orthogonal set of non-zero vectors in an inner product space $\mathbf{V}$. Let $v \in \mathbf{V}$ be any vector in $\mathbf{V}$.

Define $v' = v - (k_1u_1 + k_2u_2 + k_3u_3 + \cdots + k_ru_r)$,

where $k_1 = \frac{\langle v, u_1 \rangle}{\langle u_1, u_1 \rangle}, k_2 = \frac{\langle v, u_2 \rangle}{\langle u_2, u_2 \rangle}, \ldots, k_r = \frac{\langle v, u_r \rangle}{\langle u_r, u_r \rangle}$.

Then v' is orthogonal to $u_1, u_2, \ldots, u_r$.

Note:

Each k_i in the above theorem is the component (Fourier coefficient) of v along the given u_i.

Remark 1:

The projection of a vector $v \in \mathbf{V}$ along a subspace $\mathbf{U}$ of $\mathbf{V}$ is defined as $v = \mathbf{U} \oplus \mathbf{U}^{\perp}$.

Hence, v may be expressed uniquely in the form $v = u + u'$, where $u \in \mathbf{U}$ and $u' \in \mathbf{U}^{\perp}$.

Remark 2:

We define v to be the projection of v along $\mathbf{U}$ and denote it by $\operatorname{Proj}(v, \mathbf{U})$. If $\mathbf{U} = span\,(u_1, u_2, \ldots, u_r)$, where $u_i's$ form an orthogonal set, then $\operatorname{Proj}(v, \mathbf{U}) = k_1 u_1 + k_2 u_2 + \ldots + k_r u_r$, where k_i is the component of v along u_i.

Example 6.16:

Show that the set $\left\{(1,0,0), \left(0, \frac{3}{5}, \frac{4}{5}\right), \left(0, \frac{4}{5}, -\frac{3}{5}\right)\right\}$ is orthonormal.

Solution:

First, to show that each pair of vectors in this set is orthogonal.

$$(1,0,0) \,.\, \left(0, \frac{3}{5}, \frac{4}{5}\right) = 0$$

$$(1,0,0) \,.\, \left(0, \frac{4}{5}, -\frac{3}{5}\right) = 0$$

$$\left(0, \frac{3}{5}, \frac{4}{5}\right) \cdot \left(0, \frac{4}{5}, -\frac{3}{5}\right) = 0.$$

This implies that the vectors $\left\{(1,0,0), \left(0, \frac{3}{5}, \frac{4}{5}\right), \left(0, \frac{4}{5}, -\frac{3}{5}\right)\right\}$ are orthogonal.

Next, to show that each vector in the set is a unit vector.

Now,

$$\|(1,0,0)\| = \sqrt{1^2 + 0^2 + 0^2} = 1$$

$$\left\|\left(0, \frac{3}{5}, \frac{4}{5}\right)\right\| = \sqrt{0^2 + \left(\frac{3}{5}\right)^2 + \left(\frac{4}{5}\right)^2} = 1$$

$$\left\|\left(0, \frac{4}{5}, -\frac{3}{5}\right)\right\| = \sqrt{0^2 + \left(\frac{4}{5}\right)^2 + \left(-\frac{3}{5}\right)^2} = 1.$$

Thus, the vectors are unit and orthogonal.

Hence, the set $\left\{(1,0,0), \left(0,\frac{3}{5},\frac{4}{5}\right), \left(0,\frac{4}{5},-\frac{3}{5}\right)\right\}$ thus forms an orthonormal set.

Theorem 6.8:

Let us consider an inner product space $\mathbf{V}$. An orthogonal set of non-zero vectors in $\mathbf{V}$ is linearly independent.

Proof:

Let $\{v_1, v_2, \ldots, v_m\}$ be an orthogonal set of non-zero vectors in an inner product space $\mathbf{V}$.

Let us examine the identity $k_1v_1 + k_2v_2 + \cdots + k_mv_m = 0_v$.

We shall show that $k_1 = 0, k_2 = 0, \ldots, k_m = 0$ implies the vectors are linearly independent.

Let v_i be the vector of the orthogonal set.

Let us take the dot product of each side of this equation and use the dot products' properties. We get

$$(k_1v_1 + k_2v_2 + \cdots + k_mv_m) . v_i = 0_v . v_i$$

$$\Rightarrow (k_1v_1.v_i + k_2v_2.v_i + \cdots + k_mv_m.v_i) = 0_v.$$

Since the vectors $v_1, v_2, \ldots, v_m$ are mutually orthogonal,

i.e., $v_j.v_i = 0$ unless $j = i$.

Thus, $k_iv_i.v_i = 0$.

Now since v_i is a non-zero vector, $v_i.v_i \neq 0$.

Thus, $k_i = 0$.

Letting $i = 1, 2, \ldots, m$, we get $k_1 = 0$, $k_2 = 0, \ldots, k_m = 0$, which proves that the vectors are linearly independent.

6.9 Orthogonal Matrix Theorem

Theorem 6.9:

Let us consider an $n \times n$ square matrix A. Then the following statements are equivalent:

(1) The matrix A is orthogonal.
(2) The column vectors of the matrix A forms an orthonormal set.
(3) The row vectors of the matrix A forms an orthonormal set.

Proof:

Assume that the matrix A is orthogonal.

Thus, $A^{-1} = A^T$ implies $A^T A = I$ and $AA^T = I$.

Let $P = A^T A$. Then P_{ij} is the dot product of row vector i of A^T and column vector j of A.

Since row i of A^T is column i of A, this means that P_{ij} is the product of column vectors i and j of A.

Since P is the identity matrix

$$\begin{cases} P_{ij} = 1\ , if\ i = j \\ P_{ij} = 0\ , if\ i \neq j \end{cases},$$

it implies that the column vectors of the matrix A thus form an orthonormal set.

Similarly, if we start with $P = AA^T$, we can show that the row vectors of the matrix A form an orthonormal set. By using these arguments in the reverse direction, it can be shown that the converse is true.

Orthonormality of rows or columns implies that A is orthogonal.

6.10 Properties of the Orthogonal Matrix

Theorem 6.10:

If A is an orthogonal matrix, then $|A| = \pm 1$ and A^{-1} is an orthonormal matrix.

Proof:

Since $A^{-1} = A^T$, we have $AA^T = I$.

By the properties of determinants,

$$\begin{aligned} &\left|AA^T\right| = 1 \\ &\Rightarrow |A| \,.\, \left|A^T\right| = 1 \\ &\Rightarrow |A| \,.\, |A| = 1 \\ &\Rightarrow |A|^2 = 1 \end{aligned}.$$

Thus, $|A| = \pm 1$.

The row vectors of A form an orthonormal set and also the columns of A^T form an orthonormal set.

Since $A^{-1} = A^T$, this means that the columns of A^{-1} form an orthonormal set.

The previous theorem now implies that A^{-1} is an orthogonal set.

6.11 The Gram–Schmidt Orthogonalization Process

Theorem 6.11:

Let **V** be an inner product space. Let $\{v_1, v_2, \ldots, v_n\}$ be a basis for the inner product space **V**. Then the set of vectors $\{u_1, u_2, \ldots, u_n\}$ defined as follows are orthogonal, as shown in Figure 6.2.

To obtain an orthonormal basis for V normalizing each of the vectors $u_1, u_2, \ldots, u_n$.

$$
\begin{aligned}
u_1 &= v_1 \\
u_2 &= v_2 - \mathrm{Proj}_{u_1} v_1 \\
u_3 &= v_3 - \mathrm{Proj}_{u_1} v_3 - \mathrm{Proj}_{u_2} v_3 \\
&\ldots\ldots\ldots\ldots\ldots\ldots\ldots\ldots\ldots \\
u_n &= v_n - \mathrm{Proj}_{u_1} v_n - \mathrm{Proj}_{u_2} v_n - \cdots - \mathrm{Proj}_{u_{n-1}} v_n
\end{aligned}
$$

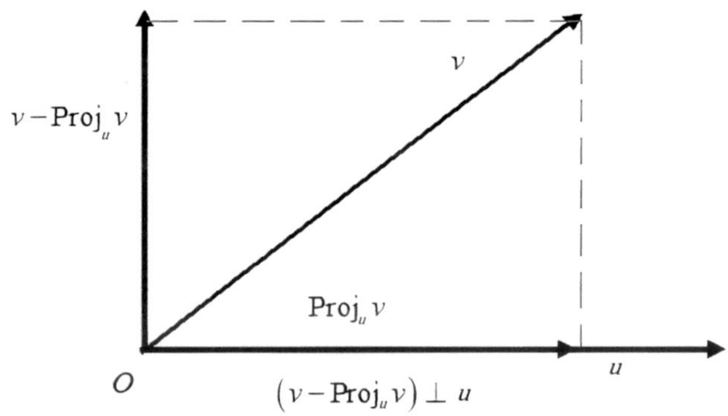

Figure 6.2 Gram–Schmidt Orthogonalization Process.

Example 6.17:

Show that the set of vectors $\{(1, 2, 0, 3), (4, 0, 5, 8), (8, 1, 5, 6)\}$ are linearly independent and form a basis for a three-dimensional subspace **V** of R^4. Hence, construct an orthonormal basis for **V**.

Solution:

Let $v_1 = (1, 2, 0, 3)$, $v_2 = (4, 0, 5, 8)$, $v_3 = (8, 1, 5, 6)$.

We now use the Gram–Schmidt process to construct an orthogonal set $\{u_1, u_2, u_3\}$ from these vectors.

Let $u_1 = v_1 = (1, 2, 0, 3)$.

Let

$$\begin{aligned} u_2 &= v_2 - \text{Proj}_{u_1} v_2 \\ &= v_2 - \frac{(v_2, u_1)}{(u_1, u_1)} u_1 \\ &= (4,0,5,8) - \frac{(4,0,5,8)\,(1,2,0,3)}{(1,2,0,3)\,(1,2,0,3)}\,(1,2,0,3) \\ &= (4,0,5,8) - 2\,(1,2,0,3) \\ &= (2,-4,5,2) \end{aligned}$$

Let

$$\begin{aligned} u_3 &= v_3 - \text{Proj}_{u_1} v_3 - \text{Proj}_{u_2} v_3 = v_3 - \frac{(v_3, u_1)}{(u_1, u_1)} u_1 - \frac{(v_3, u_2)}{(u_2, u_2)} u_2 \\ &= (8,1,5,6) - \frac{(8,1,5,6)\,(1,2,0,3)}{(1,2,0,3)\,(1,2,0,3)}\,(1,2,0,3) \\ &\quad - \frac{(8,1,5,6)\,(2,-4,5,2)}{(2,-4,5,2)\,(2,-4,5,2)}\,(2,-4,5,2) \\ &= (8,1,5,6) - 2\,(1,2,0,3) - 1\,(2,-4,5,2) \\ &= (4,1,0,-2) \end{aligned}$$

Thus, the set $\{(1,2,0,3), (2,-4,5,2), (4,1,0,-2)\}$ is an orthogonal basis for **V**.

Let us now compute the norm of each vector and then normalize the vectors to get an orthonormal basis. We get

$$\|(1,2,0,3)\| = \sqrt{1^2 + 2^2 + 0^2 + 3^2} = \sqrt{14}$$

$$\|(2,-4,5,2)\| = \sqrt{2^2 + (-4)^2 + 5^2 + 2^2} = 7$$

$$\|(4,1,0,-2)\| \sqrt{4^2 + 1^2 + 0^2 + (-2)^2} = \sqrt{21}.$$

Upon using these values, we can normalize the vectors to arrive at the following orthonormal basis for V:

$$\left\{ \left(\frac{1}{\sqrt{14}}, \frac{2}{\sqrt{14}}, 0, \frac{3}{\sqrt{14}}\right), \left(\frac{2}{7}, \frac{-4}{7}, \frac{5}{7}, \frac{2}{7}\right), \left(\frac{1}{\sqrt{21}}, \frac{1}{\sqrt{21}}, 0, \frac{1}{\sqrt{21}}\right) \right\}.$$

6.12 Projection of a Vector onto a Subspace

We have defined the projection of a vector onto another vector. We now extend this concept to the projection of a vector onto a subspace. The projection of a vector onto a subspace tells us "how much" of the vector lies in the subspace.

Let $O\vec{A}$ be the projection of a vector v onto a vector u in R^n. See Figure 6.3(a).

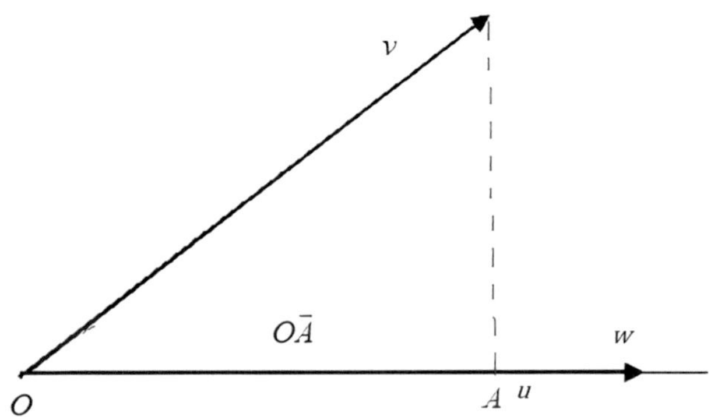

Figure 6.3(a) Orthogonal subspace.

Let **W** be the one-dimensional subspace of R^n consisting of all vectors that lie on the line defined by u. Note that the projection of v onto any vector u that lies in this subspace is $O\vec{A}$. Thus, we can determine the vector projection of v onto the subspace **W**, written $\text{proj}_w v$, to be $O\vec{A}$. The simplest expression for $O\vec{A}$ is obtained by taking u to be a unit vector. We then get

$\text{proj}_w v = O\vec{A} = (v.u)\, u$ (u is a unit vector).

Definition 6.7:

If **W** is a subspace of dimension, we extend this projection concept by expressing the projection of v onto **W** as a linear combination of the vectors of an orthonormal basis of **W** as follows.

Let **W** be a subspace of R^n. Let $\{u_1, u_2, \ldots, u_m\}$ be an orthonormal basis for **W**. If v is a vector in R^n, the projection of v onto **W** is denoted by $\text{proj}_w v$ and is defined by $\text{proj}_w v = (v.u_1)\, u_1 + (v.u_2)\, u_2 + \cdots + (v.u_m)\, u_m$.

See Figure 6.3(b)

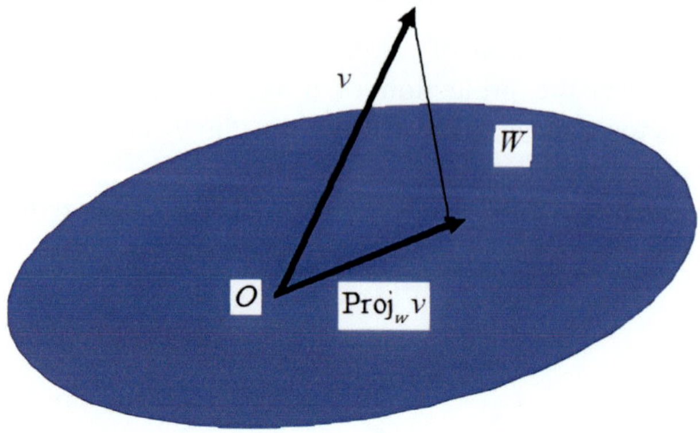

Figure 6.3(b) Orthogonal subspace.

We say that a vector v is *orthogonal to subspace* $\mathbf{W}$ if v is orthogonal to every vector in $\mathbf{W}$. The following theorem tells us that if we have a vector v in R^n and if $\mathbf{W}$ is a subspace of R^n, then we can *decompose* v into a vector that lies in the subspace $\mathbf{W}$ and a vector that is orthogonal to $\mathbf{W}$.

Theorem 6.12:

Let $\mathbf{W} \subset R^n$ be a subspace of R^n. Every vector $v \in R^n$ can be written uniquely in the form $v = w + w^{\perp}$, where $w \in \mathbf{W}$ and $w^{\perp}$ is orthogonal to $\mathbf{W}$. Then the vector w and $w^{\perp}$ are $w = \text{Proj}_w v$ and $w^{\perp} = v - \text{Proj}_w v$.

See Figure 6.3(c)

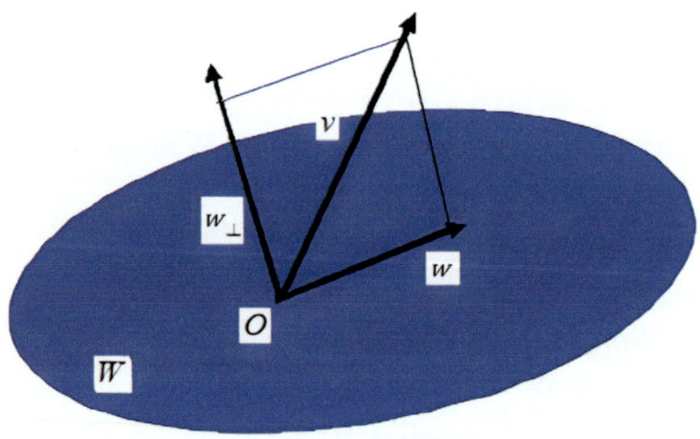

Figure 6.3(c) Orthogonal subspace.

Proof:

Observe that $w + w_{\perp} = \text{proj}_w v + (v - \text{proj}_w v) = v$.

Further $\text{proj}_w v$ is in **W**. Let us now show that $(v-\text{proj}_w v)$ is orthogonal to **W**. Let $\{u_1, u_2, \ldots, u_m\}$ be an orthonormal basis for **W**. We first show that $(v-\text{proj}_w v)$ is orthogonal to each of these base vectors of **W**. We will then be able to show that $(v-\text{proj}_w v)$ is orthogonal to an arbitrary vector in **W**. We get

$$\begin{aligned}
u_i.\,(v-\text{proj}_w v) &= u_i.\,(v-(v.u_1)\,u_1-(v.u_2)\,u_2-\cdots-(v.u_m)\,u_m) \\
&= u_i.v-u_i.\,(v.u_1)\,u_1-u_i.\,(v.u_2)\,u_2-\cdots-u_i.\,(v.u_m)\,u_m \\
&= u_i.v-(v.u_1)\,(u_i.u_1)-(v.u_2)\,(u_i.u_2)-\cdots-(v.u_m)\,(u_i.u_m) \\
&= u_i.v-(v.u_i)\,(u_i.u_i)\ \text{, since } u_i.u_j=0 \text{ unless } i=j \\
&= u_i.v-v.u_i, \text{since } u_i.u_i=1 \\
&= 0
\end{aligned}$$

Thus, $(v-\text{proj}_w v)$ is orthogonal to each of the base vectors of **W**.

Let $w' = c_1u_1 + c_2u_2 + \cdots + c_mu_m$ be an arbitrary vector in **W**. We get

$$\begin{aligned}
w'.\,(v-\text{proj}_w v) &= (c_1u_1+c_2u_2+\cdots+c_mu_m)\,.\,(v-\text{proj}_w v) \\
&= c_1u_1.\,(v-\text{proj}_w v)+\cdots+c_mu_m.\,(v-\text{proj}_w v) \\
&= 0
\end{aligned}$$

Thus, $(v-\text{proj}_w v)$ is orthogonal to **W**.

The proof of uniqueness is omitted.

Example 6.18:

Let us consider a vector $v = (3,2,6) \in R^3$. Let $\mathbf{W} \subset R^3$ be the subspace of R^3 consisting of all vectors of the form (a,b,b), i.e., $\mathbf{W} = \{(a,b,b) : a, b \in R\mathbb{R}\}$. Decompose the vector $v \in R^3$ into the sum of a vector that lies in **W** and a vector orthogonal to **W**.

Solution:

Let us consider an orthonormal basis for **W**.

With the choice of arbitrary vectors of **W**, the vector (a,b,b) can be written as

$$(a,b,b) = a\,(1,0,0) + b\,(0,1,1)\,.$$

It implies that the set $\{(1,0,0)\,,(0,1,1)\}$ spans **W** and is linearly independent.

Thus, it forms a basis for **W**, and hence the vectors are orthogonal.

Now normalize each vector $(1,0,0)$ and $(0,1,1)$ to get an orthonormal basis $\{u_1, u_2\}$ for **W**, where $u_1 = (1,0,0)\ , u_2 = \left(0, \frac{1}{\sqrt{2}}, \frac{1}{\sqrt{2}}\right)$

We get

$$\begin{aligned} w = \mathrm{Proj}_w v &= (v.u_1)\, u_1 + (v.u_2)\, u_2 \\ &= ((3,2,6)\,,(1,0,0))\,(1,0,0) + \left((3,2,6)\left(0,\frac{1}{\sqrt{2}},\frac{1}{\sqrt{2}}\right)\right)\left(0,\frac{1}{\sqrt{2}},\frac{1}{\sqrt{2}}\right) \\ &= (3,0,0) + (0,4,4) \\ &= (3,4,4) \end{aligned}$$

and

$$\begin{aligned} w^{\perp} &= v - \mathrm{Proj}_w v \\ &= (3,2,6) - (3,4,4) \\ &= (0,-2,2) \end{aligned}$$

Thus, the desired decomposition of v can be expressed as

$$(3,2,6) = (3,4,4) + (0,-2,2)\,.$$

Thus, it shows that the vector $(3,4,4)$ lies in $\mathbf{W}$, and the vector $(0,-2,2)$ is orthogonal to $\mathbf{W}$.

6.13 Distance of a Point from a Subspace

Here we discuss the distance of a point from a subspace in R^n.

Let $X = (x_1, x_2, \ldots, x_n)$ be the point in R^n, $\mathbf{W}$ be a subspace of R^n, and $\mathbf{Y} = (y_1, y_2, \ldots, y_n)$ be a point in $\mathbf{W}$.

It is natural to define the distance of X from $\mathbf{W}$, denoted to be $d\,(X, \mathbf{W})$, i.e., the minimum of the distance from X to the points of $\mathbf{W}$ and it can be expressed as $d\,(X, \mathbf{W}) = \min\{d\,(x,y)\}$, for all points y in $\mathbf{W}$.

Now we will find $d\,(\mathbf{X}, \mathbf{W})$ in terms of its projection.

Next, let us write $X - \mathbf{Y} = (X - \mathrm{Proj}_w X) + (\mathrm{Proj}_w X - \mathbf{Y})$, a decomposition of the vector $X - \mathbf{Y}$ into the sum of a vector $(X - \mathrm{Proj}_w X)$ that is orthogonal to $\mathbf{W}$, and a vector $(\mathrm{Proj}_w X - \mathbf{Y})$ that lies in $\mathbf{W}$.

Thus, the Pythagorean theorem gives

$$\|X - \mathbf{Y}\|^2 = \|X - \mathrm{Proj}_w X\|^2 + \|\mathrm{Proj}_w X - \mathbf{Y}\|^2\,.$$

Therefore, $\|X - \mathbf{Y}\|$ (which is equal to $d\,(X, \mathbf{Y})$) has a minimum value $\|X - \mathrm{Proj}_w X\|$; when $\mathrm{Proj}_w X - \mathbf{Y} = 0$; i.e., when $\mathbf{Y} = \mathrm{Proj}_w X$.

Hence, $d\,(X, \mathbf{W}) = \|X - \mathrm{Proj}_w X\|$, as shown in Figure 6.4.

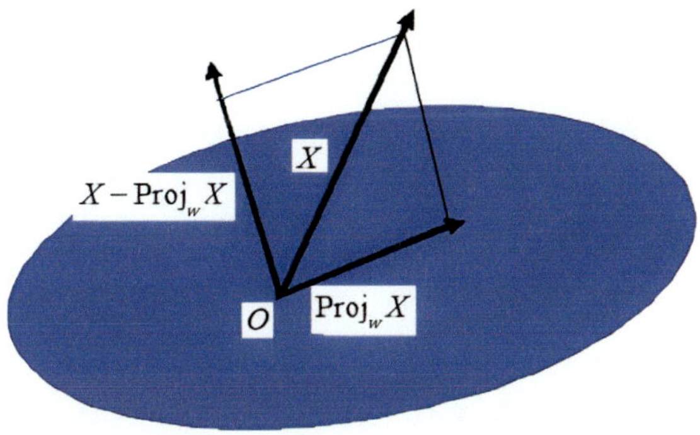

Figure 6.4 Distance of a Point from a Subspace.

Note:

A point from a subspace is the distance of the point from its projection in the subspace.

Remark:

$\text{Proj}_w X$ is the closest point to X in the subspace **W**.

Example 6.19:

Determine the distance of the point $X = (4, 1, -7)$ of R^3 from the subspace **W** consisting of all vectors of the form (a, b, b).

Solution:

From the previous example, it is found that the set $\{u_1, u_2\}$ is an orthonormal basis for **W**, where $u_1 = (1, 0, 0)\,, u_2 = \left(0, \frac{1}{\sqrt{2}}, \frac{1}{\sqrt{2}}\right)$.

Next, we compute $\text{Proj}_w \mathbf{X}$.

$$\begin{aligned}\text{Proj}_w X &= (X.u_1)\,u_1 + (X.u_2)\,u_2 \\ &= ((4,1,-7)\,.\,(1,0,0))\,(1,0,0) \\ &\quad + \left((4,1,-7)\,.\left(0, \frac{1}{\sqrt{2}}, \frac{1}{\sqrt{2}}\right)\right)\left(0, \frac{1}{\sqrt{2}}, \frac{1}{\sqrt{2}}\right) \\ &= (4,0,0) + (0,-3,-3) = (4,-3,-3)\end{aligned}$$

Thus,

$$\begin{aligned}\|X - \text{Proj}_w X\| &= \|(4,1,-7) - (4,-3,-3)\| \\ &= \|(0,4,-4)\| = \sqrt{32}\end{aligned}$$

Hence, the distance from **X** to **W** is $\sqrt{32}$.

6.14 QR-Factorization

Theorem 6.13:

Let us consider a real $m \times n$ matrix A with rank n. Then the matrix A can be written as a product, i.e., $A = QR$, where Q is a real orthogonal matrix and R is a real $n \times n$ upper triangular matrix having positive entries on its principal diagonal.

Proof:

Let $\mathbf{V}$ denote the column space of the matrix A, forming a subspace of the Euclidean inner product space R^m. Since the matrix $\mathbf{A}$ has a rank n, it implies that the n column vectors, i.e., $X_1, X_2, \ldots, X_n$ of the matrix $\mathbf{A}$ are linearly independent and form a basis of the column space $\mathbf{V}$.

Hence, the Gram–Schmidt process can be applied to these basis vectors $X_1, X_2, \ldots, X_n$ to produce an orthonormal basis of $\mathbf{V}$ say $\mathbf{Y}_1, \mathbf{Y}_2, \ldots, \mathbf{Y}_n$.

Now upon using the Gram–Schmidt procedure, these vectors $\mathbf{Y}_i$ have the form

$$\begin{aligned} \mathbf{Y}_1 &= b_{11}X_1 \\ \mathbf{Y}_2 &= b_{12}X_1 + b_{22}X_2 \\ \cdots & \cdots \quad \cdots \\ \mathbf{Y}_n &= b_{1n}X_1 + b_{2n}X_2 + \cdots + b_{nn}X_n\,, \end{aligned} \tag{6.1}$$

where b_{ij} are real numbers with $b_{ii} > 0$, i.e., positive.

Upon solving eqn (6.1) for $X_1, X_2, \ldots, X_n$ by back-substitution procedure, we get a linear system of the same general form:

$$\begin{aligned} X_1 &= r_{11}\mathbf{Y}_1 \\ X_2 &= r_{12}\mathbf{Y}_1 + r_{22}\mathbf{Y}_2 \\ \cdots & \cdots \quad \cdots \\ X_n &= r_{1n}\mathbf{Y}_1 + r_{2n}\mathbf{Y}_2 + \cdots + r_{mn}\mathbf{Y}_n\,, \end{aligned} \tag{6.2}$$

where r_{ij} are real numbers with $r_{ii} > 0$, i.e., positive.

Thus, eqn (6.2) can be expressed in matrix form as

$$\begin{aligned} A &= [X_1, X_2, \ldots, X_n] \\ &= [\mathbf{Y}_1, \mathbf{Y}_2, \ldots, \mathbf{Y}_n] \begin{bmatrix} r_{11} & r_{12} & \cdots & r_{1n} \\ 0 & r_{22} & \cdots & r_{2n} \\ \cdots & \vdots & \cdots & \vdots \\ 0 & 0 & \cdots & r_{mn} \end{bmatrix}. \end{aligned}$$

The columns of the $m \times n$ matrix, i.e., $Q = [\mathbf{Y}_1, \mathbf{Y}_2, \ldots, \mathbf{Y}_n]$ form an orthonormal basis since they constitute an orthonormal basis of $\mathbb{R}^m$ while the matrix $\mathbb{R} = \{r_{ij}\}_{m,n}$ is an upper triangular matrix.

Similarly, the most important case of this theorem is when the matrix $\mathbf{A}$ is a non-singular square matrix, then the matrix Q is $n \times n$ matrix, and equivalently it has the property

$$Q^T Q = I_n.$$

It just to say that $Q^T = Q^{-1}$.

A square matrix $\mathbf{A}$ such that $A^T = A^{-1}$ is called an *orthogonal matrix*.

Remark:

Orthogonal matrices play an essential role in the study of the canonical form of matrices.

Example 6.20:

Determine the factorization of the following matrix in the QR factorized form $A = \begin{bmatrix} 1 & 1 & 2 \\ 1 & 2 & 3 \\ 1 & 1 & 1 \end{bmatrix}$.

The method is to apply the Gram–Schmidt process to the columns X_1, X_2, X_3 of the matrix A, which are linearly independent and form a basis for the column space of A.

This yields an orthonormal basis $\{\mathbf{Y}_1, \mathbf{Y}_2, \mathbf{Y}_3\}$, where

$$\mathbf{Y}_1 = \frac{1}{\sqrt{3}} \begin{bmatrix} 1 \\ 1 \\ 1 \end{bmatrix} = \frac{1}{\sqrt{3}} X_1$$

$$\mathbf{Y}_2 = \frac{1}{\sqrt{6}} \begin{bmatrix} -1 \\ 2 \\ -1 \end{bmatrix} = \frac{2\sqrt{6}}{3} X_1 + \frac{\sqrt{6}}{2} X_2$$

$$\mathbf{Y}_3 = \frac{1}{\sqrt{2}} \begin{bmatrix} 1 \\ 0 \\ -1 \end{bmatrix} = -\frac{3\sqrt{2}}{2} X_2 + \sqrt{2} X_3 .$$

Upon back-substitution, we obtain the equations

$$X_1 = \sqrt{3}\mathbf{Y}_1$$

$$X_2 = \frac{4\sqrt{3}}{3}\mathbf{Y}_1 + \frac{\sqrt{6}}{3}\mathbf{Y}_2$$

$$X_3 = \frac{2}{\sqrt{3}}\mathbf{Y}_1 + \frac{\sqrt{6}}{2}\mathbf{Y}_2 + \frac{\sqrt{2}}{2}\mathbf{Y}_3.$$

Therefore, $A = QR$, where

$$Q = \begin{bmatrix} \frac{1}{\sqrt{3}} & -\frac{1}{\sqrt{6}} & \frac{1}{\sqrt{2}} \\ \frac{1}{\sqrt{3}} & \frac{2}{\sqrt{6}} & 0 \\ \frac{1}{\sqrt{3}} & -\frac{1}{\sqrt{6}} & -\frac{1}{\sqrt{2}} \end{bmatrix} \text{ and } R = \begin{bmatrix} \sqrt{3} & \frac{4}{\sqrt{3}} & 2\sqrt{3} \\ 0 & \frac{\sqrt{6}}{3} & \frac{\sqrt{6}}{2} \\ 0 & 0 & \frac{\sqrt{2}}{2} \end{bmatrix}.$$

Example 6.21:

Consider the matrix $\mathbf{A} = \begin{bmatrix} 1 & 1 & 0 \\ 1 & 0 & 1 \\ 0 & 1 & 1 \end{bmatrix}$ with the vectors

$X_1 = \begin{bmatrix} 1 \\ 1 \\ 0 \end{bmatrix}$, $X_2 = \begin{bmatrix} 1 \\ 0 \\ 1 \end{bmatrix}$, $X_3 = \begin{bmatrix} 0 \\ 1 \\ 1 \end{bmatrix}$, where X_1, X_2, X_3 are column vectors.

$$u_1 = X_1 = \begin{bmatrix} 1 \\ 1 \\ 0 \end{bmatrix},$$

$$\mathbf{Y}_1 = \frac{u_1}{\|u_1\|} = \frac{X_1}{\|X_1\|}$$

$$= \frac{1}{\sqrt{2}} \begin{bmatrix} 1 \\ 1 \\ 0 \end{bmatrix} = \begin{bmatrix} \frac{1}{\sqrt{2}} \\ \frac{1}{\sqrt{2}} \\ 0 \end{bmatrix}$$

$$u_2 = X_2 - (X_2.e_1)\, e_1$$

$$= \begin{bmatrix} 1 \\ 0 \\ 1 \end{bmatrix} - \frac{1}{\sqrt{2}} \begin{bmatrix} \frac{1}{\sqrt{2}} \\ \frac{1}{\sqrt{2}} \\ 0 \end{bmatrix} = \begin{bmatrix} \frac{1}{2} \\ -\frac{1}{2} \\ 1 \end{bmatrix}$$

$$\mathbf{Y}_2 = \frac{u_2}{\|u_2\|}$$

$$= \frac{1}{\sqrt{\frac{3}{2}}} \begin{bmatrix} \frac{1}{2} \\ -\frac{1}{2} \\ 1 \end{bmatrix} = \begin{bmatrix} \frac{1}{\sqrt{6}} \\ -\frac{1}{\sqrt{6}} \\ \frac{2}{\sqrt{6}} \end{bmatrix}$$

$$u_3 = X_3 - (X_3.e_1)\, e_1 - (X_3.e_2)\, e_2$$

$$= \begin{bmatrix} 0 \\ 1 \\ 1 \end{bmatrix} - \frac{1}{\sqrt{2}} \begin{bmatrix} \frac{1}{\sqrt{2}} \\ \frac{1}{\sqrt{2}} \\ 0 \end{bmatrix} - \frac{1}{\sqrt{6}} \begin{bmatrix} \frac{1}{\sqrt{6}} \\ -\frac{1}{\sqrt{6}} \\ \frac{2}{\sqrt{6}} \end{bmatrix} = \begin{bmatrix} -\frac{1}{\sqrt{3}} \\ \frac{1}{\sqrt{3}} \\ \frac{1}{\sqrt{3}} \end{bmatrix}$$

$$\mathbf{Y}_3 = \frac{u_3}{\|u_3\|} = \begin{bmatrix} -\frac{1}{\sqrt{3}} \\ \frac{1}{\sqrt{3}} \\ \frac{1}{\sqrt{3}} \end{bmatrix}.$$

Thus,

$$\begin{aligned} Q &= \left[e_1 \vdots e_2 \vdots \dots\dots\dots \vdots e_n \right] \\ &= \begin{bmatrix} \frac{1}{\sqrt{2}} & \frac{1}{\sqrt{6}} & -\frac{1}{\sqrt{3}} \\ \frac{1}{\sqrt{2}} & -\frac{1}{\sqrt{6}} & \frac{1}{\sqrt{3}} \\ 0 & \frac{2}{\sqrt{6}} & \frac{1}{\sqrt{3}} \end{bmatrix} \end{aligned}$$

and

$$\begin{aligned} R &= \begin{bmatrix} X_1.e_1 & X_2.e_1 & X_3.e_1 \\ 0 & X_2.e_2 & X_3.e_2 \\ 0 & 0 & X_3.e_3 \end{bmatrix} \\ &= \begin{bmatrix} \frac{2}{\sqrt{2}} & \frac{1}{\sqrt{2}} & \frac{1}{\sqrt{2}} \\ 0 & \frac{3}{\sqrt{6}} & \frac{1}{\sqrt{6}} \\ 0 & 0 & \frac{2}{\sqrt{3}} \end{bmatrix}. \end{aligned}$$

Example 6.22:

Orthonormalize the set of linearly independent vectors $(1, 0, 1, 1)$ $(-1, 0, -1, 1)$ $(0, -1, 1, 1)$ of $\mathbf{V}_4$.

Let $v_1 = (1, 0, 1, 1)$. Then

$$\begin{aligned} v_2 &= (-1, 0, -1, 1) - \frac{(-1, 0, -1, 1) \,.\, (1, 0, 1, 1)}{3} (1, 0, 1, 1) \\ &= \left(-\frac{2}{3}, 0, -\frac{2}{3}, \frac{4}{3} \right) \\ v_3 &= (0, -1, 1, 1) - \frac{(0, -1, 1, 1) \,.\, (1, 0, 1, 1)}{3} (1, 0, 1, 1) \\ &\quad - \frac{(0, -1, 1, 1) \,.\, \left(-\frac{2}{3}, 0, -\frac{2}{3}, \frac{4}{3}\right)}{\frac{24}{9}} \left(-\frac{2}{3}, 0, -\frac{2}{3}, \frac{4}{3} \right) \\ &= \left(-\frac{1}{2}, -1, -\frac{1}{2}, 0 \right) \end{aligned}$$

The resulting orthogonal set is

$$(1,0,1,1)\,,\left(-\frac{2}{3},0,-\frac{2}{3},\frac{4}{3}\right),\left(-\frac{1}{2},-1,-\frac{1}{2},0\right).$$

The corresponding orthonormal set is

$$\left(\frac{1}{\sqrt{3}},0,\frac{1}{\sqrt{3}},\frac{1}{\sqrt{3}}\right),\left(-\frac{1}{\sqrt{6}},0,-\frac{1}{\sqrt{6}},\frac{2}{\sqrt{6}}\right),\left(-\frac{1}{\sqrt{3}},-\frac{2}{\sqrt{3}},-\frac{1}{\sqrt{3}},0\right).$$

Example 6.23:

Find an orthonormal basis of $P_3[-1,1]$ starting from the basis $\{1,x,x^2,x^3\}$ and use the inner product defined by $f.g=\int_{-1}^{1}f(t)\,g(t)\,dt$.

We take $v_1 = 1$.

Then

$$\begin{aligned}
v_2 &= x-\frac{x.1}{2}1\\
&= x-\left(\frac{1}{2}\int_{-1}^{1}tdt\right)=x\\
v_3 &= x^2-\frac{x^2.1}{2}1-\frac{x^2.x}{\frac{2}{3}}x\\
&= x^2-\left(\frac{1}{2}\int_{-1}^{1}t^2dt\right)-\left(\frac{3}{2}x\int_{-1}^{1}t^3dt\right)\\
&= x^3-\frac{1}{3}\\
v_4 &= x^3-\frac{x^3.1}{2}1-\frac{x^3.x}{\frac{2}{3}}x-\frac{x^3.\left(x^2-\frac{1}{3}\right)}{\frac{2}{5}}\left(x^2-\frac{1}{3}\right)\\
&= x^3-\frac{3}{5}x
\end{aligned}$$

Thus, the orthogonal basis is $\{1,x,x^2-\frac{1}{3},x^3-\frac{3}{5}x\}$.

To get the corresponding orthonormal basis, we divide these by the respective norm and get

$$\left\{\frac{1}{\sqrt{2}},\frac{\sqrt{3}}{\sqrt{2}}x,\frac{3\sqrt{5}}{2\sqrt{2}}\left(x^2-\frac{1}{3}\right),\frac{5\sqrt{7}}{2\sqrt{2}}\left(x^3-\frac{3}{5}x\right)\right\}.$$

Exercises

1. Determine the inner product $u.v$ in the following cases:

   (a) $u = (1, -1), v = (2, 3)$
   (b) $u = (1, 2, 3), v = (3, 0, 2)$
   (c) $u = (-1, 1, 2, 4), v = (1, 2, -1, 1)$.

2. In an inner product space, show that

   (a) If $v.u = 0$ for all $u \in V$, then $v = 0$.
   (b) If $v.u = w.u$ for all $u \in V$, then $v = w$.

3. Show that an inner product can be defined on V_2 by

$$(x_1, x_2) . (y_1, y_2) = \frac{(x_1 - x_2) . (y_1 - y_2)}{4} + \frac{(x_1 + x_2) . (y_1 + y_2)}{4}.$$

   In this inner product space, calculate
   (a) $e_1.e_2$ (b) $(1, -1) . (1, 1)$.

4. Prove that all the eigenvalues of a symmetric matrix are real.
5. Let the set $\{v_1, v_2, \ldots, v_n\}$ be L.D. What happens when the Gram–Schmidt process of orthogonalization is applied to it?

7

Matrix Representation of Linear Transformations

Matrices form a vital tool in the study of finite-dimensional vector spaces. Hereafter though, for convenience, we deal with real vectors and the real vector space V_n, all our definitions, unless otherwise restricted, will also apply to complex vectors and the complex vector space V_n^C.

We have developed much mathematics around the concept of a basis. In this chapter, we discuss the relationship between coordinate systems and bases. We have found that a linear transformation can be represented by a matrix relative to a standard basis. We shall find a matrix representation relative to every basis. If possible, it will be of interest to find diagonal representations and determine the basis (or coordinate systems) to which this applies. Eigenvalues and eigenvectors play an essential role in these discussions. This technique will enable us to find the most suitable coordinate systems for discussing physical situations such as vibrating strings.

We shall define remarkable structure-preserving transformations between various types of vector spaces. We shall find that even though the elements of specific vector spaces such as R^n and $\mathbf{P}_n$ differ in appearance, their mathematical properties have much in common. It means that any results we develop for R^n can be applied to all such vector spaces.

7.1 Matrix Representation of Linear Transformations

Previously, we have seen that a matrix A can define a linear transformation $\mathbf{T} : R^n \to R^m$.

This section introduces a way of representing a linear transformation between general vector spaces by a matrix.

Theorem 7.1:

Let $\mathbf{T} : \mathbf{U} \to \mathbf{V}$ be a linear transformation. Let $\{u_1, u_2, \cdots, u_n\}$ be a basis for $\mathbf{U}$. The linear transformation $\mathbf{T}$ is defined by its effects on the base vectors, namely by $\mathbf{T}(u_1), \mathbf{T}(u_2), \cdots, \mathbf{T}(u_n)$ and the range of $\mathbf{T}$ is spanned by $\{\mathbf{T}(u_1), \mathbf{T}(u_2), \cdots, \mathbf{T}(u_n)\}$.

Proof:

Let u be an element of $\mathbf{U}$. Then, since $\{u_1, u_2, \cdots, u_n\}$ is a basis for $\mathbf{U}$, there exist scalars $a_1, a_2, \cdots, a_n$ such that $u = a_1u_1 + a_2u_2 + \cdots + a_nu_n$.

The linearity of $\mathbf{T}$ gives

$$\begin{aligned}\mathbf{T}(u) &= \mathbf{T}(a_1u_1 + a_2u_2 + \cdots + a_nu_n) \\ &= a_1\mathbf{T}(u_1) + a_2\mathbf{T}(u_2) + \cdots + a_n\mathbf{T}(u_n).\end{aligned}$$

Therefore, $\mathbf{T}(u)$ is known,
if $\{\mathbf{T}(u_1), \mathbf{T}(u_2), \cdots, \mathbf{T}(u_n)\}$ are known.

Further, $\mathbf{T}(u)$ may be interpreted as an arbitrary element in the range of $\mathbf{T}$ and can be expressed as a linear combination of $\{\mathbf{T}(u_1), \mathbf{T}(u_2), \cdots, \mathbf{T}(u_n)\}$.

Thus, $\{\mathbf{T}(u_1), \mathbf{T}(u_2), \cdots, \mathbf{T}(u_n)\}$ spans the range of $\mathbf{T}$.

Example 7.1:

Consider the linear transformation $\mathbf{T} : R^3\mathbb{R} \rightarrow \ R^2$ defined as follows on basis vectors of R^3. Find $\mathbf{T}(1, -2, 3)$.

$$\begin{array}{l}\mathbf{T}(1,0,0) = (3,-1) \\ \mathbf{T}(0,1,0) = (2,1) \\ \mathbf{T}(0,0,1) = (3,0)\end{array}.$$

Solution:

Since $\mathbf{T}$ is defined on basis vectors of R^3, it is specified on the whole space.

To find $\mathbf{T}(1, -2, 3)$, express the vector $(1, -2, 3)$ as a linear combination of the basis vectors and use the linearity of $\mathbf{T}$.

$$\begin{aligned}\mathbf{T}(1,-2,3) &= \mathbf{T}(1(1,0,0) - 2(0,1,0) + 3(0,0,1)) \\ &= 1.\mathbf{T}(1,0,0) - 2.\mathbf{T}(0,1,0) + 3.\mathbf{T}(0,0,1) \\ &= 1.(3,-1) - 2.(2,1) + 3.(3,0) \\ &= (8,-3).\end{aligned}$$

We have seen that a linear transformation $\mathbf{T} : R^n \rightarrow \ R^m$ can be defined by a matrix A as $\mathbf{T}(\mathbf{U}) = A\mathbf{U}$.

Here $A = \{\mathbf{T}(e_1), \mathbf{T}(e_2), \cdots, \mathbf{T}(e_n)\}$.

Note that the matrix A is constructed by finding the effects of $\mathbf{T}$ in each of these standard basis vectors R^n.

These ideas can be extended to a linear transformation $\mathbf{T} : \mathbf{U} \to \mathbf{V}$ between general vector spaces.

We shall represent the element of $\mathbf{U}$ and $\mathbf{V}$ by coordinate vectors and $\mathbf{T}$ by a matrix A that defines a transformation of coordinate vectors.

As for R^n, the matrix A is constructed by finding the effects of $\mathbf{T}$ on basis vectors.

Theorem 7.2:

Let $\mathbf{U}$ and $\mathbf{V}$ be two vector spaces with bases $B = \{u_1, u_2, \cdots, u_n\}$ and $B_r = \{v_1, v_2, \cdots, v_m\}$ and $\mathbf{T} : \mathbf{U} \to \mathbf{V}$ a linear transformation. If $v \in \mathbf{U}$ is a vector in $\mathbf{U}$ with an image $\mathbf{T}(v)$, having coordinate vectors $\bar{a}$ and $\bar{b}$ relative to these bases, then $\bar{b} = A\bar{a}$, where $A = \{\mathbf{T}(u_1)_{B_r}, \mathbf{T}(u_2)_{B_r}, \cdots, \mathbf{T}(u_n)_{B_r}\}$.

The matrix A has thus defined a transformation of coordinate vectors of $\mathbf{U}$ in the same way as $\mathbf{T}$ transforms the vectors of $\mathbf{U}$ (as shown in Figure 7.1).

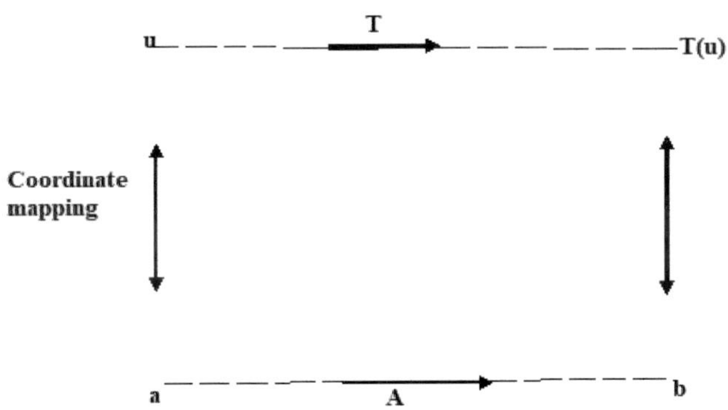

Figure 7.1 Matrix Representation of Linear Transformations.

The matrix A is called the matrix representation of $\mathbf{T}$ (or matrix of $\mathbf{T}$) concerning the bases B and B_r.

Proof:

Let $u = a_1u_1 + a_2u_2 + \cdots + a_nu_n$.

Using the linearity of $\mathbf{T}$, we can write

$$\begin{aligned}\mathbf{T}(u) &= \mathbf{T}(a_1u_1 + a_2u_2 + \cdots + a_nu_n) \\ &= a_1\mathbf{T}(u_1) + a_2\mathbf{T}(u_2) + \cdots + a_n\mathbf{T}(u_n).\end{aligned}$$

Let the effect of $\mathbf{T}$ on the basis vectors of $\mathbf{U}$ be

$$\begin{aligned}
\mathbf{T}(u_1) &= c_{11}v_1 + c_{12}v_2 + \cdots + c_{1m}v_m \\
\mathbf{T}(u_2) &= c_{21}v_1 + c_{22}v_2 + \cdots + c_{2m}v_m \\
&\vdots \\
\mathbf{T}(u_n) &= c_{n1}v_1 + c_{n2}v_2 + \cdots + c_{nm}v_m
\end{aligned}.$$

Thus,

$$\begin{aligned}
\mathbf{T}(u) = {} & a_1(c_{11}v_1 + c_{12}v_2 + \cdots + c_{1m}v_m) \\
& + a_2(c_{21}v_1 + c_{22}v_2 + \cdots + c_{2m}v_m) + \cdots \\
& + a_n(c_{n1}v_1 + c_{n2}v_2 + \cdots + c_{nm}v_m) \\
= {} & (a_1c_{11} + a_2c_{21} + \cdots + a_nc_{n1})\,v_1 \\
& + (a_1c_{12} + a_2c_{22} + \cdots + a_nc_{n2})\,v_2 + \cdots \\
& + (a_1c_{1m} + a_2c_{2m} + \cdots + a_nc_{nm})\,v_m.
\end{aligned}$$

The coordinate vector of $\mathbf{T}(u)$ is therefore

$$\bar{b} = \begin{bmatrix} (a_1c_{11} + a_2c_{21} + \cdots + a_nc_{n1}) \\ (a_1c_{12} + a_2c_{22} + \cdots + a_nc_{n2}) \\ \vdots \\ (a_1c_{1m} + a_2c_{2m} + \cdots + a_nc_{nm}) \end{bmatrix} = \begin{bmatrix} c_{11}\, c_{21} \,.....c_{n1} \\ c_{12}\, c_{22} \cdots c_{n2} \\ \vdots \\ c_{1m}c_{2m} \cdots c_{nm} \end{bmatrix} \begin{bmatrix} a_1 \\ a_2 \\ \vdots \\ a_n \end{bmatrix}$$

$$= \left[\mathbf{T}(u_1)_{B_r} \cdots \mathbf{T}(u_n)_{B_r}\right].$$

Example 7.2:

Let $\mathbf{T} : \mathbf{U} \to \mathbf{V}$ be a linear transformation. $\mathbf{T}$ and is defined relative to bases $\mathrm{B} = \{u_1, u_2, u_3\}$ and $\mathrm{B}_r = \{v_1, v_2\}$ of $\mathbf{U}$ and $\mathbf{V}$ as follows:

$$\begin{aligned}
\mathbf{T}(u_1) &= 2v_1 - v_2 \\
\mathbf{T}(u_2) &= 3v_1 + 2v_2 \\
\mathbf{T}(u_3) &= v_1 - 4v_2
\end{aligned}.$$

Find the matrix representation of $\mathbf{T}$ concerning these bases and use this matrix to determine the image of the vector $u = 3u_1 + 2u_2 - u_3$.

Solution:

The coordinate vectors of $\mathbf{T}(u_1)$, $\mathbf{T}(u_2)$, and $\mathbf{T}(u_3)$ are

$$\begin{bmatrix} 2 \\ -1 \end{bmatrix}, \begin{bmatrix} 3 \\ 2 \end{bmatrix} \text{ and } \begin{bmatrix} 1 \\ -4 \end{bmatrix}.$$

These vectors make up the columns of the matrix of the transformation **T**.

$$A = \begin{bmatrix} 2 & 3 & 1 \\ -1 & 2 & -4 \end{bmatrix}.$$

Let us now find the image of the vector $u = 3u_1 + 2u_2 - u_3$ using this matrix.

The coordinate vector of **U** is $\bar{a} = \begin{bmatrix} 3 \\ 2 \\ -1 \end{bmatrix}$.

We get $A\bar{a} = \begin{bmatrix} 2 & 3 & 1 \\ -1 & 2 & -4 \end{bmatrix} \begin{bmatrix} 3 \\ 2 \\ 1 \end{bmatrix} = \begin{bmatrix} 11 \\ 5 \end{bmatrix}$.

It implies that $\mathbf{T}(u)$ has a coordinate vector $\begin{bmatrix} 11 \\ 5 \end{bmatrix}$.

Thus, $\mathbf{T}(u) = 11v_1 + 5v_2$.

Example 7.3:

Consider the linear transformation $\mathbf{T} : R^3 \rightarrow R^2$ defined by $\mathbf{T}(x_1, x_2, x_3) = (x_1 + x_2, 2x_3)$. Then, find the matrix of the transformation **T** concerning the bases $\{u_1, u_2, u_3\}$ and $\left\{u_1{'}, u_2{'}\right\}$ of R^3 and R^2.

Here $u_1 = (1,1,0)\ , u_2 = (0,1,4)\ , u_3 = (1,2,3)$

$$u_1' = (1,0)\ , u_2' = (0,2)\,.$$

Use this matrix to find the image of the vector $u = (2,3,5)$.

Solution:

We find the effect of **T** on the basis vector of R^3.

$$\begin{aligned}
\mathbf{T}(u_1) &= \mathbf{T}(1,1,0) \\
&= (2,0) \\
&= 2.(1,0) + 0.(0,2) \\
&= 2.u_1' + 0.u_2' \\
\mathbf{T}(u_2) &= \mathbf{T}(0,1,4) \\
&= (1,8) \\
&= 1.(1,0) + 4.(0,2) \\
&= 1.u_1' + 4.u_2' \\
\mathbf{T}(u_3) &= \mathbf{T}(1,2,3)
\end{aligned}$$

$$\begin{aligned} &= (3,6) \\ &= 3.\,(1,0)\;+3.\,(0,2) \\ &= 3.u_1' + 3.u_2'. \end{aligned}$$

The coordinate vectors of $\mathbf{T}(u_1)$, $\mathbf{T}(u_2)$, and $\mathbf{T}(u_3)$ are thus

$$\begin{bmatrix} 2 \\ 0 \end{bmatrix}, \begin{bmatrix} 1 \\ 4 \end{bmatrix} \text{ and } \begin{bmatrix} 3 \\ 3 \end{bmatrix}.$$

These vectors form the column of the matrix of $\mathbf{T}$, i.e., $A = \begin{bmatrix} 2 & 1 & 3 \\ 0 & 4 & 3 \end{bmatrix}$.

Let us now use A to find the image of the vector $u = (2,3,5)$.

We determine the coordinate vector of $\mathbf{U}$.

It can be shown that

$$\begin{aligned} u = (2,3,5) &= 3\,(1,1,0)\;+2\,(0,1,4)\;-\,(1,2,3) \\ &= 3u_1 + 2u_2 + (-1)\,u_3. \end{aligned}$$

The coordinate vector of $\mathbf{U}$ is thus $\bar{a} = \begin{bmatrix} 3 \\ 2 \\ -1 \end{bmatrix}$.

The coordinate vector of $\mathbf{T}(u)$ is

$$\bar{b} = A\bar{a} = \begin{bmatrix} 2 & 1 & 3 \\ 0 & 4 & 3 \end{bmatrix} \begin{bmatrix} 3 \\ 2 \\ -1 \end{bmatrix} = \begin{bmatrix} 5 \\ 5 \end{bmatrix}.$$

Therefore,

$$\begin{aligned} \mathbf{T}(u) &= 5.u_1' + 5.u_2' \\ &= 5\,(1,0) + 5\,(0,2) \\ &= (5,10)\,. \end{aligned}$$

We can check this result directly using the definition

$$\mathbf{T}(x,y,z) = (x+y, 2z)\,.$$

For $u = (2,3,5)$, it gives $\mathbf{T}(u) = \mathbf{T}(2,3,5) = (5,10)$.

7.2 Importance of Matrix Representation

We saw that every real finite-dimensional vector space is isomorphic to R^n. This means that any such vector space can be discussed in terms of the appropriate vector space R^n. Moreover, a matrix can now represent every linear transformation, i.e., all the theoretical mathematics of these vector spaces and their linear transformation can be undertaken in vector spaces R^n and matrices.

A second reason is a computational one. The elements of R^n and matrices can be manipulated on computers. Thus, general vector spaces and their linear transformation can be discussed on computers through these representations.

Example 7.4:

Consider the linear transformation $\mathbf{T} : \mathbf{P}_2 \rightarrow \mathbf{P}_1$ defined by

$$\mathbf{T}\left(ax^2 + bx + c\right) = (a + b)\, x - c.$$

Find the matrix of $\mathbf{T}$ concerning the bases $\{u_1, u_2, u_3\}$ and $\{u_1', u_2'\}$ on $\mathbf{P}_2$ and $\mathbf{P}_1$, where
$u_1 = x^2\,, u_2 = x\,, u_3 = 1$ and $u_1' = x, u_2' = 1$.
Use this matrix to find the image of $u = 3x^2 + 2x - 1$.

Solution:

Consider the effect of $\mathbf{T}$ on each basis vector of $\mathbf{P}_2$.

$$\begin{aligned}
\mathbf{T}(u_1) &= \mathbf{T}\left(x^2\right) = x = 1.x + 0.(1) = 1.u_1' + 0.u_2' \\
\mathbf{T}(u_2) &= \mathbf{T}(x) = x = 1.x + 0.(1) = 1.u_1' + 0.u_2' \\
\mathbf{T}(u_3) &= \mathbf{T}(1) = -1 = 0.x + (-1).(1) = 0.u_1' + (-1).u_2'.
\end{aligned}$$

The coordinate vectors of $\mathbf{T}\left(x^2\right)$,$\mathbf{T}(x)$, and $\mathbf{T}(1)$ are

$$\begin{bmatrix} 1 \\ 0 \end{bmatrix}, \begin{bmatrix} 1 \\ 0 \end{bmatrix} \text{ and } \begin{bmatrix} 0 \\ -1 \end{bmatrix}.$$

The matrix of $\mathbf{T}$ is thus $A = \begin{bmatrix} 1 & 1 & 0 \\ 0 & 0 & -1 \end{bmatrix}$.

Let us now use $\mathbf{T}$ to find the image of $u = 3x^2 + 2x - 1$.

The coefficient vector of $\mathbf{U}$ relative to the basis $\left\{x^2, x, 1\right\}$ is $\bar{a} = \begin{bmatrix} 3 \\ 2 \\ -1 \end{bmatrix}$.

So, we get $\bar{b} = A\bar{a} = \begin{bmatrix} 1 & 1 & 0 \\ 0 & 0 & -1 \end{bmatrix} \begin{bmatrix} 3 \\ 2 \\ -1 \end{bmatrix} = \begin{bmatrix} 5 \\ 1 \end{bmatrix}$.

Therefore, $\mathbf{T}(u) = 5u_1' + 1u_2' = 5x + 1$.

We visualize the way this matrix representation works in Figure 7.2. The top half of the figure shows the linear transformation $\mathbf{T}$ of $\mathbf{P}_2$ to $\mathbf{P}_1$. The bottom half is analogous to the top half, with A defining a transformation of the coordinate vectors of $\mathbf{P}_2$ into the coordinate vectors of $\mathbf{P}_1$ according to

$$\begin{bmatrix} 1 & 1 & 0 \\ 0 & 0 & -1 \end{bmatrix} \begin{bmatrix} a \\ b \\ c \end{bmatrix} = \begin{bmatrix} a+b \\ -c \end{bmatrix}.$$

The bottom half is a *coordinate representation* of the top half.

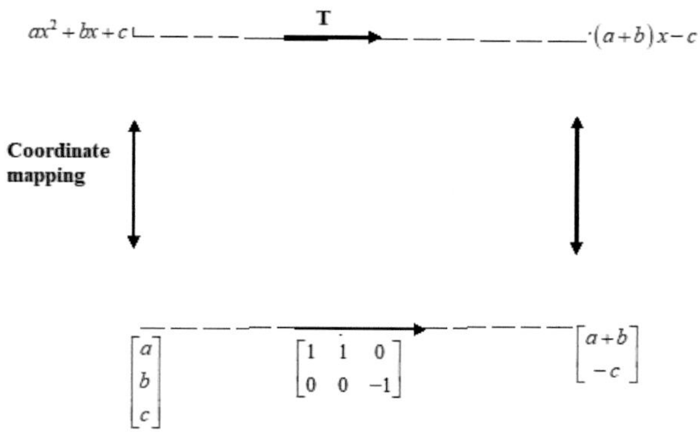

Figure 7.2 Matrix Representation of Linear Transformations.

7.3 Visualization of the Matrix Representation

The top half of Figure 7.2 shows the linear transformation $\mathbf{T}$ of $\mathbf{P}_2$ to $\mathbf{P}_1$. The bottom half is analogous to the top half with A' defining a transformation of the coordinate vectors of $\mathbf{P}_2$ into the coordinate vectors of $\mathbf{P}_1$ according to

$$\begin{bmatrix} 1 & 1 & 0 \\ 0 & 0 & -1 \end{bmatrix} \begin{bmatrix} a \\ b \\ c \end{bmatrix} = \begin{bmatrix} a+b \\ -c \end{bmatrix}.$$

The bottom half is a coordinate representation of the top half.

Example 7.5:

Let $D = \frac{d}{dx}$ be the operation of taking the derivative. D is a linear operator on $\mathbf{P}_2$. Find the matrix of D concerning the basis $\{x^2, x, 1\}$ of $\mathbf{P}_2$.

Solution:

We examine the effects of D on the basis vectors.

$$D\left(x^2\right) = 2x = 0.x^2 + 2.x + 0.(1)$$
$$D\left(x\right) = 1 = 0.x^2 + 0.x + 1.(1)$$
$$D\left(1\right) = 0 = 0.x^2 + 0.x + 0.(1).$$

The matrix of D is thus

$$A = \begin{bmatrix} 0 & 0 & 0 \\ 2 & 0 & 0 \\ 0 & 1 & 0 \end{bmatrix}.$$

The matrix A defines a linear operator that is analogous to D on $\mathbf{P}_2$, as shown in Figure 7.3

We have $D\left(ax^2 + bx + c\right) = 2ax + b$

and $\begin{bmatrix} 0 & 0 & 0 \\ 2 & 0 & 0 \\ 0 & 1 & 0 \end{bmatrix} \begin{bmatrix} a \\ b \\ c \end{bmatrix} = \begin{bmatrix} 0 \\ 2a \\ b \end{bmatrix}.$

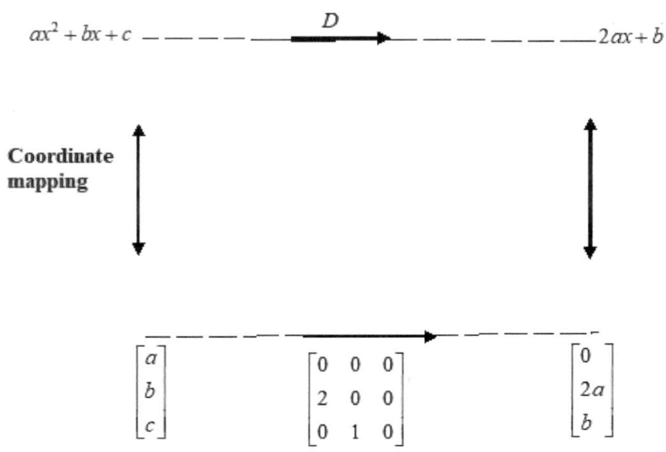

Figure 7.3 Matrix Representation of Linear Transformations.

7.4 Relation between Matrix Representation

At this time, we discuss how the matrix representation of linear operators relative to different bases is related. A transformation called *similarity transformation* plays a crucial role in this discussion.

Definition 7.1:

Let there be two square matrices A and B of the same size. The matrix B is then said to be similar to the matrix A if there is an invertible matrix C such that $B = C^{-1}AC$.

The transformation of the matrix A into the matrix B in this manner is called *similarity transformation*. We now find that the matrix representations of a linear operator relative to two bases are similar matrices.

Theorem 7.3:

Let $\mathbf{V}$ be a vector space with bases B and B_r, and $\mathbf{P}$ be the transition matrix from B_r to B. If $\mathbf{T}$ is a linear operator on $\mathbf{V}$ having matrices A and A_r concerning the first and second bases, then $\mathbf{A}' = \mathbf{P}^{-1}\mathbf{AP}$.

Proof:

Consider a vector $u \in \mathbf{V}$. Let its coordinate vectors relative to B and B_r be a and a_r.

The coordinate vectors of $\mathbf{T}(u)$ are $A\bar{a}$ and $A_r\bar{a}_r$.

As $\mathbf{P}$ is the transition matrix from B_r to B,

$\bar{a} = \mathbf{P}\bar{a}'$ and $A\bar{a} = \mathbf{P}(A'\bar{a}')$.

The second equation may be rewritten as

$$\mathbf{P}^{-1}A\bar{a} = A'\bar{a}'.$$

Substituting $\bar{a} = \mathbf{P}\bar{a}'$ into this equation gives

$$\mathbf{P}^{-1}A\mathbf{P}\bar{a}' = \mathbf{A}'\bar{a}'.$$

The effect of the matrices $\mathbf{P}^{-1}\mathbf{AP}$ and $\mathbf{A}'$ as transformations on an arbitrary coordinate vector $\bar{a}_r$ is the same.

Thus, these matrices are equal.

Example 7.6:

Consider the linear operator $\mathbf{T}(x, y) = (2x, x + y)$ on R^2. Then, find the matrix of $\mathbf{T}$ concerning the standard basis $\mathrm{B} = \{(1, 0), (0, 1)\}$ of R^2.

Use the transformation $\mathbf{A}' = \mathbf{P}^{-1}A\mathbf{P}$ to determine the matrix $\mathbf{A}'$ concerning the basis $\mathrm{B}' = \{(-2, 3), (1, -1)\}$.

Solution:

The effect of the linear transformation $\mathbf{T}$ on the vectors of the standard basis is

$$\mathbf{T}(1,0) = (2,1) = 2(1,0) + 1(0,1)$$
$$\mathbf{T}(0,1) = (0,1) = 0(1,0) + 1(0,1).$$

The matrix of the transformation $\mathbf{T}$ relative to the standard basis is

$$A = \begin{bmatrix} 2 & 0 \\ 1 & 1 \end{bmatrix}.$$

We now find that $\mathbf{P}$ is the transition matrix from B' to B. Write the vectors of B′ in terms of those of B.

$$\begin{array}{l} (-2,3) = -2\,(1,0) + 3\,(0,1) \\ (1,-1) = 1\,(1,0) - 1\,(0,1) \end{array}.$$

The transition matrix is $\mathbf{P} = \begin{bmatrix} -2 & 1 \\ 3 & -1 \end{bmatrix}$.

Therefore,

$$\begin{aligned} A' = \mathbf{P}^{-1}\mathbf{AP} &= \begin{bmatrix} -2 & 1 \\ 3 & -1 \end{bmatrix}^{-1} \begin{bmatrix} 2 & 0 \\ 1 & 1 \end{bmatrix} \begin{bmatrix} -2 & 1 \\ 3 & -1 \end{bmatrix} \\ &= \begin{bmatrix} 1 & 1 \\ 3 & 2 \end{bmatrix} \begin{bmatrix} 2 & 0 \\ 1 & 1 \end{bmatrix} \begin{bmatrix} -2 & 1 \\ 3 & -1 \end{bmatrix} \\ &= \begin{bmatrix} -3 & 2 \\ -10 & 6 \end{bmatrix}. \end{aligned}$$

Exercises

Determine the matrix $(\mathbf{T} : B_1, B_2)$ for the given linear transformation $\mathbf{T}$ and the bases B_1 and B_2.

1. $\mathbf{T} : V_2 \to V_2$ defined by $\mathbf{T}(x,y) = (-x,-y)$
   (a) $B_1 = \{e_1, e_2\}$ $B_2 = \{(1,1), (1,-1)\}$
   (b) $B_1 = \{(1,1), (1,0)\}$ $B_2 = \{(2,3), (4,5)\}$
2. $\mathbf{T} : V_3 \to V_2$ $\mathbf{T}(x,y,z) = (x+y, y+z)$
   (a) $B_1 = \{(1,1,1), (1,0,0), (1,1,0)\}$, $B_2 = \{e_1, e_2\}$
   (b) $B_1 = \left\{\left(1,1,\frac{2}{3}\right), (-1,2,-1), \left(2,3,\frac{1}{2}\right)\right\}$, $B_2 = \left\{(1,3), \left(\frac{1}{2},1\right)\right\}$.
3. $\mathbf{T} : V_4 \to V_5$ defined by

$$\mathbf{T}(x_1,x_2,x_3,x_4) = (2x_1 + x_2, x_2 - x_3, x_3 + x_4, x_1, x_1 + x_2 + 3x_3 + x_4)$$

$$\begin{aligned} B_1 &= \{(1,2,3,1), (1,0,0,1), (1,1,0,0), (0,1,1,1)\} \\ B_2 &= \{e_1, e_2, e_3, e_4, e_5\} \end{aligned}$$

4. $D : \mathbf{P}_n \to \mathbf{P}_n$ defined by $D(p) = p'$

$$B_1 = B_2 = \left\{1, x, x^2, x^3, \cdots, x^n\right\}$$

5. $\mathbf{T} : \mathbf{P}_4 \to \mathbf{P}_4$ defined by $\mathbf{T}(p)(x) = \int_1^x p'(t)\,dt$
   (a) $B_1 = B_2 = \{1, x, x^2, x^3, x^4\}$.
   (b) $B_1 = \{1, x, x^2, x^3, x^4\}$,
   $B_2 = \{x - 1, x + 1, x^2 - x^4, x^3 + x^4, x^2 + x\}$.

6. $\mathbf{T} : \mathbf{P}_2 \to \mathbf{P}_3$ defined by $\mathbf{T}(p)(x) = xp(x)$
   (a) $B_1 = \{1, 1 + x, 1 - x + x^2\}, B_2 = \{1, 1 + x, x^2, 2x - x^3\}$
   (b) $B_1 = \{1, x, x^2\}, B_2 = \left\{1 + x, (1 + x)^2, (1 + x)^3, 1 - x\right\}$.

The linear map associated with a matrix:
For each given matrix A and bases B_1 and B_2, determine a linear transformation $\mathbf{T} : V_n \to V_m$ such that $A = (\mathbf{T} : B_1, B_2)$.

7.
$$A = \begin{bmatrix} 1 & 1 & 2 & 3 \\ 1 & 0 & 1 & -1 \\ 1 & 2 & 0 & 0 \end{bmatrix}.$$

(a) B_1 and B_2 are standard bases for V_4 and V_3 respectively.
(b) $B_1 = \{(1, 1, 1, 2)\ (1, -1, 0, 0)\ (0, 0, 1, 1)\ (0, 1, 0, 0)\}$

$$B_2 = \{(1, 2, 3)\ (1, -1, 1)\ (2, 1, 1)\}$$

8.
$$A = \begin{bmatrix} 1 & 0 & 0 \\ 0 & 1 & 0 \\ 0 & 0 & 1 \end{bmatrix}.$$

(a) $B_1 = B_2 = \{e_1, e_2, e_3\}$.
(b) $\begin{array}{l} B_1 = \{(1, 1, 1)\ (1, 0, 0)\ (0, 1, 0)\} \\ B_2 = \{(1, 2, 3)\ (1, -1, 1)\ (2, 1, 1)\} \end{array}.$
(c) $\begin{array}{l} B_1 = \{(1, 2, 3)\ (1, -1, 1)\ (2, 1, 1)\} \\ B_2 = \{(1, 1, 1)\ (1, 0, 0)\ (0, 1, 0)\} \end{array}.$

9.
$$A = \begin{bmatrix} 1 & -1 & 2 \\ 3 & 1 & 0 \end{bmatrix}$$

(a) B_1 and B_2 are standard bases for V_3 and V_2 respectively.
(b) $\begin{array}{l} B_1 = \{(1, 1, 1)\ (1, 2, 3)\ (1, 0, 0)\} \\ B_2 = \{(1, 1)\ (1, -1)\} \end{array}$
(c) $\begin{array}{l} B_1 = \{(1, -1, 1)\ (1, 2, 0)\ (0, -1, 0)\} \\ B_2 = \{(1, 0)\ (2, -1)\} \end{array}.$

10.

$$A = \begin{bmatrix} 1 & 2 \\ 0 & 1 \\ -1 & 3 \end{bmatrix}$$

(a) B_1 and B_2 are standard bases for V_2 and V_3 respectively.

(b) $\begin{matrix} B_1 = \{(1,1)\,(-1,1)\} \\ B_2 = \{(1,1,1)\,(1,-1,1)\,(0,0,1)\} \end{matrix}$.

(c) $\begin{matrix} B_1 = \{(1,2)\,(-2,1)\} \\ B_2 = \{(1,-1,-1)\,(1,2,3)\,(-1,0,2)\} \end{matrix}$.

11. If $\begin{bmatrix} \cos\theta & -\sin\theta \\ \sin\theta & \cos\theta \end{bmatrix}$ is the matrix of a linear map $\mathbf{T} : V_2 \to V_2$ relative to the standard bases, then find the matrix of $\mathbf{T}^{-1}$ relative to the standard bases.

Linear operations in $M_{m \times n}$:

In Problems 12–15, determine $\alpha A + \beta B$ for the given scalars α and β and the matrices A and B.

12.

$$A = \begin{bmatrix} 1 & 2 & 3 \\ -1 & 1 & 1 \end{bmatrix}, B = \begin{bmatrix} 1 & 0 & 1 \\ 2 & 1 & -1 \end{bmatrix}.$$

(a) $\alpha = 2, \beta = 7$ (b) $\alpha = 3, \beta = -2$.

13.

$$A = \begin{bmatrix} 1 & 2 \\ 3 & 1 \end{bmatrix}, B = \begin{bmatrix} 1 & 0 \\ 2 & -1 \end{bmatrix}.$$

(a) $\alpha = 3, \beta = 5$ (b) $\alpha = 2, \beta = -3$.

14.

$$A = \begin{bmatrix} 1 & 2 & 3 & 4 \\ -1 & 1 & 1 & 1 \\ 3 & 1 & 2 & 0 \end{bmatrix}, B = \begin{bmatrix} 3 & -1 & 2 & 0 \\ 1 & 5 & 7 & 3 \\ 2 & 1 & 0 & 1 \end{bmatrix}$$

(a) $\alpha = 2, \beta = -6$ (b) $\alpha = 3, \beta = 5$

(c) $\alpha = -7, \beta = 3$.

15.

$$A = \begin{bmatrix} 1 & -1 & 1 \\ 2 & -3 & -1 \\ 3 & 2 & 0 \end{bmatrix}, B = \begin{bmatrix} 1 & 0 & 2 \\ 0 & 1 & 1 \\ 3 & 1 & -1 \end{bmatrix}$$

(a) $\alpha = 7, \beta = -5$ (b) $\alpha = \frac{1}{2}, \beta = \frac{2}{3}$

(c) $\alpha = \frac{1}{3}, \beta = \frac{4}{5}$.

16. Let $\mathbf{T}_1, \mathbf{T}_2 : \mathbf{U} \to \mathbf{V}$ be two linear maps. Let B_1, B_2 be ordered bases for $\mathbf{U}$ and $\mathbf{V}$, respectively. Then prove that

$$(\alpha_1 \mathbf{T}_1 + \alpha_2 \mathbf{T}_2 : B_1, B_2) = \alpha_1 (\mathbf{T}_1 : B_1, B_2) + \alpha_2 (\mathbf{T}_2 : B_1, B_2).$$

17. Let $S, \mathbf{T} : V_3 \to V_4$ be defined as

$$\begin{array}{l} S(x_1, x_2, x_3) = (x_1 + x_2, x_1 - 2x_2 + x_3, x_2 + 3x_3, x_1 + x_3) \\ \mathbf{T}(x_1, x_2, x_3) = (x_1 + 2x_2, x_1 - x_2, 3x_2 + x_3, x_1 + x_2 + x_3) \end{array}.$$

Determine the matrix of $3S - 4\mathbf{T}$ relative to the standard bases by two different methods.

18. Let $S, \mathbf{T} : \mathbf{P}_3 \to \mathbf{P}_4$ be defined as

$$\begin{array}{l} (S(p))(x) = (x^2 - 1)\, p'(x) \\ (\mathbf{T}(p))(x) = (3x + 2)\, p(x) - \int_1^x p'(t)\, dt. \end{array}$$

Determine the matrix of $4S + 2\mathbf{T}$ relative to the ordered bases $B_1 = \{1, x, x^2, x^3\}$ and $B_2 = \left\{(1-x), (1+x), (1-x)^2, (1-x)^3, \frac{x^4}{2}\right\}$ by two different methods.

19. Let $B_1 = \{u_1, u_2, \ldots.., u_n\}$ and $B_2 = \{v_1, v_2, \ldots.., v_n\}$ be the ordered bases for the vector spaces $\mathbf{U}$ and $\mathbf{V}$, respectively. Define $\mathbf{T}_{i,j} : \mathbf{U} \to \mathbf{V}$,$1 \le i \le m,\ 1 \le j \le n$ such that

$$\mathbf{T}_{i,j}(u_k) = \begin{cases} 0 & if\, k \ne i \\ v_j & if\, k = i \end{cases}.$$

Then prove that
(a) $(\mathbf{T}_{i,j} : B_1, B_2) = E_{i,j}$
(b) $\{\mathbf{T}_{i,j}\}$ is a basis of $L(\mathbf{U}, \mathbf{V})$.

20. Define $\mathbf{T} : M_{2,2} \to M_{2,3}$ such that

$$\mathbf{T}\begin{bmatrix} \alpha_{11} & \alpha_{12} \\ \alpha_{21} & \alpha_{22} \end{bmatrix} = \begin{bmatrix} \alpha_{11} + \alpha_{12} & 0 & \alpha_{12} + \alpha_{22} \\ \alpha_{12} & \alpha_{21} + \alpha_{22} & 0 \end{bmatrix}.$$

Prove that $\mathbf{T}$ is linear and determine its matrix relative to the standard bases for $M_{2,2}$ and $M_{2,3}$.

21. Repeat Problem 9 for $\mathbf{T} : M_{2,3} \to M_{2,2}$ defined as

$$\mathbf{T}\begin{bmatrix} \alpha_{11} & \alpha_{12} \\ \alpha_{21} & \alpha_{22} \end{bmatrix} = \begin{bmatrix} \alpha_{11} + \alpha_{12} & 0 & \alpha_{12} + \alpha_{22} \\ \alpha_{12} & \alpha_{21} + \alpha_{22} & 0 \end{bmatrix}.$$

22. Let $\mathbf{V}$ be the subspace of $C^m(-\infty,\infty)$ spanned by the functions $\sin x, \cos x, \sin x\cos x, \sin^2 x, \cos^2 x$. Determine the dimension of $\mathbf{V}$ and prove that the differential operator D^n maps $\mathbf{V}$ into itself for every positive integer $n \le m$.
    Determine the matrix of (a) $2D+3$ (b) $3D^2 - D + 4$ relative to the basis of $\mathbf{V}$ obtained from the given spanning set of $\mathbf{V}$.

23. Find the range, kernel, rank, and nullity of the following matrices:

    (a) $\begin{bmatrix} 1 & 3 & 2 \\ -1 & 7 & 2 \\ 1 & 0 & 1 \end{bmatrix}$ (b) $\begin{bmatrix} 1 & -1 & 2 \\ 3 & -2 & 5 \end{bmatrix}$

    (c) $\begin{bmatrix} 2 & 0 & 1 \\ 7 & 1 & 2 \\ 3 & -1 & 1 \end{bmatrix}$ (d) $\begin{bmatrix} 2 & 3 & 1 & 2 & 0 \\ 0 & 3 & -1 & 2 & 1 \\ 1 & -3 & 2 & 4 & 3 \\ 2 & 3 & 0 & 3 & 0 \end{bmatrix}$

    (e) $\begin{bmatrix} 1 & -1 & 1 & 0 \\ 2 & 3 & -1 & 1 \\ 1 & 5 & 2 & 0 \\ 0 & 0 & 1 & 1 \end{bmatrix}$ (f) $\begin{bmatrix} -1 & 1 & 1 \\ 3 & 1 & -1 \\ 2 & 2 & 1 \end{bmatrix}$

24. Prove that the following matrices are non-singular and find their inverses:

    (a) $\begin{bmatrix} 1 & 2 \\ -1 & 0 \end{bmatrix}$ (b) $\begin{bmatrix} 1 & 0 & 0 \\ 2 & 1 & 0 \\ 3 & 4 & 2 \end{bmatrix}$

    (c) $\begin{bmatrix} 1 & -1 & 2 \\ 3 & 0 & 1 \\ 0 & 1 & -1 \end{bmatrix}$ (d) $\begin{bmatrix} 1 & -1 & 2 & 3 \\ 0 & 2 & 1 & 1 \\ 1 & 3 & 1 & -1 \\ 1 & 1 & -1 & 0 \end{bmatrix}$

    (e) $\begin{bmatrix} 2 & 2 & 1 & 1 \\ 0 & 2 & 1 & 1 \\ 0 & 0 & 1 & 1 \\ 0 & 0 & 1 & 0 \end{bmatrix}$

25. Find the values of α and β for which the following matrix is invertible. Find the inverse when it exists.

$$\begin{bmatrix} \alpha & \beta & 0 \\ 0 & \alpha & \beta \\ \beta & 0 & \alpha \end{bmatrix}$$

26. Prove the following:
    (a) If two rows of a matrix are interchanged, then the rank does not change.
    (b) If a row of a matrix is multiplied by a non-zero scalar, then the rank does not change.

8

Diagonalizations

This chapter discusses the diagonalization process that includes minimal polynomials, Cayley–Hamilton theorem, and diagonal matrix representation of a linear operator. It also consists of the diagonalization of matrices, diagonalization of symmetric matrices, and orthogonal diagonalization.

A self-contained illustration of the role of linear transformation in computer graphics is presented. The bases of the space define these isomorphisms. Different bases also lead to other matrix representations of linear transformation. The vital aspect of eigenvalues and eigenvectors and minimal polynomials in finding diagonal representations is discussed. These techniques are used to arrive at the normal modes of oscillating systems.

8.1 Minimal Polynomials

Let us consider an $n{\times}n$ square matrix $\mathbf{A}$. Let $I(\mathbf{A})$ denote the collection of all polynomials $f(t)$ for which the matrix $\mathbf{A}$ is a root, i.e., for which $f(\mathbf{A}) = 0$.

The Cayley–Hamilton theorem states that, the characteristic polynomial $\Delta(\lambda)$ of $\mathbf{A}$ belongs to the non-empty set $I(\mathbf{A})$. Let $m(\lambda)$ denote the monic polynomial of the lowest degree in $I(\mathbf{A})$. (Such a polynomial $m(\lambda)$ exists and is unique.) We shall call $m(\lambda)$ the minimal polynomial of the matrix $\mathbf{A}$, provided $m(\lambda)$ will satisfy specific properties of the minimal polynomial.

Monic polynomial: If the polynomial's leading coefficient equals one, then the polynomial $f(\lambda)$ is monic.

Definition 8.1:

Let us consider an $n \times n$ square matrix $\mathbf{A}$ defined over a field F. Then the *monic polynomial* is a *minimal polynomial* of the matrix $\mathbf{A}$, if the *monic polynomial* $m(\lambda)$ of least degree satisfies its characteristic equation, i.e., $m(\mathbf{A}) = 0$.

Theorem 8.1:

Let $f(\lambda)$ be a polynomial of the matrix $\mathbf{A}$ such that $f(\mathbf{A}) = 0$. Then the *minimal polynomial* $m(\lambda)$ divides $f(\lambda)$.

Proof:

By Euclidean division, let there be polynomials $q(\lambda)$ and $r(\lambda)$ such that

$$f(\lambda) = m(\lambda) \,.\, q(\lambda) + r(\lambda), \text{ with } r(\lambda) = 0$$

or

$$\deg r(\lambda) < \deg m(\lambda).$$

Now by hypothesis, we know that $f(\mathbf{A}) = 0$, and by definition $m(\mathbf{A}) = 0$, it implies that $r(\mathbf{A}) = 0$.

Consequently, let us assume $r(\mathbf{A}) = 0$.

Then by the definition of $m(\lambda)$, $\deg r(\lambda)$ cannot be smaller than $\deg m(\lambda)$; so we must have $r(\lambda) = 0$.

It follows that

$f(\lambda) = m(\lambda)\, q(\lambda)$ implies $m(\lambda)$ divides $f(\lambda)$.

Corollary 8.1:

The *minimum polynomial* $m(\lambda)$ of a matrix $\mathbf{A}$ divides the *characteristic polynomial*$\Delta(\lambda)$.

Note: It is immediate from the above analogy that every zero of the minimal polynomials $m(\lambda)$ is zero of the *characteristic polynomials* $\Delta(\lambda)$. The *converse* is also true.

Theorem 8.2:

The *minimum polynomial* $m(\lambda)$ and the *characteristic polynomial* $\Delta(\lambda)$ have the same zeros.

Proof:

Suppose that λ is a zero of $\Delta(\lambda)$. Then λ is an eigenvalue, and a non-zero vector $\mathbf{X}$ such that $\mathbf{AX} = \lambda\mathbf{X}$.

Let $g(\lambda) = a_0 + a_1\lambda + \cdots + a_k\lambda^k$.

We then have

$$\begin{aligned} g(\mathbf{A})\, x &= a_0 x + a_1 \mathbf{A} x + \cdots + a_k \mathbf{A}^k x \\ &= a_0 x + a_1 \lambda x + \cdots + a_k \lambda^k x \\ &= \left(a_0 + a_1\lambda + \cdots + a_k\lambda^k\right) x \\ &= g(\lambda)\, x. \end{aligned}$$

It happens whence $g(\lambda)$ is an eigenvalue of $g(\mathbf{A})$.

Thus, $g(\lambda)$ is a zero of the characteristic equation for $g(\mathbf{A})$.

Now, if we take $g(\lambda)$ to be $m(\lambda)$, then for every zero λ of $\Delta(\lambda)$, we have that $m(\lambda)$ is a zero of the characteristic polynomial of $m(\mathbf{A})$, i.e.,

$$\begin{aligned}\Delta_{m(\mathbf{A})}(\lambda) &= C_0(\lambda)\\ &= \det(-\lambda I_n)\\ &= (-1)^n \lambda^n.\end{aligned}$$

Since the only zeros of this characteristic polynomial are 0, we have $m_{\mathbf{A}}(\lambda) = 0$ a zero of $m_{\mathbf{A}}(\lambda)$.

8.2 Cayley–Hamilton Theorem

Theorem 8.3 (Cayley—Hamilton Theorem):

Let us consider an $n \times n$ square matrix $\mathbf{A}$ over the field of complex numbers. If $\Delta(\lambda)$ is the characteristic polynomial of the matrix $\mathbf{A}$, then $\Delta(\mathbf{A}) = 0$ and hence the minimal polynomials of the matrix $\mathbf{A}$ divide the characteristic polynomial of the matrix $\mathbf{A}$.

Proof:

Let us consider an $n \times n$ matrix $\mathbf{A}$ over the field of complex numbers and let the matrix be similar to an upper triangular matrix U.

Since the matrix $\mathbf{A}$ is similar to an upper triangular matrix U, by definition, we have $C^{-1}AC = U$, where C is an invertible matrix.

Since the matrices $\mathbf{A}$ and U are similar, they have the same minimal polynomial and the same characteristic polynomials.

Therefore, for the proof of the *Cayley—Hamilton theorem*, it is sufficient to prove the hypothesis of the theorem for the triangular matrix U.

As we know, the characteristic polynomial of a triangular matrix U is $(a_{11} - \lambda)(a_{22} - \lambda)\cdots(a_{nn} - \lambda)$, and hence the direct matrix multiplication gives

$$(a_{11}I - U).(a_{22}I - U)\cdots(a_{nn}I - U) = 0.$$

That satisfies the characteristic equation.

Again, the upper triangular matrices have the same characteristic matrix and the same minimal polynomials since the matrices $\mathbf{A}$ and T are equivalent matrices. So, it can be concluded that the minimal polynomial of the matrix $\mathbf{A}$ divides the characteristic polynomial of the matrix $\mathbf{A}$.

Hence proved.

Note: Certain essential features of the matrix can be obtained from the minimal polynomial, but that cannot be obtained from the characteristic polynomial.

Example 8.1:

Find the *minimal polynomial* of the matrix

$$\mathbf{A} = \begin{bmatrix} 1 & 0 & 1 \\ 0 & 2 & 1 \\ -1 & 0 & 3 \end{bmatrix}.$$

Solution:

The characteristic polynomial of the matrix

$$\mathbf{A} = \begin{bmatrix} 1 & 0 & 1 \\ 0 & 2 & 1 \\ -1 & 0 & 3 \end{bmatrix} \text{ is}$$

$$\Delta(\lambda) = (\lambda - 2)^3 .$$

Since $\mathbf{A} - 2I_3 \neq 0$ and $(\mathbf{A} - 2I_3)^2 \neq 0$,
we have $m(\lambda) = \Delta(\lambda)$, i.e., the *minimum polynomial* $m(\lambda)$ and the *characteristic polynomial* $\Delta(\lambda)$ have the same zeros.

Example 8.2:

Find the *minimal polynomial* of the matrix

$$\mathbf{A} = \begin{bmatrix} 5 & -6 & -6 \\ -1 & 4 & 2 \\ 3 & -6 & -4 \end{bmatrix}.$$

Solution:

For the matrix

$$\mathbf{A} = \begin{bmatrix} 5 & -6 & -6 \\ -1 & 4 & 2 \\ 3 & -6 & -4 \end{bmatrix},$$

we have the characteristic polynomial

$$\Delta(\lambda) = (\lambda - 1)(\lambda - 2)^2 .$$

By Theorem 8.3, the *minimum polynomial* is therefore either
$(\lambda - 1)$, $(\lambda - 2)$, $(\lambda - 1)(\lambda - 2)$ or $(\lambda - 1)(\lambda - 2)^2$.

Since $(\mathbf{A} - I_3)(\mathbf{A} - 2I_3) = 0$, it follows that the minimal polynomial of the matrix $\mathbf{A}$ is

$$m(\lambda) = (\lambda - 1)(\lambda - 2) .$$

Example 8.3:

Find the *minimal polynomial* $m(\lambda)$ of the matrix

$$\mathbf{A} = \begin{bmatrix} 2 & 2 & -5 \\ 3 & 7 & -15 \\ 1 & 2 & -4 \end{bmatrix}.$$

Solution:

Here we have

$$tr(\mathbf{A}) = 2 + 7 - 4 = 5$$
$$\mathbf{A}_{11} + \mathbf{A}_{22} + \mathbf{A}_{33} = 2 - 3 + 8 = 7$$
$$|\mathbf{A}| = 3.$$

Hence, the characteristic polynomial of $\mathbf{A}$ is

$$\begin{aligned} \Delta(\lambda) &= \lambda^3 - 5\lambda^2 + 7\lambda - 3 \\ &= (\lambda - 1)^2 (\lambda - 3). \end{aligned}$$

Since the *minimal polynomial* $m(\lambda)$ must divide the characteristic polynomial $\Delta(\lambda)$, this implies that each irreducible factor of $\Delta(\lambda)$, i.e., $(\lambda - 3)$ and $(\lambda - 1)$ must also be a factor of the *minimal polynomial* $m(\lambda)$.

Thus, the minimal polynomial $m(\lambda)$ of $\mathbf{A}$ is precisely one of the following:

$f(\lambda) = (\lambda - 3)(\lambda - 1)$ and $g(\lambda) = (\lambda - 3)(\lambda - 1)^2$.

Since λ is a root of the matrix $\mathbf{A}$, by the Cayley–Hamilton theorem, we find

$$g(\mathbf{A}) = \Delta(\mathbf{A}) = 0.$$

Hence, we need the only test $f(\lambda)$. We have

$$\begin{aligned} f(\mathbf{A}) &= (\mathbf{A} - I)(\mathbf{A} - 3I) \\ &= \begin{bmatrix} 1 & 2 & -5 \\ 3 & 6 & -15 \\ 1 & 2 & -5 \end{bmatrix} \begin{bmatrix} -1 & 2 & 5 \\ 3 & 4 & -15 \\ 1 & 2 & -7 \end{bmatrix}. \\ &= \begin{bmatrix} 0 & 0 & 0 \\ 0 & 0 & 0 \\ 0 & 0 & 0 \end{bmatrix} \end{aligned}$$

Thus,

$$f(\lambda) = m(\lambda) = (\lambda - 3)(\lambda - 1)$$

$$= \lambda^2 - 4\lambda + 3$$

is the *minimal polynomial* of the matrix $\mathbf{A}$.

Example 8.4:

Determine the minimal polynomial of the following matrix:

$$\mathbf{A} = \begin{bmatrix} 2 & 1 & 1 \\ 0 & 2 & 0 \\ 0 & 0 & 2 \end{bmatrix}.$$

Solution:

The characteristic equation of the matrix $\mathbf{A}$ is

$$(\mathbf{A} - 2I_3)^2 = 0$$

and the characteristic polynomial of $\mathbf{A}$ is $(\lambda - 2)^2$.

Since the minimal polynomial $m(\lambda)$ must divide the characteristic polynomial $(\lambda - 2)^2$, there arise two possibilities, i.e., either $m(\lambda) = \lambda - 2$ or $m(\lambda) = (\lambda - 2)^2$.

However, since $\mathbf{A} - 2I \neq 0$, the polynomial $m(\lambda)$ cannot be equal to $\lambda - 2$.

Thus, the minimal polynomial of the matrix $\mathbf{A}$ is, therefore, $m(\lambda) = (\lambda - 2)^2$, which is the characteristic polynomial of $\mathbf{A}$.

Example 8.5:

Determine the minimal polynomial of a diagonal matrix D.

Solution:

Let $d_{11}, d_{22}, \cdots, d_{rr}$ be the distinct diagonal entries of the diagonal matrix D.

Again, the diagonal matrix D satisfies the characteristic equation, i.e.,

$$(D - d_{11}I) \cdots (D - d_{rr}I) = 0.$$

This implies that the characteristic polynomial of D

$$(\lambda - d_{11})(\lambda - d_{22}) \cdots (\lambda - d_{rr}).$$

Since the minimal polynomial divides the characteristic polynomial, i.e., the product

$$(\lambda - d_{11})(\lambda - d_{22}) \cdots (\lambda - d_{rr}),$$

it must be a factor of the product of certain of the factors $\lambda - d_{ii}$, $(i = 1, 2, \cdots, r)$. However, we cannot discard that every one of these factors is from the products of the form $D - d_{jj}I$ for $j \neq i$ is not zero since $d_{ii} \neq d_{jj}$, if $j \neq i$. Thus, it observes that the minimal polynomial of the diagonal matrix D is the product of all the factors that are

$$(\lambda - d_{11})(\lambda - d_{22}) \cdots (\lambda - d_{rr}) .$$

The following statements discuss some more about the *minimal polynomials.*

Lemma 8.1:

Similar matrices have the same minimal polynomials.

Proof:

Since we know that similar matrices describe the same linear operator, it implies that their minimal polynomial also equalizes the linear operator's minimal polynomial.

Note: Using this result, we can find the minimal polynomial of any diagonalizable complex matrix.

Example 8.6:

Determine the minimal polynomial of the matrix

$$\mathbf{A} = \begin{bmatrix} 1 & 2 \\ 2 & 1 \end{bmatrix}.$$

Solution:

Here the matrix $\mathbf{A}$ is similar to a diagonal matrix, i.e.,

$D = \begin{bmatrix} 3 & 0 \\ 0 & -1 \end{bmatrix}$ and the characteristic polynomial of the matrix $\mathbf{A}$ is $(\lambda - 3)(\lambda + 1)$.

Thus, the minimal polynomial of the given matrix $\mathbf{A}$ is

$$(\lambda - 3)(\lambda + 1) .$$

Note: The characteristic polynomial alone cannot tell us that if a matrix is diagonalizable or not. Next, we will discuss that there is a connection between the characteristic polynomial and the minimal polynomials.

Example 8.7:

Consider the matrices I_2 and $\mathbf{A} = \begin{bmatrix} 1 & 1 \\ 0 & 1 \end{bmatrix}.$

It can be seen that both the matrices I_2 and **A** have the characteristic polynomial $(\lambda - 1)^2$, but the first matrix I_2 is diagonalizable while the second matrix **A** is not.

Thus, it implies that the characteristic polynomial alone cannot express to us whether a matrix is diagonalizable or not.

Similarly, if we consider the minimal polynomial of both matrices, then the two matrices have different minimal polynomials, i.e., $(\lambda - 1)$ and $(\lambda - 1)^2$ respectively.

Thus, it can be concluded from this example that the minimum polynomial can determine whether a matrix is diagonalizable or not. In contrast, the characteristic polynomial alone cannot tell us about the diagonalizability of a matrix.

Theorem 8.4:

Let us consider an $n \times n$ square matrix **A** defined over the field of complex numbers C. Then the matrix **A** is said to be *diagonalizable* if and only if the *minimal polynomial* of the matrix **A** splits into a product of n-distinct linear factors.

Proof:

Necessary part:

Suppose the matrix **A** is diagonalizable. So by definition, there exists an invertible matrix C such that $C^{-1}AC = D$ is a diagonal matrix.

Since the matrices **A** and D are similar, they have the same minimal polynomials.

Let $d_{11}, d_{22}, \cdots, d_{rr}$ be the diagonal entries of the diagonal matrix D.

Then it shows that the minimal polynomial of the matrix D is $(\lambda - d_{11})(\lambda - d_{22}) \cdots (\lambda - d_{rr})$, which is a product of distinct linear factors.

Sufficient part:

Suppose that the matrix **A** has a minimal polynomial, i.e.,

$$m(\lambda) = (\lambda - d_{11})(\lambda - d_{22}) \cdots (\lambda - d_{rr}),$$

where $d_{11}, d_{22}, \cdots, d_{rr}$ are the distinct diagonal elements of the matrix D.

Now let us define $g_i(\lambda)$ to be the polynomial obtained from the minimal polynomial $m(\lambda)$ by deleting the factor $\lambda - d_{ii}$; thus,

$$g_i(\lambda) = \frac{1}{\lambda - d_{ii}}.$$

Upon using the method of partial fractions and with the use of constants $b_1, b_2, \cdots, b_r$, we write

$$\frac{1}{m(\lambda)} = \sum_{i=1}^{r} \frac{b_i}{\lambda - d_{ii}}.$$

Upon multiplying both sides of this equation by $m(\lambda)$, we obtain

$$1 = b_1 g_1 + b_2 g_2 + \cdots + b_r g_r \text{ by the definition of } g_i(\lambda).$$

Upon introducing the linear operator $T : R^n \rightarrow R^n$ on the complex vector space R^n defined by $T(\mathbf{X}) = \mathbf{AX}$, it follows from the above equation that

$b_1 g_1(T) + b_2 g_2(T) + \cdots + b_r g_r(T) = I$, where I is the identity function and hence

$\mathbf{X} = b_1 g_1(T)\mathbf{X} + b_2 g_2(T)\mathbf{X} + \cdots + b_r g_r(T)\mathbf{X}$, for any vector $\mathbf{X}$.

Let U_i denote the set of all elements of the form $g_i(T)\mathbf{X}$ with $\mathbf{X} \in R^n$.

Then U_i forms a subspace, and the above equation $\mathbf{X}$ tells us that

$$\mathbf{X} = b_1 U_1 + b_2 U_2 + \cdots + b_r U_r\,, \text{ Where } \mathbf{X} \in R^n,$$

which implies that $R^n = U_1 \oplus U_2 \oplus \cdots \oplus U_r$.

Next, since the vector space R^n is the direct sum of the subspace U_i, it implies that the intersection of a U_i and the sum of the remaining U_j with $j \neq i$ is zero.

To examine why this is valid, let $\mathbf{X}$ be a vector in the intersection.

We need to observe that $g_i(T)\, g_j(T) = 0$ if $i \neq j$.

Each factor $\lambda - d_{kk}$ is existing in the polynomial $g_i g_j$.

Thus, $g_k(T)\mathbf{X} = 0$, for all k.

Now since $\mathbf{X} = \sum_{k=1}^{n} b_k g_k(T)\mathbf{X}$

it follows that $\mathbf{X} = 0$.

Hence, the vector space R^n is the direct sum

$$R^n = U_1 \oplus U_2 \oplus \cdots \oplus U_r.$$

Now the impact of T on vectors in v_i is simply to multiply them by d_{ii}, since

$$(T - d_{ii})\, g_i(T) = f(T) = 0\,.$$

Thus, if we select the bases for each subspace $U_1, U_2, \cdots, U_r$ and combine them to form a basis, then a diagonal matrix will be represented.

Consequently, the matrix $\mathbf{A}$ is similar to a diagonal matrix.

Example 8.8:

Show that the matrix $\mathbf{A} = \begin{bmatrix} 2 & 1 & 1 \\ 0 & 2 & 0 \\ 0 & 0 & 2 \end{bmatrix}$ is not diagonalizable.

Solution:

The matrix $\mathbf{A} = \begin{bmatrix} 2 & 1 & 1 \\ 0 & 2 & 0 \\ 0 & 0 & 2 \end{bmatrix}$ has a minimal polynomial $(\lambda - 2)^2$.

Since the minimal polynomial is not a product of distinct linear factors, the matrix cannot be diagonalized.

Example 8.9:

Consider an $n \times n$ upper triangular matrix.

The matrix

$$\mathbf{A} = \begin{bmatrix} \mu & 1 & 0 & \cdots & 0 & 0 \\ 0 & \mu & 1 & \cdots & 0 & 0 \\ 0 & 0 & \mu & \cdots & 0 & 0 \\ \cdots & \cdots & \cdots & \cdots & \cdots & \cdots \\ 0 & 0 & 0 & \cdots & \mu & 1 \\ 0 & 0 & 0 & \cdots & 0 & \mu \end{bmatrix}$$

has minimum polynomial $(\lambda - \mu)^n$. This is because $(\mathbf{A} - \lambda I)^n = 0$, but $(\mathbf{A} - \lambda I)^{n-1} \neq 0$.

Hence, the matrix $\mathbf{A}$ is diagonalizable if and only if $n = 1$, whereas the characteristic polynomial of the matrix $\mathbf{A}$ equals $(\mu - \lambda)^n$.

Theorem 8.5:

A square matrix $\mathbf{A}$ is *invertible* if and only if the constant term in its characteristic polynomial $\Delta(\lambda)$ is not zero.

Proof:

Necessary part:

Let $\mathbf{A}$ be a square matrix.

By definition, we know that a scalar λ is an eigenvalue of $\mathbf{A}$ if and only if $\det(\mathbf{A} - \lambda I_n) = 0$.

If the matrix $\mathbf{A}$ is invertible, then 0 is not an eigenvalue of the matrix $\mathbf{A}$, which implies that 0 is not a zero of the characteristic polynomials $\Delta(\lambda)$.

Thus, the constant term in the characteristic polynomial $\Delta(\lambda)$ is then non-zero.

Hence proved.

Sufficient part:

Suppose that the constant term of the characteristic polynomial $\Delta(\lambda)$ of the matrix $\mathbf{A}$ is non-zero.

Then by the *Cayley–Hamilton theorem*, we have that $\Delta(\mathbf{A}) = 0$, which, by theorem, can be reported in the form $\mathbf{A}\,[P(\mathbf{A})] = I_n$ for some polynomial $P(\mathbf{A})$.

Hence, the matrix $\mathbf{A}$ is invertible.

Example 8.10:

Show that the matrix $\mathbf{A} = \begin{bmatrix} 1 & 1 & 1 \\ 0 & 1 & 1 \\ 0 & 0 & 1 \end{bmatrix}$ is invertible and hence find its inverse, i.e., $\mathbf{A}^{-1}$.

Solution:

The characteristic polynomial of the matrix

$$\mathbf{A} = \begin{bmatrix} 1 & 1 & 1 \\ 0 & 1 & 1 \\ 0 & 0 & 1 \end{bmatrix}$$

is $\Delta(\lambda) = (\lambda - 1)^3$.

Since the constant term in the characteristic polynomial of $\mathbf{A}$ is not zero, the matrix $\mathbf{A}$ is invertible.

Thus, upon applying the *Cayley–Hamilton theorem*, we have

$$0 = (\mathbf{A} - I_3)^3 = \mathbf{A}^3 - 3\mathbf{A}^2 + 3\mathbf{A} - I_3\,,$$

which gives

$$\mathbf{A}\left(\mathbf{A}^2 - 3\mathbf{A} + 3I_3\right) = I_3,$$

which implies

$$\mathbf{A}^{-1} = \mathbf{A}^2 - 3\mathbf{A} + 3I_3 = \begin{bmatrix} 1 & -1 & 0 \\ 0 & 1 & -1 \\ 0 & 0 & 1 \end{bmatrix}.$$

Remark 8.1:

Let $\mathbf{A} = [a_{ij}]$ be a triangular matrix of order n. Then the *characteristic polynomial* $|\mathbf{A} - \lambda I|$ of the matrix $\mathbf{A}$ is a *triangular matrix* with diagonal entries $a_{ii} - \lambda$ and hence

$$\begin{aligned} \Delta(\lambda) &= \det(\mathbf{A} - \lambda I) \\ &= (a_{11} - \lambda)(a_{22} - \lambda) \ldots\ldots\ldots (a_{nn} - \lambda) \end{aligned}.$$

It implies that the roots of $\Delta(\lambda)$are the diagonal elements of $\mathbf{A}$. □

Similar matrices: The matrices P and Q are said to be *similar* if there is an *invertible matrix* C such that $Q = CPC^{-1}$.

The following result indicates some properties of *similar matrices.*

Theorem 8.6:

Similar matrices $\mathbf{A}$ and $\mathbf{B}$ have the same eigenvalues.

Proof:

Let $\mathbf{A}$ and $\mathbf{B}$ be similar matrices.

Thus, by definition, there exists an *invertible matrix* C such that

$$\mathbf{B} = C^{-1}AC.$$

The characteristic polynomial of the matrix $\mathbf{B}$ is $\det(\mathbf{B} - \lambda I)$.

Substituting for $\mathbf{B}$ and using the multiplicative properties of determinants, we get

$$\begin{aligned}
\det(\mathbf{B} - \lambda I) &= \det\left(C^{-1}AC - \lambda I\right) \\
&= \det\left(C^{-1}(\mathbf{A} - \lambda I)\, C\right) \\
&= \det\left(C^{-1}\right)\det(\mathbf{A} - \lambda I)\det(C) \\
&= \det(\mathbf{A} - \lambda I)\det\left(C^{-1}\right)\det(C) \\
&= \det(\mathbf{A} - \lambda I)\,(\det(C))^{-1}\det(C) \\
&= \det(\mathbf{A} - \lambda I)\ \det\left(C^{-1}C\right) \\
&= \det(\mathbf{A} - \lambda I)\det(I) \\
&= \det(\mathbf{A} - \lambda I)\,.
\end{aligned}$$

This implies that the characteristic polynomial of the matrices $\mathbf{A}$ and $\mathbf{B}$ is identical.

Hence, the eigenvalues of $\mathbf{A}$ and $\mathbf{B}$ are the same, i.e., the matrices $\mathbf{A}$ and $\mathbf{B}$ have the same eigenvalues.

Note 1: On the other hand, one cannot expect that similar matrices have the same eigenvectors.

Generally, the condition of a column vector $\mathbf{X}$ to be an eigenvector of $\mathbf{B} = C^{-1}AC$ with eigenvalue λ is $\left(CAC^{-1}\right)\mathbf{X} = \lambda\mathbf{X}$, which is equivalent to $\mathbf{A}\left(C^{-1}\mathbf{X}\right) = \lambda\left(C^{-1}\mathbf{X}\right)$.

Thus, $\mathbf{X}$ is an eigenvector of $\mathbf{B} = C^{-1}AC$ if and only if $C^{-1}\mathbf{X}$ is an eigenvector of $\mathbf{A}$.

Recall that over the field C of complex numbers, this equation has $n-$roots, some of which may be repeated.

Note 2: If $\lambda_1, \lambda_2, \cdots, \lambda_k$ are the distinct roots (eigenvalues) of a matrix **A**, then the characteristic polynomial of the matrix **A**can be factorized in the form

$(-1)^n (\lambda - \lambda_1)^{r_1} (\lambda - \lambda_2)^{r_2} \cdots (\lambda - \lambda_k)^{r_k}$, where we call $r_1, r_2, \cdots,$ r_k are the algebraic multiplicities of the eigenvalues $\lambda_1, \lambda_2, \cdots, \lambda_k$.

Example 8.11:

Consider the linear mapping $T : R^3 \rightarrow R^3$, defined by $T(\mathbf{X}) = \mathbf{AX}$, where the function T is relative to the *natural ordered basis* of R^3.

The matrix of the transformation $T(\mathbf{X}) = \mathbf{AX}$ is defined by

$$\mathbf{A} = \begin{bmatrix} 0 & 1 & 1 \\ 1 & 0 & 1 \\ 1 & 1 & 0 \end{bmatrix}.$$

Determine the algebraic multiplicity of the matrix **A**.

Solution:

The characteristic polynomial of the matrix **A** is

$$\begin{aligned} \det(\mathbf{A} - \lambda I_3) &= \det \begin{bmatrix} -\lambda & 1 & 1 \\ 1 & -\lambda & 1 \\ 1 & 1 & -\lambda \end{bmatrix}. \\ &= -(\lambda + 1)^2 (\lambda - 2) \end{aligned}$$

This implies that the eigenvalues of the matrix **A** are -1 and 2.

Hence, it implies that the algebraic multiplicity of -1 and 2 are 2 and 1, respectively.

Example 8.12:

Find the eigenvalues, and their *algebraic multiplicities* of the linear mapping $T : R^3 \rightarrow R^3$ are given by

(1) $T(x_1, x_2, x_3) = (x_1 + 2x_2 + 2x_3, 2x_2 + x_3, -x_1 + 2x_2 + 2x_3)$.
(2) $T(x_1, x_2, x_3) = (x_2 + x_3, 0, x_1 + x_2)$.

Example 8.13:

Consider a 2×2 matrix $\mathbf{A} = \begin{bmatrix} 0 & 1 \\ -1 & 0 \end{bmatrix}$. Determine the algebraic multiplicity of the matrix **A** over R and over C.

Solution:

We have

$$\det(\mathbf{A} - \lambda I_2) = \det \begin{bmatrix} -\lambda & 1 \\ -1 & -\lambda \end{bmatrix} = \lambda^2 + 1.$$

Since the characteristic polynomial $\lambda^2 + 1$ has no real roots, it implies that the matrix $\mathbf{A}$ has no real eigenvalues over R.

Still, if we consider the matrix $\mathbf{A}$ as a matrix over the complex field C, then the matrix $\mathbf{A}$ has two eigenvalues, namely i and $-i$, each being of algebraic multiplicity 1.

Example 8.14:

Determine the algebraic multiplicity of the matrix

$$\mathbf{A} = \begin{bmatrix} -3 & 1 & -1 \\ -7 & 5 & -1 \\ -6 & 6 & -2 \end{bmatrix} \text{ over } C.$$

Solution:

The characteristic polynomial of the matrix $\mathbf{A}$ can be computed as follows:

$$\det(\mathbf{A} - \lambda I_3) = -(2+\lambda)(4-\lambda)(-2-\lambda) = (2+\lambda)^2(4-\lambda).$$

This shows that the eigenvalues of the matrix $\mathbf{A}$ are 4 and -2, which are of *algebraic multiplicity* 1 and 2, respectively.

Definition 8.2:

A square matrix $\mathbf{A}$ is diagonalizable if there is an invertible matrix C such that $D = C^{-1}AC$ is a *diagonal matrix*.

Theorem 8.7:

Let $\mathbf{A}$ be an $n \times n$ matrix.

(1) If the matrix $\mathbf{A}$ has $n-$linearly independent eigenvectors, then the matrix $\mathbf{A}$ is diagonalizable.
(2) If the matrix $\mathbf{A}$ is diagonalizable, then the matrix $\mathbf{A}$ has $n-$linearly independent eigenvectors.

Proof:

(1) Let the matrix $\mathbf{A}$ have $n-$eigenvalues, i.e., $\lambda_1, \lambda_2, \lambda_3, \cdots, \lambda_n$ (which need not be distinct) with corresponding linearly independent eigenvectors $\mathbf{X}_1, \mathbf{X}_2, \mathbf{X}_3, \cdots, \mathbf{X}_n$.

Let C be the matrix having $\mathbf{X}_1, \mathbf{X}_2, \mathbf{X}_3, \cdots, \mathbf{X}_n$ as column vectors, i.e., $C = [\mathbf{X}_1, \mathbf{X}_2, \mathbf{X}_3, \cdots, \mathbf{X}_n]$.
Since

$$\begin{array}{l} \mathbf{AX}_1 = \lambda_1 \mathbf{X}_1 \\ \mathbf{AX}_2 = \lambda_2 \mathbf{X}_2 \\ ,\cdots, \\ \mathbf{AX}_n = \lambda_n \mathbf{X}_n \end{array}.$$

Matrix multiplication in terms of columns gives

$$\begin{aligned} AC &= \mathbf{A}\left[\mathbf{X}_1, \mathbf{X}_2, \mathbf{X}_3, \cdots, \mathbf{X}_n\right] \\ &= \left[\mathbf{AX}_1, \mathbf{AX}_2, \mathbf{AX}_3, \cdots, \mathbf{AX}_n\right] \\ &= \left[\lambda_1 \mathbf{X}_1, \lambda_2 \mathbf{X}_2, \lambda_3 \mathbf{X}_3, \cdots, \lambda_n \mathbf{X}_n\right] \\ &= \left[\mathbf{X}_1, \mathbf{X}_2, \mathbf{X}_3, \cdots, \mathbf{X}_n\right] \begin{bmatrix} \lambda_1 & 0 & 0 \\ 0 & \ddots & 0 \\ 0 & 0 & \lambda_n \end{bmatrix}. \\ &= C \begin{bmatrix} \lambda_1 & 0 & 0 \\ 0 & \ddots & 0 \\ 0 & 0 & \lambda_n \end{bmatrix} \end{aligned}$$

Since the columns of C are linearly independent, it implies that the matric C is non-singular.

Thus, $C^{-1}AC = \begin{bmatrix} \lambda_1 & 0 & 0 \\ 0 & \ddots & 0 \\ 0 & 0 & \lambda_n \end{bmatrix}.$

Therefore, if a matrix $\mathbf{A}$ has $n-$linearly independent eigenvectors, then the eigenvectors can be used as the column of a matrix C that diagonalizes the matrix $\mathbf{A}$.
The diagonal matrix D has the eigenvalues of $\mathbf{A}$ as its diagonal elements.

(2) The converse is proved by retracing the above steps.
Commence with the assumption that C is a matrix $[\mathbf{X}_1, \mathbf{X}_2, \mathbf{X}_3, \cdots, \mathbf{X}_n]$ that diagonalizes the matrix $\mathbf{A}$.
Thus, there exist scalars $\gamma_1, \gamma_2, \cdots, \gamma_n$ such that

$$C^{-1}AC = \begin{bmatrix} \gamma_1 & 0 & \cdots & 0 \\ 0 & \gamma_2 & \cdots & 0 \\ \vdots & \vdots & \ddots & \vdots \\ 0 & 0 & \cdots & \gamma_n \end{bmatrix}.$$

Retracing these steps, we arrive at a conclusion that

$$\begin{aligned} \mathbf{AX}_1 &= \gamma_1 X_1 \\ \mathbf{AX}_2 &= \gamma_2 \mathbf{X}_2 \\ &, \cdots, \\ \mathbf{AX}_n &= \gamma_n \mathbf{X}_n \end{aligned} .$$

Thus, $[\mathbf{X}_1, \mathbf{X}_2, \mathbf{X}_3, \cdots, \mathbf{X}_n]$ are the eigenvectors of $\mathbf{A}$.
Since the matrix C is non-singular, these vectors (column vectors of C) are linearly independent.
Thus, if an $n \times n$ matrix $\mathbf{A}$ is diagonalizable, then it has $n-$linearly independent eigenvectors.

Example 8.15:

(1) Show that the following matrix $\mathbf{A}$ is diagonalizable.
(2) Find the diagonal matrix D that is similar to $\mathbf{A}$.
(3) Determine the similarity transformation that diagonalizes $\mathbf{A}$, where

$$\mathbf{A} = \begin{bmatrix} -4 & -6 \\ 3 & 5 \end{bmatrix}.$$

Note: If the matrix $\mathbf{A}$ is similar to a diagonal matrix D under the transformation $C^{-1}AC$, then it can be shown that $\mathbf{A}^k = CD^kC^{-1}$.

These results can be used to compute the power of a matrix, i.e., $\mathbf{A}^k$.

Let us derive this result and then apply it.

Since the matrix $\mathbf{A}$ is similar to the diagonal matrix D, $D = C^{-1}AC$, which implies

$$\begin{aligned} D^k &= \left(C^{-1}AC\right)^k \\ &= \left(C^{-1}AC\right)\left(C^{-1}AC\right), \cdots, \left(C^{-1}AC\right) \\ &= C^{-1}A^kC \end{aligned} .$$

This leads to $\mathbf{A}^k = CD^kC^{-1}$.

Example 8.16:

Compute $\mathbf{A}^9$ for the following matrix $\mathbf{A}$

$$\mathbf{A} = \begin{bmatrix} -4 & -6 \\ 3 & 5 \end{bmatrix}.$$

This technique is used in solving an equation called the *difference equation*.

Example 8.17:

Show that the following matrix $\mathbf{A} = \begin{bmatrix} 5 & -3 \\ 3 & -1 \end{bmatrix}$ is not *diagonalizable.*

Remark 8.2:

Not every matrix is *diagonalizable.*

Remark 8.3:

The eigenspace of the above 2×2 matrix $\mathbf{A}$ is one-dimensional, but it does not have two linearly independent eigenvectors. So, it implies the matrix $\mathbf{A}$ is not diagonalizable.

Example 8.18:

Find $\mathbf{A}^n$, where $\mathbf{A} = \begin{bmatrix} \frac{1}{4} & \frac{1}{20} \\ \frac{3}{4} & \frac{19}{20} \end{bmatrix}$.

Solution:

The eigenvalues of $\mathbf{A}$ are the roots of the equation

$$\left(\frac{1}{4} - \lambda\right)\left(\frac{19}{20} - \lambda\right) - \frac{3}{80} = 0.$$

It can easily be checked that this reduces to

$$(5\lambda - 1)(\lambda - 1) = 0,$$

which implies that the eigenvalues of matrix $\mathbf{A}$ are $\frac{1}{5}$ and 1.

Hence, it follows that the matrix $\mathbf{A}$ is diagonalizable and that an eigenvector associated with the eigenvalue $\lambda_1 = \frac{1}{5}$ is

$$\begin{bmatrix} \frac{1}{20} \\ \frac{1}{5} - \frac{1}{4} \end{bmatrix} = \begin{bmatrix} \frac{1}{20} \\ -\frac{1}{20} \end{bmatrix}$$

and that an eigenvector associated with the eigenvalue $\lambda_2 = 1$ is

$$\begin{bmatrix} \frac{1}{20} \\ 1 - \frac{1}{4} \end{bmatrix} = \begin{bmatrix} \frac{1}{20} \\ \frac{3}{4} \end{bmatrix}.$$

Thus, it can assert that the matrix $C = \begin{bmatrix} 1 & 1 \\ -1 & 15 \end{bmatrix}$ is invertible and is such that

$$C^{-1}AC = \begin{bmatrix} \frac{1}{5} & 0 \\ 0 & 1 \end{bmatrix},$$

which implies

$$C^{-1} = \frac{1}{16}\begin{bmatrix} 15 & -1 \\ 1 & 1 \end{bmatrix}.$$

Next, we can compute $\mathbf{A}^n$

i.e., $\mathbf{A}^n = C\begin{bmatrix} \frac{1}{5} & 0 \\ 0 & 1 \end{bmatrix}C^{-1}$.

We have

$$\begin{aligned} \mathbf{A}^n &= \tfrac{1}{16}\begin{bmatrix} 1 & 1 \\ -1 & 15 \end{bmatrix}\begin{bmatrix} \frac{1}{5^n} & 0 \\ 0 & 1 \end{bmatrix}\begin{bmatrix} 15 & -1 \\ 1 & 1 \end{bmatrix} \\ &= \tfrac{1}{16}\begin{bmatrix} 1 & 1 \\ -1 & 15 \end{bmatrix}\begin{bmatrix} \frac{15}{5^n} & -\frac{1}{5^n} \\ 1 & 1 \end{bmatrix} \\ &= \tfrac{1}{16}\begin{bmatrix} 1+\frac{15}{5^n} & 1-\frac{1}{5^n} \\ 15\left(1-\frac{1}{5^n}\right) & 15+\frac{1}{5^n} \end{bmatrix} \end{aligned}.$$

Note: If there is a 2×2 matrix $\mathbf{A}$ (non-diagonal) whose eigenvalues are not distinct, then in this case, we can say that the matrix $\mathbf{A}$ is not diagonalizable.

Similarly, suppose there is only one eigenvalue for a 2×2 matrix $\mathbf{A}$. In that case, the characteristic equation $(\mathbf{A} - \lambda I_2)\,\mathbf{X} = 0$ of the matrix reduces to a single equation, which results in the dimension of the solution space $2 - 1 = 1$, which implies there cannot exist two linearly independent eigenvectors.

Next, o find the high powers of a matrix $\mathbf{A}$.

8.3 Power of a Matrix

For this case, we can proceed differently,

If $\mathbf{A} = \begin{bmatrix} a & b \\ c & d \end{bmatrix}$, then the characteristic polynomial of $\mathbf{A}$ is

$$\Delta(\lambda) = \lambda^2 - (a+d)\,\lambda + ad - bc.$$

Observe now that

$$\begin{aligned} \mathbf{A}^2 &= \begin{bmatrix} a^2+bc & b\,(a+d) \\ c\,(a+d) & bc+d^2 \end{bmatrix} \\ &= (a+d)\begin{bmatrix} a & b \\ c & d \end{bmatrix} - (ad-bc)\begin{bmatrix} 1 & 0 \\ 0 & 1 \end{bmatrix}, \\ &= (a+d)\,\mathbf{A} - (ad-bc)\,I_2 \end{aligned}$$

and we see that $\Delta(\mathbf{A}) = 0$.

Next for $n \geq 2$, consider the *Euclidean division* of λ^n by $\Delta(\lambda)$.

Since the polynomial $\Delta(\lambda)$ is of degree 2, we have

$$\lambda^n = \Delta(\lambda) . q(\lambda) + \alpha_1 \lambda + \alpha_2. \tag{8.1}$$

By substituting the matrix $\mathbf{A}$ for λ in this polynomial identity, it obtains, by the above observation,

$$\mathbf{A}^n = \alpha_1 \mathbf{A} + \alpha_2 I_2.$$

Now we can determine α_1 and α_2 as follows.

Upon differentiating eqn (8.1) and substituting the value of λ (the single eigenvalue of $\mathbf{A}$) for $\mathbf{X}$,

we obtain $n\lambda^{n-1} = \alpha_1$ (since $\Delta(\lambda) = 0$).

Also, substituting λ for $\mathbf{X}$ in eqn (8.1) and again using $f(\lambda) = 0$, we obtain

$$\lambda^n = \alpha_1 \lambda + \alpha_2 = n\lambda^n + \alpha_2$$

which implies $\alpha_2 = (1 - n)\lambda^n$.

It now follows that $\mathbf{A}^n = n\lambda^{n-1}\mathbf{A} + (1 - n)\lambda^n I_2$.

Example 8.19:

Consider the $n \times n$ trigonal matrix

$$\mathbf{A}_n = \begin{bmatrix} 2 & 1 & 0 & 0 & \cdots & 0 & 0 \\ 1 & 2 & 1 & 0 & \cdots & 0 & 0 \\ 0 & 1 & 2 & 1 & \cdots & 0 & 0 \\ \vdots & \vdots & \vdots & \vdots & \vdots & \vdots & \vdots \\ 0 & 0 & 0 & 0 & \cdots & 2 & 1 \\ 0 & 0 & 0 & 0 & \cdots & 1 & 2 \end{bmatrix}$$

Let $a_n = \det \mathbf{A}_n$.

Upon using Laplace expansion along the first row, it obtains

$$\begin{aligned} a_n &= 2a_{n-1} - \det \begin{bmatrix} 1 & 1 & 0 & 0 & \cdots & 0 \\ 0 & 2 & 1 & 0 & \cdots & 0 \\ 0 & 1 & 2 & 1 & \cdots & 0 \\ \vdots & \vdots & \vdots & \vdots & \cdots & \vdots \\ 0 & 0 & 0 & 0 & \cdots & 2 \end{bmatrix}. \\ &= 2a_{n-1} - a_{n-2} \end{aligned}$$

Expressing this recurrence relation in the usual way as a system of difference equations, we find

$$\begin{aligned} a_n &= 2a_{n-1} - b_{n-1} \\ b_n &= a_{n-1} \,. \end{aligned}$$

Here we consider the system as $\mathbf{X}_n = \mathbf{A}\mathbf{X}_{n-1}$,
where $\mathbf{X} = \begin{bmatrix} a_n \\ b_n \end{bmatrix}$ and $\mathbf{A} = \begin{bmatrix} 2 & -1 \\ 1 & 0 \end{bmatrix}$.

Now $\det(\mathbf{A} - \lambda I_2) = \lambda(\lambda - 2) + 1 = (\lambda - 1)^2$
and so the matrix $\mathbf{A}$ has the single eigenvalue 1 of algebraic multiplicity 2.

Next, we can compute $\mathbf{A}^n$ as in the above:

$$\mathbf{A}^n = n\mathbf{A} + (1-n)\, I_2 = \begin{bmatrix} n+1 & -n \\ n & 1-n \end{bmatrix}.$$

Consequently, we have

$$\begin{bmatrix} a_n \\ b_n \end{bmatrix} = \mathbf{A}^{n-2} \begin{bmatrix} a_2 \\ b_2 \end{bmatrix} = \begin{bmatrix} n-1 & -n+2 \\ n-2 & 3-n \end{bmatrix} \begin{bmatrix} 3 \\ 2 \end{bmatrix} = \begin{bmatrix} n+1 \\ n \end{bmatrix}.$$

Hence, we see that $\det \mathbf{A}_n = a_n = n + 1$.

8.4 Diagonal Matrix Representation of a Linear Operator

Let us now see how to find a diagonal matrix representation of a linear operator T, if one exists.

A *diagonal matrix* representation is usually the representation that provides most information in applications.

Let T be a linear operator on a vector space V of dimension n.

Let $\mathbf{B}$ be a basis for V and let $\mathbf{A}$ be the matrix representation of T relative to the basis $\mathbf{B}$.

The matrix representation of T relative to another basis B′ can be obtained using a similarity transformation $\mathbf{A}' = P^{-1}\mathbf{A}P$, where P is the transition matrix from the basis $\mathbf{B}'$ to the basis $\mathbf{B}$.

Suppose the matrix $\mathbf{A}$ has n eigenvalues $\lambda_1, \lambda_2, \cdots, \lambda_n$ with n-corresponding linearly independent eigenvectors $\mathbf{X}_1, \mathbf{X}_2, \cdots, \mathbf{X}_n$.

If $P = [\mathbf{X}_1, \mathbf{X}_2, \cdots, \mathbf{X}_n]$, then we know that $\mathbf{A}'$ is the diagonal matrix,

$$\mathbf{A}' = \begin{bmatrix} \lambda_1 & 0 & 0 \\ 0 & \ddots & 0 \\ 0 & 0 & \lambda_n \end{bmatrix}.$$

The coordinate vectors $\mathbf{B}'$ relative to $\mathbf{B}$ are the eigenvectors $\mathbf{X}_1, \mathbf{X}_2, \cdots, \mathbf{X}_n$.

For a reason stated earlier, it is desirable to use an orthogonal transformation, if possible, to arrive at this basis $\mathbf{B}'$, which provides the diagonal matrix $\mathbf{A}'$.

Example 8.20:

Consider the linear operator $T(x, y) = (3x + y, x + 3y)$ onR^2. Find a *diagonal matrix* representation of T. Determine the *basis* for this representation and give a *geometrical representation* of T.

Solution:

Let us start by finding the matrix representation $\mathbf{A}$ relative to the standard basis $\mathbf{B} = \{(1, 0), (0, 1)\}$ of R^2.

We get

$$T(1, 0) = (3, 1) = 3(1, 0) + 1(0, 1)$$
$$T(0, 1) = (1, 3) = 1(1, 0) + 3(0, 1).$$

The coordinate vectors of $T(1, 0)$ and $T(0, 1)$ relative to $\mathbf{B}$ are

$$\begin{bmatrix} 3 \\ 1 \end{bmatrix} \text{ and } \begin{bmatrix} 1 \\ 3 \end{bmatrix}.$$

The matrix representation of T relative to the standard basis $\mathbf{B}$ is thus

$$\mathbf{A} = \begin{bmatrix} 3 & 1 \\ 1 & 3 \end{bmatrix}.$$

The matrix $\mathbf{A}$ has the following eigenvalues and eigenvectors:

$$\lambda_1 = 4\ , \mathbf{X}_1 = r\begin{bmatrix} 1 \\ 1 \end{bmatrix} \text{ and } \lambda_2 = 2\ , \mathbf{X}_2 = s\begin{bmatrix} -1 \\ 1 \end{bmatrix}.$$

The following matrix $\mathbf{A}_r$ is thus a diagonal matrix representation of T

$$\mathbf{A}_r = \begin{bmatrix} \lambda_1 & 0 \\ 0 & \lambda_2 \end{bmatrix} = \begin{bmatrix} 4 & 0 \\ 0 & 2 \end{bmatrix}.$$

Let us now find the basis $\mathbf{B}_r$, which gives this representation. Observe that $\mathbf{A}$ is a *symmetric matrix*. Then select the *unit orthogonal eigenvectors* for the coordinate vectors of $\mathbf{B}_r$ relative to B.

The transition matrix from $\mathbf{B}$ to $\mathbf{B}_r$ will then be orthogonal, and the geometry will be preserved.

Let $\mathbf{B}_r = \left\{\left(\frac{1}{\sqrt{2}}, \frac{1}{\sqrt{2}}\right), \left(-\frac{1}{\sqrt{2}}, \frac{1}{\sqrt{2}}\right)\right\}$.

Note that the basis $\mathbf{B}_r$ is obtained from the basis $\mathbf{B}$ by rotation through $\frac{\pi}{4}$.

Geometrical representation:

The standard basis $\mathbf{B}$ defines an XY-coordinate system. Let the basis $\mathbf{B}'$ illustrate a $X'Y'P'Q'R'$ coordinate system.

The figure shows that the matrix $\mathbf{A}'$ tells us that T is scaling in the $X'Y'$ coordinate system with factor 4 in the direction and factor 2 in the Y'-direction, as shown in Figure 8.1.

Thus, for example, T maps the square $PQRO$ into the rectangle $P'Q'R'O$.

This example illustrates a situation that frequently arises in physics and engineering.

Theorem 8.8:

The eigenvectors corresponding to distinct eigenvalues are linearly independent.

Proof:

The proof can be done by induction.

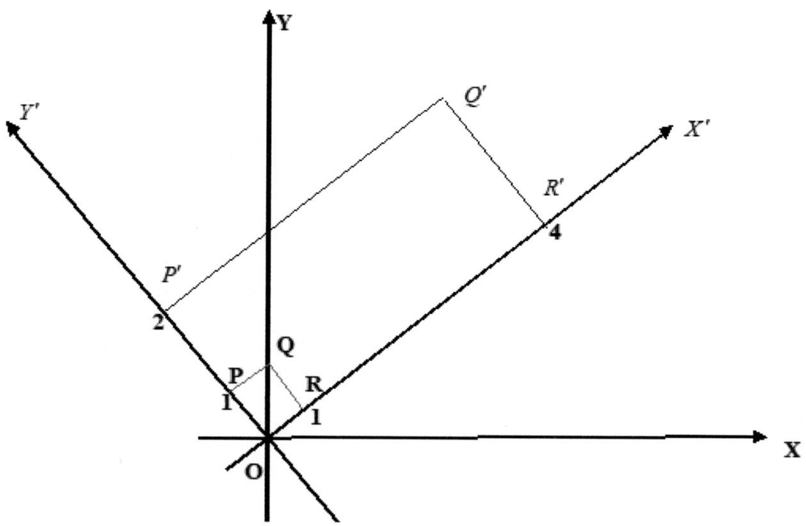

Figure 8.1 Coordinate Representations.

If a linear transformation $T : V \to V$ has only one eigenvalue λ, and if $\mathbf{X}$ is a corresponding eigenvector, then since $\mathbf{X} \neq 0$, we know that $\{\mathbf{X}\}$ is linearly independent.

For the inductive step, assume that every set of n-eigenvectors that corresponds to n-distinct eigenvalues is linearly independent.

Let $\mathbf{X}_1, \mathbf{X}_2, \cdots, \mathbf{X}_{n+1}$ be the eigenvectors that correspond to distinct eigenvalues $\lambda_1, \lambda_2, \cdots, \lambda_{n+1}$.

For any n scalars $a_1, a_2, \cdots, a_n, a_{n+1}$, if we have

$$a_1 x_1 + a_2 x_2 + \cdots + a_n x_n + a_{n+1} x_{n+1} = 0_v \tag{8.2}$$

Then applying the linear transformation T and using the fact that

$$T(\mathbf{X}_i) = \lambda_i \mathbf{X}_i,$$

we obtain

$$a_1 \lambda_1 \mathbf{X}_1 + a_2 \lambda_2 \mathbf{X}_2 + \cdots + a_n \lambda_n \mathbf{X}_n + a_{n+1} \lambda_{n+1} \mathbf{X}_{n+1} = 0_v. \tag{8.3}$$

Now take eqn 8.2 and 8.3 to get

$$a_1 (\lambda_1 - \lambda_{n+1}) \mathbf{X}_1 + a_2 (\lambda_2 - \lambda_{n+1}) \mathbf{X}_2 + \cdots + a_n (\lambda_n - \lambda_{n+1}) \mathbf{X}_n = 0_v.$$

By the induction hypothesis and the fact that $\lambda_1, \lambda_2, \cdots, \lambda_{n+1}$ are distinct, we deduce that $a_1 = a_2 = \cdots = a_n = 0$.

It now *follows* that $a_{n+1} \mathbf{X}_{n+1} = 0$, whence since $\mathbf{X}_{n+1} \neq 0$, we also have $a_{n+1} = 0$. Hence, the eigenvectors $\mathbf{X}_1, \mathbf{X}_2, \cdots, \mathbf{X}_{n+1}$ are linearly independent, and the result follows.

8.5 Diagonalization of Matrices

Definition 8.3:

A linear transformation $T : V \to V$ said to be *diagonalizable* if there is an ordered basis $(v_i)_n$ of V concerning which the matrix of T is diagonal. Thus, the linear transformation T is diagonalizable if and only if there exists an ordered basis $(v_i)_n$ of V such that

$$\begin{aligned} T(v_1) &= \lambda_1 v_1 \\ T(v_2) &= \lambda_2 v_2 \\ &\vdots \\ T(v_n) &= \lambda_n v_n \end{aligned}$$

In this case, the $\lambda_i's$ are the eigenvalues of the mapping T.

We can, thus, assert the following theorem.

Theorem 8.9:

A linear transformation $T : V \to V$ is *diagonalizable* if and only if the vector space V has a basis consisting of eigenvectors of T. □

Note: Here, it can be considered matrices that are not *diagonalizable*.

For example: Consider a 2×2 matrix $\mathbf{A} = \begin{bmatrix} 1 & 1 \\ 0 & 1 \end{bmatrix}$.

Here if the matrix $\mathbf{A}$ is *diagonalizable*, then it would be similar to the identity matrix I_2.

Since both the eigenvalues of the matrix $\mathbf{A}$ are equal to 1 and $C^{-1}AC = I_2$, for some, invertible matrix C implies that the last equation gives $\mathbf{A} = CI_2C^{-1} = I_2$, which is not true.

An exciting feature of the above theorem is that it finds an invertible matrix C that diagonalizes the matrix $\mathbf{A}$.

For this, it is enough to find a set of linearly independent eigenvectors of $\mathbf{A}$ that can be considered to form the columns of the invertible matrix C.

Example 8.21:

Determine an invertible matrix C that diagonalizes the matrix

$$\mathbf{A} = \begin{bmatrix} 2 & -1 \\ 2 & 4 \end{bmatrix}.$$

Solution:

The eigenvalues of the matrix $\mathbf{A}$ are $3 + \sqrt{-1}$ and $3 - \sqrt{-1}$ and hence the eigenvectors of $\mathbf{A}$ are

$$C = \begin{bmatrix} \frac{(-1+\sqrt{-1})}{2} & \frac{-(1+\sqrt{-1})}{2} \\ 1 & 1 \end{bmatrix}.$$

That diagonalizes the matrix $\mathbf{A}$, i.e., in the form of an invertible matrix C.

Then by the previous theorem, it can be written as

$$C^{-1}AC = \begin{bmatrix} 3 + \sqrt{-1} & 0 \\ 0 & 3 - \sqrt{-1} \end{bmatrix}. \square$$

The following section shows that the linear transformation $T : V \to V$ is *diagonalizable* if the geometric and algebraic multiplicities coincide for every eigenvalue.

Theorem 8.10:

Let V be an n-dimensional vector space. If $\lambda_1, \lambda_2, \cdots, \lambda_m$, the eigenvalues of a linear mapping $T : V \to V$ and their geometric multiplicities $d_1, d_2, \cdots, d_m$ satisfy the inequality $d_1 + d_2 + \cdots + d_m \leq n$, if and only if the linear mapping T is diagonalizable.

Proof:

Let $\mathbf{B}_i$ be a basis of the eigenspace E_{λ_i}, for each i.

Let us observe first that

$v_1 + v_2 + \cdots + v_m = 0_v$, where each $v_i \in \mathbf{B}_\iota$.

Then certainly each $v_i = 0$, since by Theorem 8.8, the eigenvectors corresponding to distinct eigenvalues are linearly independent.

We next observe that $\bigcup_{i=1}^{m} \mathbf{B}_\iota$ is linearly independent.

If $\mathbf{B} = \{e_{i1}, e_{i2}, \cdots, e_{id_i}\}$ and $v_i = \mu_{i1}e_{i1} + \mu_{i2}e_{i2} + \cdots + \mu_{id_i}e_{id_i} \in \mathbf{B}_\iota$, Then $\sum_{i=1}^{k} v_i = 0$ gives each $v_i = 0$, whence all the coefficient $\mu_{ij} = 0$.

Since the $\mathbf{B}_\iota' s$ are pairwise disjoint, it observes that

$$d_1 + d_2 + \cdots + d_m = \left| \bigcup_{i=1}^{m} \mathbf{B}_\iota \right| \leq n.$$

Finally, equality occurs if the vector space V has n-linearly independent eigenvectors, i.e., by Theorem 8.9, if and only if the transformation T is diagonalizable. □

Theorem 8.11:

Let $T : V \to V$ be a linear mapping and λ be an eigenvalue of T. Then the *geometric multiplicity* of the eigenvalue λ is less than or equal to the *algebraic multiplicity* of λ, i.e., $G.M \leq \mathbf{A}.M$.

Proof:

Let $T : V \to V$ be a linear transformation.

Let $\{e_1, e_2, \cdots, e_k\}$ be a basis of E_λ and extend this to a basis $\mathbf{B} = \{e_1, e_2, \cdots, e_n\}$ of V.

Thus, the matrix of the linear transformation T relative to the basis $\mathbf{B}$ is of the form

$$M = \begin{bmatrix} \lambda I_k & C \\ 0 & D \end{bmatrix}.$$

This implies that the characteristic polynomial of the matrix M is of the form $(\lambda - \mathbf{X})^d \, p(\mathbf{X})$, where $p(\mathbf{X})$ is a polynomial of degree $n - k$. Thus, it follows that the value of k is less than or equal to the algebraic multiplicity of λ. □

It can be deduced from the above that the following necessary and sufficient condition for a linear transformation $T : V \to V$ is said to be *diagonalizable*.

Theorem 8.12:

The following statements are equivalent:

(1) The linear transformation $T : V \to V$ is diagonalizable.
(2) For every eigenvalue λ of the linear transformation T, the geometric multiplicity (G.M) of the eigenvalue λ coincides with the algebraic multiplicity (A.M) of λ.

Proof:

As we know, the sum of the algebraic multiplicities of the eigenvalues is the degree of the characteristic polynomial, namely $n = \dim V$. The result, therefore, follows from *Theorems 8.10 and 8.11*. □

Example 8.22:

Show that the matrix $\mathbf{A} = \begin{bmatrix} -3 & 1 & -1 \\ -7 & 5 & -1 \\ -6 & 6 & -2 \end{bmatrix}$ is not *diagonalizable*.

Solution:

The matrix $\mathbf{A}$ has only two distinct eigenvalues, namely 4 and -2, where the algebraic multiplicity of -2 is 2.

To determine the *eigenspace* E_{-2} for $\lambda = -2$, we solve
$(\mathbf{A} + 2I_3)\,\mathbf{X} = 0_v$, i.e.,

i.e., $$\begin{bmatrix} -1 & 1 & -1 \\ -7 & 7 & -1 \\ -6 & 6 & 0 \end{bmatrix}\begin{bmatrix} x_1 \\ x_2 \\ x_3 \end{bmatrix} = \begin{bmatrix} 0 \\ 0 \\ 0 \end{bmatrix}.$$

Upon using the reduced echelon form of the matrix, the corresponding system of the linear equation reduces to

$$\begin{aligned} x_1 - x_2 &= 0 \\ x_3 &= 0 \end{aligned},$$

which gives the rank of the coefficient matrix $\mathbf{A}$ is 2. Consequently, the solution space of the matrix is of dimension $3 - 2 = 1$.

Thus, the eigenvalue -2 is of *geometric multiplicity* 1.

Hence, by *Theorem 8.10*, it follows that the matrix $\mathbf{A}$ is not diagonalizable. □

Example 8.23:

Show that the matrix $\mathbf{A} = \begin{bmatrix} 1 & -3 & 3 \\ 3 & -5 & 3 \\ 6 & -6 & 4 \end{bmatrix}$ is diagonalizable.

Solution:

The characteristic polynomial of the matrix $\mathbf{A}$ is

$$|\mathbf{A} - \lambda I_3| = (4 - \lambda)(\lambda + 2)^2 .$$

This implies that the eigenvalues of the matrix $\mathbf{A}$ are 4 and -2, having respective algebraic multiplicities of 1 and 2.

Next, to determine the eigenspace E_{-2} for $\lambda = -2$, for which we solve

$$(\mathbf{A} + 2I_3)\,\mathbf{X} = 0,$$

i.e.,

$$\begin{bmatrix} 3 & -3 & 3 \\ 3 & -3 & 3 \\ 6 & -6 & 6 \end{bmatrix} \begin{bmatrix} x_1 \\ x_2 \\ x_3 \end{bmatrix} = \begin{bmatrix} 0 \\ 0 \\ 0 \end{bmatrix} .$$

Upon using the reduced echelon form of the matrix, the corresponding system of equations reduces to

$$x_1 - x_2 + x_3 = 0,$$

which implies that the coefficient matrix $\mathbf{A}$ is of rank 1.

So, the dimension of the solution space is $3 - 1 = 2$.

Thus, the eigenvalue $\lambda = -2$ is of geometric multiplicity 2.

Similarly, for the eigenvalue $\lambda = 4$, since its algebraic multiplicity is 1, by *Theorem 8.11*, it follows that its geometric multiplicity of $\lambda = 4$ is also 1.

Hence, it follows that the matrix $\mathbf{A}$ is diagonalizable. □

Example 8.24:

Let us consider any two linearly independent eigenvectors $\begin{bmatrix} 1 \\ 1 \\ 0 \end{bmatrix} \begin{bmatrix} 1 \\ 0 \\ -1 \end{bmatrix}$, in E_{-2} that correspond to the eigenvalue -2 and constitute a basis for E_{-2}.

Similarly, let us consider any single non-zero vector

$\begin{bmatrix} 1 \\ 1 \\ 2 \end{bmatrix}$ in E_4 that fits the eigenvalue 4 and forms a basis for E_4.

Then, clearly, these three eigenvectors
$\begin{bmatrix} 1 \\ 1 \\ 0 \end{bmatrix}$, $\begin{bmatrix} 1 \\ 0 \\ -1 \end{bmatrix}$ and $\begin{bmatrix} 1 \\ 1 \\ 2 \end{bmatrix}$ are linearly independent.

Upon using these eigenvectors together, we can obtain an invertible matrix C, i.e.,

$$C = \begin{bmatrix} 1 & 1 & 1 \\ 1 & 0 & 1 \\ 0 & -1 & 2 \end{bmatrix},$$

and this is such that

$$C^{-1}AC = \begin{bmatrix} -2 & 0 & 0 \\ 0 & -2 & 0 \\ 0 & 0 & 4 \end{bmatrix}. \square$$

Example 8.25:

For each of the matrices $\mathbf{A}$ given by

(a) $\mathbf{A} = \begin{bmatrix} 1 & 0 & 1 \\ 0 & 1 & 0 \\ 1 & 0 & 1 \end{bmatrix}$ (b) $\mathbf{A} = \begin{bmatrix} -2 & 5 & 7 \\ 1 & 0 & -1 \\ -1 & 1 & 2 \end{bmatrix}$

(c) $\mathbf{A} = \begin{bmatrix} -3 & -7 & 19 \\ -2 & -1 & 8 \\ -2 & -3 & 10 \end{bmatrix}$ and (d) $\mathbf{A} = \begin{bmatrix} -4 & 0 & -3 \\ 1 & 3 & 1 \\ 4 & -2 & 3 \end{bmatrix}$.

Find an invertible matrix C such that $C^{-1}AC$ is diagonal.

Theorem 8.13:

Let $\mathbf{A}$ be a 2×2 matrix defined by

$\mathbf{A} = \begin{bmatrix} a_1 & a_2 \\ a_3 & a_4 \end{bmatrix}$, which has distinct eigenvalues λ_1, λ_2.

Then the matrix $\mathbf{A}$ is diagonalizable.

When $a_2 \neq 0$, there exists an invertible matrix C such that

$C^{-1}AC = \begin{bmatrix} \lambda_1 & 0 \\ 0 & \lambda_2 \end{bmatrix}$ is the matrix

$$C = \begin{bmatrix} a_2 & a_2 \\ \lambda_1 - a_1 & \lambda_2 - a_1 \end{bmatrix}.$$

Proof:

The first statement is immediate from Theorems 8.7 and 8.8. With Theorems 8.7 and 8.8, it can be easily proved that if a matrix $\mathbf{A}$ has distinct eigenvalues, then the matrix **A** is diagonalizable. Thus, the first statement is proved.

For the second statement, let us observe that

$$\det\begin{bmatrix} a_1-\lambda & a_2 \\ a_3 & a_4-\lambda \end{bmatrix} = \lambda^2 - (a_1+a_4)\lambda + a_1a_4 - a_2a_3,$$

where the eigenvalues of the matrix **A** are

$$\lambda_1 = \frac{1}{2}\left\{(a_1+a_4) + \sqrt{(a_1-a_4)^2 + 4a_2a_3}\right\}$$

$$\lambda_2 = \frac{1}{2}\left\{(a_1+a_4) - \sqrt{(a_1-a_4)^2 + 4a_2a_3}\right\}.$$

Let us consider the column matrix $\mathbf{X}_1$ as: $\mathbf{X}_1 = \begin{bmatrix} a_2 \\ \lambda_1 - a_1 \end{bmatrix}$, in which, by hypothesis $a_2 \neq 0$, we have

$$\begin{bmatrix} a_1 & a_2 \\ a_3 & a_4 \end{bmatrix}\begin{bmatrix} a_2 \\ \lambda_1 - a_1 \end{bmatrix} = \begin{bmatrix} a_2\lambda_1 \\ ca_2 + a_4(\lambda_1 - a_1) \end{bmatrix} = \lambda_1\begin{bmatrix} a_2 \\ \lambda_1 - a_1 \end{bmatrix}.$$

The final equality results from the fact that

$$\begin{aligned}\lambda_1(\lambda_1 - a_1) - a_3a_2 - a_4(\lambda_1 - a_1) &= {\lambda_1}^2 - (a_1+a_4)\lambda_1 + a_1a_4 - a_2a_3 \\ &= 0\end{aligned}.$$

This implies that $\mathbf{X}_1$ is an eigenvector associated with the eigenvalue $\lambda = \lambda_1$.

Similarly, one can show that

$\mathbf{X}_2 = \begin{bmatrix} b \\ \lambda_2 - a \end{bmatrix}$ is the eigenvector associated with the eigenvalue $\lambda = \lambda_2$.

Upon using these eigenvectors $\mathbf{X}_1$ and $\mathbf{X}_2$ together, the required invertible matrix C can be obtained. □

8.6 Diagonalization of Symmetric Matrices

The diagonalization of a matrix is related to its eigenvectors. The following theorem summarizes the properties of eigenvalues and eigenvectors of symmetric matrices and paving the way for their diagonalization results.

Theorem 8.14:

Let **A** be an $n \times n$ matrix.

(1) All the eigenvalues of $\mathbf{A}$ are real numbers.
(2) The dimension of an eigenspace of $\mathbf{A}$ is the multiplicity of the eigenvalue as a root of the characteristic equation.
(3) The eigenspaces of the matrix$\mathbf{A}$ are orthogonal.
(4) The matrix $\mathbf{A}$ has n-linearly independent eigenvectors. □

Example 8.26:

Consider a symmetric matrix $\mathbf{A} = \begin{bmatrix} 5 & 4 & 2 \\ 4 & 5 & 2 \\ 2 & 2 & 2 \end{bmatrix}$. Verify the above theorem for $\mathbf{A}$.

8.7 Orthogonal Diagonalization

If a matrix Q is orthogonal, then $Q^{-1} = Q^T$; thus, if such a matrix is used in a similarity transformation, then the transformation becomes $D = Q^T AQ$.

These types of computation, of course, is easier to compute than $Q^{-1}AQ$.

This is important if one is performing computations by hand, but not essential when using a computer.

Note: There is a role of *similarity transformation* in going from one coordinate system to another. Similarity transformation involving an orthogonal matrix is the transformation that is used to relate orthogonal coordinate systems (coordinate systems where the axes are at right angles).

Definition 8.4:

An $n \times n$ square matrix $\mathbf{A}$ is orthogonally diagonalizable, if there is an orthogonal matrix Q such that $Q^T AQ = D$ is a diagonal matrix.

Remark 8.4:

The set of orthogonally diagonalizable matrices is, in fact, the set of symmetric matrices.

Theorem 8.15:

Let $\mathbf{A}$ be an $n \times n$ square matrix. Then the matrix $\mathbf{A}$ is *orthogonally diagonalizable* if and only if the matrix $\mathbf{A}$ is *symmetric.*

Proof:

Suppose that the matrix $\mathbf{A}$ is symmetric. The following steps can be taken to construct an *orthogonal matrix* Q such that $Q^T AQ$ is *orthogonal.*

From the previous theorem, it can be ensured that this algorithm can be carried out.

(1) First, to find a basis for each eigenspace of the matrix $\mathbf{A}$.
(2) Then to find an *orthonormal basis* for each eigenspace by using the *Gram–Schmidt process*.
(3) Let Q be an orthogonal matrix whose columns are orthogonal vectors.
(4) The matrix $Q^T AQ = D$ will be the diagonal matrix.

Conversely:

Assume that the matrix $\mathbf{A}$ is *orthogonally diagonalizable*. Thus, there exists an orthogonal matrix Q such that $\mathbf{A} = QDQ^T$.

Upon using the properties of transpose, we get

$$\begin{aligned}\mathbf{A}^T &= \left(QDQ^T\right)^T \\ &= \left(Q^T\right)^T (QD)^T \\ &= QDQ^T \\ &= \mathbf{A}.\end{aligned}$$

Hence, the matrix $\mathbf{A}$ is symmetric.

Example 8.27:

Orthogonally diagonalize the following symmetric matrix $\mathbf{A}$:

$$\mathbf{A} = \begin{bmatrix} 1 & -2 \\ -2 & 1 \end{bmatrix}.$$

Solution:

The eigenvalue and the corresponding eigenspaces of this matrix are as follows:

$$\lambda_1 = -1 \ , v_1 = \left\{ s \begin{bmatrix} 1 \\ 1 \end{bmatrix} \right\} \ \text{and} \ \lambda_2 = 3 \ , v_2 = \left\{ r \begin{bmatrix} -1 \\ 1 \end{bmatrix} \right\}.$$

Since the matrix $\mathbf{A}$ is symmetric, it can be diagonalized to give

$$D = \begin{bmatrix} -1 & 0 \\ 0 & 3 \end{bmatrix}.$$

Let us determine the transformation.

The eigenspaces v_1 and v_2 are to be expected orthogonally.

Use a unit vector in each eigenspace as columns of an *orthogonal matrix* Q.

We get $Q = \begin{bmatrix} \frac{1}{\sqrt{2}} & -\frac{1}{\sqrt{2}} \\ \frac{1}{\sqrt{2}} & \frac{1}{\sqrt{2}} \end{bmatrix}$.

The orthogonal transformation that leads to D is

$$Q^T A Q = \begin{bmatrix} \frac{1}{\sqrt{2}} & \frac{1}{\sqrt{2}} \\ -\frac{1}{\sqrt{2}} & \frac{1}{\sqrt{2}} \end{bmatrix} \begin{bmatrix} 1 & -2 \\ -2 & 1 \end{bmatrix} \begin{bmatrix} \frac{1}{\sqrt{2}} & -\frac{1}{\sqrt{2}} \\ \frac{1}{\sqrt{2}} & \frac{1}{\sqrt{2}} \end{bmatrix}.$$

Now we apply these tools to the study of matrix representations of linear operators.

9

Application to Conics and Quadrics

This chapter discusses the applications of conics and quadrics and the canonical form of matrices that include some vital discussion such as bilinear forms, Sylvester's law of inertia.

In this chapter, we shall use the concept of an orthogonal matrix to reduce a real symmetric matrix to the diagonal form by a similarity transformation. This chapter starts with the discussion of quadratic forms, conics, and quadrics. It includes some vital discussion such as bilinear forms, eigenvalues of congruent matrices, Sylvester's law of inertia, and skew-symmetric bilinear form.

9.1 Quadratic Forms

A quadratic form $\mathbf{q}$ in the set of real variables $x_1, x_2, \ldots, x_n$ is a polynomial in $x_1, x_2, \ldots, x_n$ with real coefficients in which every term has degree 2.

For example, a statement $a_1{x_1}^2 + 2a_2x_1x_2 + a_3x_2^2$ is a quadratic form in x_1 and x_2.

Quadratic forms arise in many contexts.

For example:

The *equation of a conic* in the plane and a quadratic surface in *three-dimensional space* involves quadratic forms.

We commence by noticing that a quadratic form $\mathbf{q} = a_1{x_1}^2 + 2a_2x_1x_2 + a_3x_2^2$ in x_1 and x_2 can be expressed as a product of these matrices, namely

$$\mathbf{q} = [x_1\ x_2]\begin{bmatrix} a_1 & a_2 \\ a_2 & a_3 \end{bmatrix}\begin{bmatrix} x_1 \\ x_2 \end{bmatrix} = \begin{bmatrix} x_1 \\ x_2 \end{bmatrix}^T \begin{bmatrix} a_1 & a_2 \\ a_2 & a_3 \end{bmatrix}\begin{bmatrix} x_1 \\ x_2 \end{bmatrix}.$$

Primarily, any quadratic form $\mathbf{q}$ in $x_1, x_2, \ldots, x_n$ can be expressed in this form.

Let the equation give the quadratic form

$\mathbf{q} = \sum_{i=1}^{n} \sum_{j=1}^{n} a_{ij} x_i x_j$, where a_{ij} are real numbers.

By context $\mathbf{A} = [a_{ij}]_{n,n}$ and expressing $\mathbf{X}$ for the column vector with entries $x_1, x_2, \ldots, x_n$. We look from the definition of the matrix product that the quadratic form $\mathbf{q}$ may be described in the form

$$\mathbf{q} = X^T AX.$$

This implies that the real matrix $\mathbf{A}$ can determine the quadratic form $\mathbf{q}$.

Now, we consider that the matrix $\mathbf{A}$ is symmetric. Since $X^T AX$ is scalar, the quadratic form $\mathbf{q}$ may equally well be expressed as

$$\left(X^T AX\right)^T = X^T \mathbf{A}^T X.$$

Thus,

$$\begin{aligned} \mathbf{q} &= \tfrac{1}{2}\left(X^T AX + X^T \mathbf{A}^T X\right) \\ &= X^T \left(\tfrac{1}{2}\left(\mathbf{A} + \mathbf{A}^T\right)\right) X \end{aligned}.$$

It follows that the symmetric matrix $\mathbf{A}$ can replace the matrix $\frac{1}{2}\left(\mathbf{A} + \mathbf{A}^T\right)$.

Note:

(1) The matrix combined with a quadratic form is symmetric.
(2) A quadratic form $\mathbf{q}$ can be described in terms of square only.

9.2 Conics

Let us consider a conic, which is a curve defined in the two-dimensional plane with an equation of the second degree of the form

$$a_1 x^2 + 2a_2 xy + a_3 y^2 + a_4 x + a_5 y + a_6 = 0, \tag{9.1}$$

where the coefficients a_1, a_2, a_3, a_4, a_5, and a_6 are all real numbers.

Eqn (9.1) can be expressed in matrix form as

$$X^T AX + [a_4 \ a_5] X + a_6 = 0, \tag{9.2}$$

where $\mathbf{X} = \begin{bmatrix} x \\ y \end{bmatrix}$ and $\mathbf{A} = \begin{bmatrix} a_1 & a_2 \\ a_2 & a_3 \end{bmatrix}$,
which is a quadratic form in x and y that is involved in this conic.

Next, let us apply the *spectral theorem* to examine its effect on the equation of conic (9.1).

Let Q be a real orthogonal matrix such that

$Q^T AQ = \begin{bmatrix} a_1{}' & 0 \\ 0 & a_3' \end{bmatrix}$, where entries $a_1{}'$ and a_3' are the eigenvalues of the matrix $\mathbf{A}$.

Let us put $\mathbf{X}' = \mathbf{Q}^T\mathbf{X}$ and let the entries of X' be x' and y'. Then $\mathbf{X} = \mathbf{Q}\mathbf{X}'$ and the equation of the conic takes the form:

$$(X')^T \begin{bmatrix} a_1{}' & 0 \\ 0 & a_3{}' \end{bmatrix} X' + [a_4 \;\; a_5]\, QX' + a_6 = 0. \tag{9.3}$$

Or, equivalently, the equation can be expressed as

$$a_1{}' x'^2 + a_3{}' y'^2 + a_4{}' x' + a_5{}' y' + a_6 = 0,$$

for specific real numbers, $a_4{}'$ and $a_5{}'$.

Therefore, the advantage of changing to the new variables x', y' is that there is no "cross term" $x'y'$ in the quadratic form.

Geometrical interpretation:

Geometrically, it corresponds to a rotation of axes x and y to a new set of coordinates x' and y'.

Any 2×2 real orthogonal matrix represents either a rotation or a reflection in R^2; however, no reflection will arise in the present instance.

Example 9.1:

Determine the conic of the equation defined by the quadratic equation $x^2 + 4xy + y^2 + 3x + y - 1 = 0$.

Solution:

The matrix $\mathbf{A}$ associated with the quadratic form of the equation

$$x^2 + 4xy + y^2 \text{ is } \mathbf{A} = \begin{bmatrix} 1 & 2 \\ 2 & 1 \end{bmatrix}.$$

The eigenvalues of the matrix $\mathbf{A}$ can be found to be 3 and -1, and the orthogonal matrix $Q = \frac{1}{\sqrt{2}} \begin{bmatrix} 1 & -1 \\ 1 & 1 \end{bmatrix}$ can diagonalize the matrix $\mathbf{A}$.

Putting $\mathbf{X}' = \mathbf{Q}^T\mathbf{X}$, where X' has entries x' and y', then $\mathbf{X} = \mathbf{Q}\mathbf{X}'$ and we find that

$$\begin{aligned} x &= \tfrac{1}{\sqrt{2}}(x' - y') \\ y &= \tfrac{1}{\sqrt{2}}(x' + y') \end{aligned}.$$

So we have $\theta = \frac{\pi}{4}$, and the rotation of axes found for this conic is through an angle $\frac{\pi}{4}$ in an anticlockwise direction.

Now upon substituting the value of x and y in the equation of the conic, it obtains

$$3x'^2 - y'^2 + 2\sqrt{2}x' - \sqrt{2}y' - 1 = 0,$$

which shows that the *conic* is a *hyperbola.*

Next, to find the standard form, we put the equation in a complete square in x' and y', i.e.,

$$3\left(x' + \frac{\sqrt{2}}{3}\right)^2 - \left(y' + \frac{1}{\sqrt{2}}\right)^2 = \frac{7}{6}.$$

Thus, the equation of the hyperbola in standard form is

$$3x''^2 - y''^2 = \frac{7}{6},$$

where the values of x'' and y'' are defined as

$$\begin{aligned} x'' &= x' + \tfrac{\sqrt{2}}{3} \\ y'' &= y' + \tfrac{1}{\sqrt{2}} \end{aligned}.$$

This represents a hyperbola having a center (x', y'), where $x' = -\frac{\sqrt{2}}{3}$ and $y' = -\frac{1}{\sqrt{2}}$.

Thus, the hyperbola center in xy-coordinates form is $\left(\frac{1}{6}, -\frac{5}{5}\right)$, and the axes of the hyperbola are defined by the lines $x'' = 0$ and $y'' = 0$,

i.e., $x - y = 1$ and $x + y = -\frac{2}{3}$.

9.3 Quadrics

A quadric is an equation of surface defined on a *three-dimensional* space, having degree 2 of the form

$$a_1x^2 + a_2y^2 + a_3z^2 + 2a_4xy + 2a_5yz + 2a_6zx + a_7x + a_8y + a_9z + a_{10} = 0.$$

Let us consider a symmetric matrix $\mathbf{A}$, defined by

$$\mathbf{A} = \begin{bmatrix} a_1 & a_4 & a_6 \\ a_4 & a_2 & a_5 \\ a_6 & a_5 & a_3 \end{bmatrix}.$$

Then the equation of the quadric can be expressed in the form

$$X^T AX + [a_7 \;\; a_8 \;\; a_9]\, X + a_{10} = 0,$$

where X is the column with entries x, y, z.

From analytical geometry, a quadric is one of the following surfaces, i.e., a *hyperboloid*, an *ellipsoid*, a *paraboloid*, a *cylinder*, and a *cone* (or a degenerate form).

So, just as conic, this quadric class can also be obtained by a rotation to principal axes. For finding a real orthogonal matrix Q, It must satisfies $QTAX = D$, where D is a diagonal matrix with diagonal entries $a_{1\_}, a_{2\_}, a_{3\_}$, say.

Let us substitute $X' = Q^T X$, Then it obtains $X = QX'$ and $X^T AX = (X')^T DX'$.

Thus, the quadric equation becomes

$$(X')^T DX' + [a_7 \;\; a_8 \;\; a_9]\, QX' + a_{10} = 0,$$

which is similar to

$$a_1{}' x'^2 + a_2' y'^2 + a_3' z'^2 + a_7' x' + a_8' y' + a_9' z' + a_{10} = 0.$$

Here a_1', a_2', a_3' are the eigenvalues of the matrix $\mathbf{A}$, while a_7', a_8', a_9' are specific real numbers.

By accomplishing the squares in x', y', z' as necessary, we can get the quadric equation in standard form and then it will be desirable to recognize its type and position. The last step illustrates the translation of axes.

Example 9.2:

Determine the quadric surface of the equation

$$x^2 + y^2 + z^2 + 2xy + 2yz + 2zx - x + 2y - z = 0. \qquad (9.4)$$

The matrix $\mathbf{A}$ of the suitable quadric form of eqn (9.4) is

$$\mathbf{A} = \begin{bmatrix} 1 & 1 & 1 \\ 1 & 1 & 1 \\ 1 & 1 & 1 \end{bmatrix}.$$

If we write the equation of the quadric in matrix form, then it will take the form as

$$X^T AX + [-1 \;\; 2 \;\; -1]\, X = 0.$$

Next, we can continue to diagonalize the matrix $\mathbf{A}$ using an orthogonal matrix Q. Upon finding the eigenvalues of the matrix $\mathbf{A}$, the eigenvalues of the matrix **A** can be obtained as 0, 0, and 3 with the corresponding eigenvectors

$$\begin{bmatrix} \frac{1}{\sqrt{2}} \\ -\frac{1}{\sqrt{2}} \\ 0 \end{bmatrix}, \begin{bmatrix} 0 \\ \frac{1}{\sqrt{2}} \\ -\frac{1}{\sqrt{2}} \end{bmatrix}, \text{ and } \begin{bmatrix} \frac{1}{\sqrt{3}} \\ \frac{1}{\sqrt{3}} \\ \frac{1}{\sqrt{3}} \end{bmatrix}.$$

Here the eigenvalue $\lambda = 0$ obtains the first two eigenvectors that generate the eigenspace.

Next, to identify an orthonormal basis of this subspace, this can be done either by guessing such a basis or by applying the *Gram–Schmidt procedure*, which turns out to be

$$\frac{1}{\sqrt{2}}\begin{bmatrix} 1 \\ -1 \\ 0 \end{bmatrix}, \frac{1}{\sqrt{6}}\begin{bmatrix} 1 \\ 1 \\ -2 \end{bmatrix}.$$

Therefore, the matrix **A** can be diagonalized by the orthogonal matrix Q as

$$Q = \begin{bmatrix} \frac{1}{\sqrt{2}} & \frac{1}{\sqrt{6}} & \frac{1}{\sqrt{3}} \\ -\frac{1}{\sqrt{2}} & \frac{1}{\sqrt{6}} & \frac{1}{\sqrt{3}} \\ 0 & -\frac{2}{\sqrt{6}} & \frac{1}{\sqrt{3}} \end{bmatrix}.$$

The matrix Q illustrates a rotation of axes.

Put $\mathbf{X}' = Q^T\mathbf{X}$; then $X = QX'$ and

$$X^TAX = (X')^T (Q^TAQ) X' = (X')^T DX',$$

where D is the diagonal matrix with diagonal entries 0, 0, and 3.

Hence, the equation of the quadric is obtained as

$$(X')^T DX' + [-1 \quad 2 \quad -1]\, QX' = 0$$

or $z'^2 = \frac{1}{\sqrt{2}}x' - \frac{1}{\sqrt{6}}y' = 0.$

This represents a parabolic cylinder whose axis is in line with the equations

$$y' = \sqrt{3}x', \; z' = 0.$$

9.4 Definite Quadratic Form

Let us assume a quadratic form $\mathbf{q} = X^TAX$ in real variables $x_1, x_2, \ldots, x_n$, where **A** is a real symmetric matrix.

In some applications, it is the sign of the quadratic form $\mathbf{q}$ that is important. The quadratic form $\mathbf{q}$ is said to be *positive definite*, if $\mathbf{q} > 0$ whenever $X \neq 0$. Similarly, the quadratic form $\mathbf{q}$ is called *negative definite* if $\mathbf{q} < 0$, whenever $X \neq 0$. The quadratic form $\mathbf{q}$ is said to be *indefinite*, if $\mathbf{q}$ can have both positive and negative values.

For example:

Positive-definite quadratic form:

Here, the expression $3x^2 + 2y^2$ is *positive* unless $x = 0$ and $y = 0$; so this is a *positive definite quadratic form.*

Negative definite quadratic form:

Here, the expression $-3x^2 - 2y^2$ is *negative definite* since the term is negative unless $x = 0$ and $y = 0$. Otherwise, the form $3x^2 - 2y^2$ can take both positive and negative values; so it is an *indefinite quadratic form.*

One can quickly obtain the nature of the quadratic form if it contained only the squared term, just as done in the above examples. However, it is not possible to determine the nature of the problem by simple investigation in general. In symmetric matrices, the diagonalization process reduces the problem to a quadratic form whose matrix is diagonal and includes only squared terms. From this, it is evident that it is the signs of the eigenvalues of the matrix that are prominent. The conclusive result is as follows.

Theorem 9.1:

Let us consider a real symmetric matrix $\mathbf{A}$ and the quadratic form $\mathbf{q} = X^T AX$. Then

(1) The quadratic form $\mathbf{q}$ is *positive definite* if and only if all the eigenvalues of the matrix $\mathbf{A}$ are positive.

(2) The quadratic form $\mathbf{q}$ is said to be *negative definite* if and only if all the eigenvalues of the matrix $\mathbf{A}$ are negative.

(3) The quadratic form $\mathbf{q}$ is indefinite if and only if all the eigenvalues of the matrix $\mathbf{A}$ have both positive and negative eigenvalues.

Proof:

Let there exist a real orthogonal matrix Q such that $Q^T \mathbf{A} Q = D$ is diagonal, with diagonal entries $\lambda_1, \lambda_2, \ldots, \lambda_n$, say.

Let us put $X' = Q^T X$; then $X = QX'$ and

$$\mathbf{q} = X^T AX$$

$$= (X')^T (Q^T AQ) X'$$
$$= (X')^T DX'.$$

This implies that the quadratic form $\mathbf{q}$ takes the form $\mathbf{q} = \lambda_1 x'^2{}_1 + \lambda_2 x'^2{}_2 + \cdots + \lambda_n x'^2{}_n$, where $x'_1, x'_2, \ldots, x'_n$ are the entries of X'.

Thus, the quadratic form $\mathbf{q}$ in $x'_1, x'_2, \ldots, x'_n$ involves only squares.

Next, it can be observed that as X varies over all the non-zero vectors of R^n, so does $X' = Q^T X$.

This is since $Q^T = Q^{-1}$ is invertible. Therefore, $\mathbf{q} > 0$ for all $\mathbf{X} \neq 0$ if and only if $\mathbf{q} > 0$ for all $\mathbf{X}' \neq 0$.

Similarly, one can find that it is sufficient to discuss the behavior of the *quadratic form* $\mathbf{q}$ in $x'_1, x'_2, \ldots, x'_n$. Still, the quadratic form $\mathbf{q}$ will be *positive definite* as such a form precisely, when $\lambda_1, \lambda_2, \ldots, \lambda_n$ $\lambda'_i s > 0, \; for\, i = 1, 2, \ldots, n$ and similarly for negative definite and indefinite.

Thus, since $\lambda_1, \lambda_2, \ldots, \lambda_n$ are the eigenvalues of the matrix $\mathbf{A}$, it proves the statement of the theorem.

Hence proved. □

Next, let us discuss the crucial case of a quadratic form $\mathbf{q}$ in two variables x and y, say, $\mathbf{q} = a_1 x^2 + 2a_2 xy + a_3 y^2$.

The associated symmetric matrix is

$$\mathbf{A} = \begin{bmatrix} a_1 & a_2 \\ a_2 & a_3 \end{bmatrix}.$$

Let the eigenvalues of the symmetric matrix $\mathbf{A}$ be μ_1 and μ_2. Then, we have the relationship $\det(\mathbf{A}) = \mu_1\mu_2, tr(\mathbf{A}) = \mu_1 + \mu_2$ and hence

$\mu_1\mu_2 = a_1 a_3 - a_2{}^2$ and $\mu_1 + \mu_2 = a_1 + a_3$.

The quadratic form $\mathbf{q}$ is said to be positive definite if and only if the eigenvalues μ_1 and μ_2 are both positive. It happens precisely when $a_1 a_3 > a_2{}^2$ and $a_1 > 0$, and these conditions are indeed necessary (by Theorem 9.1).

If the eigenvalues μ_1, μ_2 are both positive, then the coefficients $a_1 > 0$, $a_3 > 0$, and $a_1 a_3 > a_2{}^2$. Then it shows that a_1 and a_3 are of the same sign.

Similarly, it can also be argued that the conditions $a_1 a_3 > a_2{}^2$ and $a_1 < 0$ will be for the matrix $\mathbf{A}$ to be *negative definite*.

Finally, the quadratic form $\mathbf{q}$ is said to be *indefinite*, if and only if $a_1 a_3 < a_2{}^2$. The condition for the quadratic form $\mathbf{q}$ to be indefinite is that μ_1 and μ_2 are of opposite signs, i.e., $\mu_1\mu_2 < 0$. Therefore, we have the following results.

Corollary 9.1:

Let us consider the quadratic form $\mathbf{q} = a_1x^2 + 2a_2xy + a_3y^2$ in x and y. Then

(1) The quadratic form $\mathbf{q}$ is positive definite if and only if $a_1a_3 > {a_2}^2$ and $a_1 > 0$.

(2) The quadratic form $\mathbf{q}$ is negative definite if and only if $a_1a_3 > {a_2}^2$ and $a_1 < 0$.

(3) The quadratic form $\mathbf{q}$ is indefinite if and only if $a_1a_3 < {a_2}^2$.

Example 9.3:

Let the quadratic form $\mathbf{q}$ be defined by $\mathbf{q} = -2x^2 + xy - 3y^2$.

Here $a_1 = -2$, $a_2 = \frac{1}{2}$, $a_3 = -3$

Since $a_1a_3 > {a_2}^2$ and $a_1 < 0$, it implies that the quadratic form Q is negative definite.

Similarly, one can consider the quadratic forms in three or more variables; for it, one must use (Theorem 7.1) to decide on definiteness.

Example 9.4:

Let $\mathbf{q} = -2x^2 - y^2 - 2z^2 + 6zx$ be a quadratic form in x, y, z.

The coefficient matrix of the quadratic form can be defined as

$$\mathbf{A} = \begin{bmatrix} -2 & 0 & 3 \\ 0 & -1 & 0 \\ 3 & 0 & -2 \end{bmatrix}.$$

Since the matrix's eigenvalues $\mathbf{A}$ are -5, -1, and 1, the quadratic form $\mathbf{q}$ is said to be indefinite.

Next, we discuss the criterion for a matrix to be positive definite.

Theorem 9.2:

Let us consider a real symmetric matrix $\mathbf{A}$. Then the symmetric matrix $\mathbf{A}$ is said to be *positive definite* if and only if there is an invertible matrix $\mathbf{A} = \mathbf{C}^T\mathbf{C}$.

Proof:

Let us consider that the matrix $\mathbf{C}$ is invertible and $\mathbf{A} = \mathbf{C}^T\mathbf{C}$.

Then the quadratic form $\mathbf{q} = X^TAX$ can be rewritten as

$$\begin{aligned}\mathbf{q} &= X^T\left(\mathbf{C}^T\mathbf{C}\right)X \\ &= X^T\mathbf{C}^TCX\end{aligned}$$

$$= (CX)^T CX$$
$$= \|CX\|^2.$$

Now, if $X \neq 0$, then it implies $CX \neq 0$ (since the matrix $\mathbf{C}$ is invertible). Thus, it means $\|\mathbf{CX}\|$ is positive, if $X \neq 0$.

Hence, it observes that the quadratic form $\mathbf{q}$ is positive implies that the real symmetric matrix $\mathbf{A}$ is positive definite.

Conversely, let us consider that the real symmetric matrix $\mathbf{A}$ is positive definite, implying that all the matrix eigenvalues $\mathbf{A}$ are positive.

Then there exists a real orthogonal matrix Q such that $Q^T\mathbf{A}Q = D$ is diagonal, having diagonal entries $d_{11}, d_{22}, \ldots, d_{nn}$ say, and these are the eigenvalues of the matrix $\mathbf{A}$ and all $d'_{ii}s$ are positive.

Now define $\sqrt{D}$ to be the real diagonal matrix having its diagonal entries as $\sqrt{d_{11}}, \sqrt{d_{22}}, \ldots, \sqrt{d_{nn}}$.

Then it has

$$\begin{aligned} \mathbf{A} &= \left(Q^T\right)^{-1} DQ^T \\ &= QDQ^T \end{aligned}.$$

Since $Q^T = Q^{-1}$, it becomes

$$\begin{aligned} \mathbf{A} &= Q\left(\sqrt{D}\sqrt{D}\right)Q^T \\ &= \left(\sqrt{D}Q^T\right)^T\left(\sqrt{D}Q^T\right) \end{aligned}.$$

Finally, let us put $\mathbf{C} = \sqrt{D}Q^T$, and it can be observed that the matrix $\mathbf{C}$ is invertible, since both Q and $\sqrt{D}$ are invertible.

9.5 Bilinear Form

A *bilinear form* can be defined as a scalar-valued linear function of two vector variables: an *inner product* described in a real vector space $\mathbf{V}$. Thus, there is a close network between the *bilinear forms* and quadratic forms.

Let us consider a vector space $\mathbf{V}$ defined over a field $\mathbf{F}$.

Let $\mathbf{V} \times \mathbf{V} = \{(u, v) : u, v \in \mathbb{R}\}$.

Then a bilinear form defined on a vector space $\mathbf{V}$ is a function $\phi : \mathbf{V} \times \mathbf{V} \to \mathbf{F}$.

That is to say, a rule that assigns to each pair of vectors (u, v), a scalar that satisfies $\phi(u, v)$ the following requirements:

(1) $\phi(u_1 + u_2, v) = \phi(u_1, v) + \phi(u_2, v)$.
(2) $\phi(u, v_1 + v_2) = \phi(u, v_1) + \phi(u, v_2)$.

(3) $\phi(ku, v) = k\phi(u, v)$.
(4) $\phi(u, kv) = k\phi(u, v)$.

These must hold for all vectors u, u_1, u_2, v, v_1, v_2 in $\mathbf{V}$ and for all scalars k in $\mathbf{F}$. If $\phi(u, v)$ satisfies all the above four properties, then $\phi(u, v)$ is said to be *linear* in both the variables u and v.

As cited, an inner product $\langle . \rangle$ on a real vector space is a *bilinear form* ϕ in which $\phi\langle u, v\rangle = \langle u, v\rangle$ and a bilinear form $\phi\langle u, v\rangle$ arises whenever a square matrix is given, which is shown by this example.

For example:

Let $\mathbf{A}$ be a $n \times n$ matrix over a field $\mathbf{F}$ and $\phi : \mathbf{F}^n \times \mathbf{F}^n \to \mathbf{F}$ defined by the rule $\phi(\mathbf{X}, \mathbf{Y}) = \mathbf{X}^T\mathbf{A}\mathbf{Y}$ is a bilinear form on $\mathbf{F}^n$.

It is a typical bilinear form on finite-dimensional vector spaces, which we will state further.

9.6 Matrix Representation of Bilinear Forms

Let us consider a bilinear form $\phi : \mathbf{V} \times \mathbf{V} \to \mathbf{F}$ defined on an n-dimensional vector space $\mathbf{V}$ over a field $\mathbf{F}$. Let $\mathbf{B} = \{v_1, v_2, \ldots.., v_n\}$ be an ordered basis of $\mathbf{V}$ and represent c_{ij} to be the scalar $\phi(v_i, v_j)$.

Thus, the function ϕ can be associated with the $n \times n$ matrix $\mathbf{A} = \{a_{ij}\}$.

Next, let $u, v \in \mathbf{V}$ and

$u = \sum_{i=1}^{n} \alpha_i v_i$ and $v = \sum_{j=1}^{n} \beta_j v_j$.

Then the coordinate vectors of the vectors u and v can be defined for the given basis $\mathbf{B}$ are

$$[u]_{\mathbf{B}} = \begin{bmatrix} b_1 \\ \vdots \\ b_n \end{bmatrix} \text{ and } [v]_{\mathbf{B}} = \begin{bmatrix} c_1 \\ \vdots \\ c_n \end{bmatrix}.$$

Upon using the linear properties of ϕ, the function $\phi(u, v)$ can be calculated in terms of the matrix $\mathbf{A}$ as

$$\begin{aligned}
\phi(u, v) &= \phi\left(\sum_{i=1}^{n} \alpha_i v_i, \sum_{j=1}^{n} \beta_j v_j\right) \\
&= \sum_{i=1}^{n} \alpha_i \phi\left(v_i, \sum_{j=1}^{n} \beta_j v_j\right) \\
&= \sum_{i=1}^{n} \sum_{j=1}^{n} \alpha_i \phi(v_i, v_j) \beta_j.
\end{aligned}$$

Since $\phi(v_i, v_j) = c_{ij}$, it becomes

$$\phi(u, v) = \sum_{i=1}^{n} \sum_{j=1}^{n} \alpha_i c_{ij} \beta_j.$$

From this, the fundamental equation can be obtained as

$$\phi(u, v) = ([u]_{\mathbf{B}})^T \mathbf{A} [v]_{\mathbf{B}}.$$

Thus, the linear form ϕ is defined concerning the basis $\mathbf{B}$ by the $n \times n$ matrix $\mathbf{A}$ whose (i, j) entry is $\phi(v_i, v_j)$, where the value of ϕ can be calculated using the above rule, especially if ϕ is a bilinear form on $\mathbf{F}^n$, with the standard basis, then $\phi(X, Y) = X^T AY$.

Conversely, let us begin with a matrix $\mathbf{A}$ and define ϕ by

$$\phi(u, v) = ([u]_{\mathbf{B}})^T \mathbf{A} [v]_{\mathbf{B}}.$$

Then it can be easily verified that a function ϕ is a bilinear form on the vector space $\mathbf{V}$ and that the matrix representing the function ϕ concerning the basis $\mathbf{B}$ is $\mathbf{A}$.

Now let us consider another ordered basis $\mathbf{B}'$, and we will check the effect of $\mathbf{B}'$ upon the matrix $\mathbf{A}$.

Let $\mathbf{C}$ be an invertible matrix that describes the change of basis from $\mathbf{B}'$ to $\mathbf{B}$, i.e., $\mathbf{B}' \to \mathbf{B}$. Thus, $[u]_{\mathbf{B}} = \mathbf{C} [u]_{\mathbf{B}'}$.

And it obtains

$$\begin{aligned}\phi(u, v) &= (\mathbf{C} [u]_{\mathbf{B}'})^T \mathbf{A} (\mathbf{C} [v]_{\mathbf{B}'}) \\ &= ([u]_{\mathbf{B}'})^T (\mathbf{C}^T AC) [v]_{\mathbf{B}'}.\end{aligned}$$

This demonstrates that the matrix $\mathbf{C}^T AC$ describes the function ϕ with respect to the basis $\mathbf{B}'$. □

Here, we find a new relationship between the matrices that have arisen. It can be defined that a matrix $\mathbf{B}$ is said to be *congruent* to a matrix $\mathbf{A}$, if there exists an invertible matrix $\mathbf{C}$ such that $\mathbf{B} = \mathbf{C}^T A\mathbf{C}$.

There is some correlation between congruence and similarity, but matrices need not be congruent in analogous general, nor are congruent matrices identical. The point that has appeared from the preceding discussion is that matrices representing the same bilinear form (concerning different vector space bases) are congruent. This result is to be correlated because the matrices representing the same linear transformation are similar.

The conclusion of the last few paragraphs outlines the following fundamental theorem.

Theorem 9.3:

Let $\mathbf{V}$ be an n-dimensional vector space and ϕ be a *bilinear form* on $\mathbf{V}$ over F.

(1) Let $\mathbf{B}$ be an ordered basis of $\mathbf{V}$ defined by $\mathbf{B} = \{v_1, v_2, \ldots, v_n\}$. Let us define an $n \times n$ matrix $\mathbf{A}$ whose (i, j) entry is $\phi(v_i, v_j)$; then the bilinear form ϕ is defined by $\phi(u, v) = ([u]_{\mathbf{B}})^T \mathbf{A} [v]_{\mathbf{B}}$.
Here, $\mathbf{A}$ is the $n \times n$ matrix representing the function ϕ concerning the ordered basis $\mathbf{B}$.

(2) If there is another ordered basis $\mathbf{B}$, then the bilinear form ϕ is represented concerning the ordered basis $\mathbf{B}'$ by the matrix $\mathbf{C}^T AC$, and $\mathbf{C}$ is an *invertible matrix* defining the change of basis $\mathbf{B}' \rightarrow \mathbf{B}$.

(3) Conversely, if $\mathbf{A}$ is any $n \times n$ matrix defined over the field $\mathbf{F}$, then a *bilinear form* ϕ on $\mathbf{V}$ is determined by the rule as

$$\phi(u, v) = ([u]_{\mathbf{B}})^T \mathbf{A} [v]_{\mathbf{B}} \,.$$

The matrix $\mathbf{A}$ represents it concerning the basis $\mathbf{B}$.

9.7 Symmetric and Skew-symmetric Bilinear Form

A bilinear form ϕ defined on a real or complex vector space $\mathbf{V}$ is *symmetric* if its values are unaltered by reversing the arguments, i.e., if $\phi(u, v) = \phi(v, u)$, for all u and v.

Similarly, a bilinear form ϕ is said to be *skew-symmetric* if $\phi(u, v) = -\phi(v, u)$ and is always true.

For example, any real inner product on an inner product space $\mathbf{V}$ is a *symmetric bilinear form*; otherwise, the structure ϕ is defined by the rule

$$\phi\left(\begin{bmatrix} x_1 \\ x_2 \end{bmatrix}, \begin{bmatrix} y_1 \\ y_2 \end{bmatrix}\right) = x_1 y_2 - x_2 y_1,$$

which is an example of a *skew-symmetric bilinear form* defined on R^2.

Note:

There are connections with symmetric and skew-symmetric matrices. The following theorem discusses it further.

Theorem 9.4:

Let $\mathbf{V}$ be a finite-dimensional vector space and ϕ be a bilinear form defined on $\mathbf{V}$. Let the matrix $\mathbf{A}$ represent the function ϕ concerning to some basis of $\mathbf{V}$. Then the function ϕ is said to be *symmetric* if and only if the matrix $\mathbf{A}$ is symmetric, and the equivalent statement is also valid for *skew-symmetric* bilinear forms.

Proof:

Let us consider the matrix $\mathbf{A}$ to be symmetric and $[u]^T \mathbf{A}[v]$ to be scalar; then we have

$$\begin{aligned}\phi(u,v) &= [u]^T \mathbf{A}[v] = \left([u]^T \mathbf{A}[v]\right)^T \\ &= [v]^T \mathbf{A}^T [u] = [v]^T \mathbf{A}[v] \\ &= \phi(v,u).\end{aligned}$$

Thus, ϕ is symmetric.

Conversely, assume that the function ϕ is symmetric and let $\mathbf{B} = \{v_1, v_2, \ldots, v_n\}$ be the ordered basis of $\mathbf{V}$. Then

$$a_{ij} = \phi(v_i, v_j) = \phi(v_j, v_i) = a_{ji},$$

which implies that the matrix $\mathbf{A}$ is symmetric.

The proof of the *skew-symmetric* case is also related to the *symmetric* case.

9.8 Symmetric Bilinear Forms and Quadratic Forms

Let ϕ be a bilinear form on R^n defined by

$$\phi(X,Y) = X^T AY.$$

Then the bilinear form ϕ determines a quadratic form $\mathbf{q}$, where

$$\mathbf{q} = \phi(X,X) = X^T AX.$$

Conversely, if $\mathbf{q}$ is a quadratic form in $x_1, x_2, \ldots, x_n$, then an equivalent symmetric bilinear form ϕ on R^n can be defined by means of the rule

$$\phi(X,Y) = \frac{1}{2}\{\mathbf{q}(X+Y) - \mathbf{q}(X) - \mathbf{q}(Y)\},$$

where X and Y are the column vectors consisting of $x_1, x_2, \ldots, x_n$ and $y_1, y_2, \ldots, y_n$.

To show the function $\phi(X,Y)$ is bilinear, first, we express $\mathbf{q}(X)$ as $\mathbf{q}(X) = X^T AX$ where the matrix $\mathbf{A}$ is symmetric.
Then

$$\begin{aligned}\phi(X,Y) &= \frac{1}{2}\left\{(X+Y)^T \mathbf{A}\ (X+Y) - X^T AX - Y^T AY\right\}\\ &= \frac{1}{2}\left(X^T AY + Y^T AX\right)\\ &= X^T AY.\end{aligned}$$

Since $X^T AY = Y^T AX$, this shows that $\phi(X,Y)$ is bilinear.

Theorem 9.5:

A bijection arises between the set of quadratic forms in $x_1, x_2, \ldots, x_n$ and the set of symmetric bilinear form on R^n. □

We would suppose to get important information about symmetric bilinear forms by utilizing the spectral theorem from experience. In fact, what is obtained is a canonical or standard form for such bilinear conditions.

Theorem 9.6:

Let ϕ be a symmetric bilinear form defined on an n-dimensional real vector space $\mathbf{V}$. Then there exists a basis $\mathbf{B}$ of $\mathbf{V}$ such that

$$\begin{aligned}\phi(u,v) = u_1v_1 + u_2v_2 + \cdots + u_kv_k\\ -u_{k+1}v_{k+1} - u_{k+2}v_{k+2} - \cdots - u_Lv_L\end{aligned}$$

where $u_1, u_2, \ldots, u_n$ and $v_1, v_2, \ldots, v_n$ are the entries of the coordinate vectors $[u]_{\mathbf{B}}$ and $[v]_{\mathbf{B}}$ respectively, and k and L are integers satisfying $0 \le k \le L \le n$.

Proof:

Let a matrix $\mathbf{A}$ represent the bilinear form ϕ concerning some basis $\mathbf{B}'$ of $\mathbf{V}$. Then the matrix $\mathbf{A}$ is *symmetric*. Thus, $\mathbf{Q}^T\mathbf{AQ} = \mathbf{D}$ is diagonal for some orthonormal matrix Q, having diagonal entries $d_1, d_2, \ldots, d_n$, and these are the eigenvalues of the matrix $\mathbf{A}$.

Suppose that the first k diagonal entries are non-negative, i.e., $d_1, d_2, \ldots, d_k > 0$ while $d_{k+1}, d_{k+2}, \ldots, d_L < 0$ and $d_{L+1} = d_{L+2} = \cdots = d_n = 0$ by altering the eigenvectors, if necessary.

Let E be the $n \times n$ diagonal matrix whose diagonal matrix entries are the real numbers

$$\frac{1}{\sqrt{d_1}}, \frac{1}{\sqrt{d_2}}, \ldots, \frac{1}{\sqrt{d_k}}, \frac{1}{\sqrt{(-d_{k+1})}}, \ldots, \frac{1}{\sqrt{(-d_L)}}, 1, \ldots, 1.$$

Then

$$(QE)^T \mathbf{A}(QE) = E^T (Q^T AQ) E$$

$$= EDE = \begin{bmatrix} \mathbf{I}_K & \vdots & 0 & \vdots & 0 \\ \cdots & \vdots & \cdots & \vdots & \cdots \\ 0 & \vdots & -\mathbf{I}_{L-K} & \vdots & 0 \\ \cdots & \vdots & \cdots & \vdots & \cdots \\ 0 & \vdots & 0 & \vdots & 0 \end{bmatrix} = S \, .$$

It implies that the matrix $Q\mathbf{E}$ is invertible, and its inverse finds out a change of basis from $\mathbf{B}'$ to $\mathbf{B}$ say. Then the bilinear form ϕ will be shown by the matrix S concerning the basis $\mathbf{B}$.

This implies that $\phi(u, v) = ([u]_{\mathbf{B}})^T S [v]_{\mathbf{B}}$.

Hence, on multiplying the matrices together, the result follows.

Example 9.5:

Determine the canonical form of the symmetric bilinear form on R^2 defined by $\phi(x, y) = x_1y_1 + 2x_1y_2 + 2x_2y_1 + x_2y_2$.

The matrix of the bilinear form ϕ concerning the standard basis can be expressed as

$$\mathbf{A} = \begin{bmatrix} 1 & 2 \\ 2 & 1 \end{bmatrix}.$$

The matrix $\mathbf{A}$ has eigenvalues 3 and -1 and can be diagonalized by the matrix

$$Q = \frac{1}{\sqrt{2}} \begin{bmatrix} 1 & -1 \\ 1 & 1 \end{bmatrix}.$$

Upon putting $X' = Q^T X$ and $Y' = Q^T Y$, we find

$$\begin{aligned} \phi(X, Y) &= X^T \mathbf{A} Y \\ &= (X')^T Q^T \mathbf{A} Q Y' \\ &= (X')^T \begin{bmatrix} 3 & 0 \\ 0 & -1 \end{bmatrix} Y' \end{aligned}$$

so that $\phi(X, Y) = 3x_1{}'y_1{}' - x_2{}'y_2{}'$.

Here $x_1{}' = \frac{1}{\sqrt{2}}(x_1 + x_2)$ and $x_2{}' = \frac{1}{\sqrt{2}}(-x_1 + x_2)$ with the corresponding formulas in y.

Next, to obtain the canonical form of the bilinear form ϕ,

put $x_1'' = \sqrt{3}x_1'$, $y_1'' = \sqrt{3}y_1'$ and $x_2'' = x_2'$, $y_2'' = y_2'$.
Then $\phi(\mathbf{X}, \mathbf{Y}) = x_1'' y_1'' - x_2'' y_2''$, which is the canonical form.

9.9 Eigenvalues of Congruent Matrices

Since the congruent matrices describe the same symmetric bilinear form, it expects that such congruent matrices must have some common properties as that of similar matrices. Still, similar matrices possess the same eigenvalues, whereas this is not true in congruent matrices.

For example:

The eigenvalues of the matrix

$\begin{bmatrix} 2 & 0 \\ 0 & -3 \end{bmatrix}$ are 2 and -3,

whereas the congruent matrix

$$\begin{bmatrix} 1 & 0 \\ 1 & 1 \end{bmatrix} \begin{bmatrix} 2 & 0 \\ 0 & -3 \end{bmatrix} \begin{bmatrix} 1 & 1 \\ 0 & 1 \end{bmatrix} = \begin{bmatrix} 2 & 2 \\ 2 & -1 \end{bmatrix}$$

has eigenvalues -2 and 3.

Remark:

It is well known that, although the numbers of *positive and negative eigenvalues* are alike for each matrix, the eigenvalues of the *congruent matrices* are different.

9.10 Sylvester's Law of Inertia

Theorem 9.7 (Sylvester's law of inertia):

Let us consider an $n \times n$ real symmetric $\mathbf{A}$ matrix and $\mathbf{C}$ an invertible $n \times n$ matrix. Then the matrices $\mathbf{A}$ and $\mathbf{C}^T AC$ have an equal number of eigenvalues and have the same number of negative, positive, and zero eigenvalues.

Proof:

First, let us assume that the matrix $\mathbf{C}$ is invertible. It is possible to express QR factorization, Q a real orthogonal matrix, and R a real upper triangular with positive diagonal entries obtained using the *Gram–Schmidt process*.

The objective of the criterion of the theorem is to get a continuous chain of matrices that leads $\mathbf{C}$ to an orthogonal matrix Q.

The point of this is that $Q^T\mathbf{A}Q = Q^{-1}\mathbf{A}Q$; indeed, it has the same eigenvalues as the matrix $\mathbf{A}$ (since Q is an orthogonal matrix).

Define
$\mathbf{C}(t) = tQ + (1-t)\,\mathbf{C}$, where $0 \le t \le 1$.
Thus, $\mathbf{C}(0) = \mathbf{C}$, while $\mathbf{C}(1) = Q$.
Now note $U = tI + (1-t)\,R$,
so that $\mathbf{C}(t) = QU$.

Next, as U is an upper triangular matrix, its diagonal entries are $t + (1-t)\,r_{ii}$; so these cannot be zero $0 \le t \le 1$, since $r_{ii} > 0$ and $0 \le t \le 1$.

Hence, the matrix U is invertible, while the matrix Q is absolutely invertible since it is orthogonal.

It observes that $\mathbf{C}(t) = QU$ is invertible.
Thus, $\det(\mathbf{C}(t)) \ne 0$.
Let us consider $\mathbf{A}(t) = \mathbf{C}(t)^T AC(t)$.
So, it implies that $\det(\mathbf{A}(t)) = \det(\mathbf{A}) . \det(\mathbf{C}(t))^2 \ne 0$.
(Since $\det(\mathbf{C}(t)) \ne 0$ and $\det \mathbf{C}(t)^T = \det \mathbf{C}(t)$.)
It shows that the matrix $\mathbf{A}(t)$ cannot have zero eigenvalues.

As t goes from 0 to 1, the eigenvalues of $\mathbf{A}(0) = \mathbf{C}^T AC$ gradually shift to those of $\mathbf{A}(1) = Q^T\mathbf{A}Q$, that is, to those of $\mathbf{A}$.

But in this method, no eigenvalue can alter its sign because the eigenvalues that occur are a continuous function of t, and they are never zero.

Therefore, the numbers of negative and positive eigenvalues of $\mathbf{C}^T AC$ equal to those of $\mathbf{A}$. □

Note:

To check whether $\mathbf{A}$ is singular, we have to analyze the matrix $\mathbf{A} + \varepsilon\mathbf{I}$, which may be a concept of *perturbation* of $\mathbf{A}$. Now $\mathbf{A} + \varepsilon\mathbf{I}$ will be invertible only if ε is sufficiently small and positive.

Suppose we consider $\det(\mathbf{A} + x\mathbf{I})$ as a polynomial of a degree n that vanishes for at most n-values of x.

Thus, the prior disagreement shows that the conclusion is valid for $\mathbf{A}+\varepsilon\mathbf{I}$ also. If ε is small and positive, then take the limit as $\varepsilon \to 0$ and we can deduct the results for the matrix $\mathbf{A}$. □

Remark:

The theorem shows that the bilinear form uniquely determines the number of negative and positive signs in the canonical form. Moreover, it does not depend on the particular choice of the basis.

Example 9.6:

Show that the matrices $\mathbf{A} = \begin{bmatrix} 2 & 1 \\ 1 & 2 \end{bmatrix}$ and $\mathbf{B} = \begin{bmatrix} 1 & 2 \\ 2 & 1 \end{bmatrix}$ are not congruent.

Solution:

The eigenvalues of the matrix $\mathbf{A}$ are 1 and 3, while the matrix $\mathbf{B}$ has the eigenvalues 3 and -1. So, by Theorem 9.7, the matrices $\mathbf{A}$ and $\mathbf{B}$ cannot be congruent.

9.11 Skew-symmetric Bilinear Form

We have observed that obtaining a canonical form for a symmetric bilinear form on a real vector space $V$is possible. Next, the discussion extends to *skew-symmetric bilinear forms* in the light of congruent matrices.

The following theorem yields a solution to the problem.

Theorem 9.8:

Let ϕ be a *skew-symmetric bilinear form* defined on an n-dimensional vector space $\mathbf{V}$ over the field $\mathbf{F}$, where $\mathbf{F}$ is either $\mathbb{R}$ or $\mathbb{C}$. Then there is an ordered basis $\mathbf{B}$ of the vector space $\mathbf{V}$ with the form $\{u_1, v_1, u_2, v_2, \ldots, u_k, v_k, w_1, \ldots, w_{n-2k}\}$, where $0 \leq 2k \leq n$ such that $\phi(u_i, v_i) = 1 = -\phi(v_i, u_i)$, $i = 1, 2, \ldots, k$ and ϕ vanishes on all other pairs of basis elements.

Before going to the proof of the theorem, let us first examine the consequence of this theorem. Upon using the basis as provided by the above theorem, the bilinear form ϕ can be represented by the matrix

$$\mathbf{M} = \left[\begin{array}{ccccc:ccc} 0 & 1 & \cdots & 0 & 0 & 0 & \cdots & 0 \\ -1 & 0 & \cdots & 0 & 0 & 0 & \cdots & 0 \\ \cdots & \cdots & \cdots & \cdots & \cdots & \cdots & \cdots & \cdots \\ 0 & 0 & \cdots & 0 & 1 & 0 & \cdots & 0 \\ 0 & 0 & \cdots & -1 & 0 & 0 & \cdots & 0 \\ \cdots & \cdots & \cdots & \cdots & \cdots & \cdots & \cdots & \cdots \\ 0 & 0 & \cdots & 0 & 0 & 0 & \cdots & 0 \\ \cdots & \cdots & \cdots & \cdots & \cdots & \cdots & \cdots & \cdots \\ 0 & 0 & \cdots & 0 & 0 & 0 & \cdots & 0 \end{array}\right].$$

The number of blocks of the type $\begin{bmatrix} 0 & 1 \\ -1 & 0 \end{bmatrix}$ allows us to derive an important conclusion about the *skew-symmetric matrices*.

Corollary 9.2:

Every skew-symmetric matrix $\mathbf{A}$ defined over the field $\mathbf{F}$ (where $\mathbf{F}$ is either $\mathbb{R}$ or $\mathbb{C}$) is congruent to a matrix $\mathbf{M}$ with the above form. □

The bilinear form ϕ given by $\phi(\mathbf{X}, \mathbf{Y}) = \mathbf{X}^T \mathbf{AY}$ is *skew-symmetric* and is represented concerning a suitable basis by a type matrix $\mathbf{M}$. Hence, the matrix $\mathbf{A}$ must be congruent to a matrix $\mathbf{M}$ with the above form.

Proof of Theorem 9.8:

Let $y_1, y_2, \ldots, y_n$ be any basis of a vector space $\mathbf{V}$.

If $\phi(y_i, y_j) = 0$ for all i and j, then $\phi(u, v) = 0$ for all vectors u and v.

So ϕ is the zero bilinear form, and the zero matrices represent it. This is the case for $k = 0$.

Assume that $\phi(y_i, y_j) \neq 0$ for some i and j. Since we can reorder the basis, we may suppose that $\phi(y_1, y_2) = a \neq 0$.

Then $\phi(a^{-1}y_1, y_2) = a^{-1}\phi(y_1, y_2) = a^{-1}a = 1$.

Now replace y_1 by $a^{-1}y_1$; the effect is to make $\phi(y_1, y_2) = 1$, and of course $\phi(y_2, y_1) = -1$, since ϕ is skew-symmetric.

Next put $b_i = \phi(y_1, y_i)$, where $i > 2$.

Then

$$\begin{aligned}\phi(y_1, y_i - by_2) &= \phi(y_1, y_i) - b\phi(y_1, y_2)\\ &= b - b = 0.\end{aligned}$$

It implies that the basis can also be altered by replacing y_i by $y_i - by_2$, for $i > 2$.

Since this does not disturb linear independence, we have a basis for the vector space $\mathbf{V}$. The effect of this substitution is to make

$\phi(y_1, y_i) = 0$ for $i = 3, \ldots, n$.

Next, we have to focus on the feasibility that $\phi(y_2, y_i)$ may be non-zero when $i > 2$.

Let $c = \phi(y_2, y_i)$. Then

$$\begin{aligned}\phi(y_2, y_i + cy_1) &= \phi(y_2, y_i) + c\phi(y_2, y_1)\\ &= c + (-c)\\ &= 0.\end{aligned}$$

This implies that the later step should be to replace y_i by $y_i + cy_1$, where $i > 2$. Again, we have to notice that $y_1, y_2, \ldots, y_n$ will still form a basis for the vector space $\mathbf{V}$. Also, there is an introductory observation that this replacement will not nullify what we have already been attained; the idea is that when $i > 2$,

$$\phi(y_1, y_i + cy_1) = \phi(y_1, y_i) + c\phi(y_1, y_1) = 0 \,.$$

We have now reached the stage where $\phi(y_1, y_2) = 1 = -\phi(y_2, y_1)$ and also $\phi(y_1, y_i) = 0 = \phi(y_2, y_i)$ for all $i > 2$.

Here, we rename the first two basis elements by noting $u_1 = y_1$ and $v_1 = y_2$.

Till now, the matrix defining ϕ has the form

$$\begin{bmatrix} 0 & 1 & \vdots & 0 & 0 \\ -1 & 0 & \vdots & 0 & 0 \\ \cdots & \cdots & \vdots & \cdots & \cdots \\ 0 & 0 & \vdots & & \mathbf{B} \end{bmatrix}.$$

The matrix $\mathbf{B}$ is a skew-symmetric matrix having $n-2$ rows and columns; the above argument can be repeated for the subspace with basis $\{y_3, \ldots, y_n\}$. It can be observed by induction that there is a basis for this subspace concerning which ϕ can be expressed by a matrix of the appropriate form. Indeed, let $u_2, u_3, \ldots, u_k, v_2, v_3, \ldots, v_k, w_1, w_2, \ldots, w_{n-2k}$ be this basis. A basis $\mathbf{V}$ is obtained by adjoining the basis elements u_1 and v_1, concerning which a matrix of the required form represents the bilinear form.

Example 9.7:

Determine a canonical form of the skew-symmetric matrix

$$\mathbf{A} = \begin{bmatrix} 0 & 0 & 2 \\ 0 & 0 & -1 \\ -2 & 1 & 0 \end{bmatrix}.$$

For this, we can execute the procedure signified in the proof of the theorem.

Solution:

Let$\{e_1, e_2, e_3\}$ be the standard basis of R^3. Then the matrix $\mathbf{A}$ finds out a skew-symmetric bilinear form ϕ with the properties

$$\begin{array}{l} \phi(e_1, e_3) = 2 = -\phi(e_3, e_1) \\ \phi(e_3, e_2) = 1 = -\phi(e_2, e_3) \\ \phi(e_1, e_2) = 0 = \phi(e_2, e_1) \end{array}.$$

First, let us alter the basis, i.e., $\{e_1, e_3, e_2\}$; this is necessary since $\phi(e_1, e_2) = 0$, whereas $\phi(e_1, e_3) \neq 0$.

Now change $\{e_1, e_3, e_2\}$ by $\left\{\frac{1}{2}e_1, e_3, e_2\right\}$ and note that

$$\phi\left(\frac{1}{2}e_1, e_3\right) = 1 = -\phi\left(e_3, \frac{1}{2}e_1\right).$$

Next since $\phi(e_3, e_2) = 1$, we replace e_2 by

$$e_2 + \phi(e_3, e_2)\frac{1}{2}e_1 = \frac{1}{2}e_1 + e_2.$$

Similarly, by noting that

$$\phi\left(\frac{1}{2}e_1, \frac{1}{2}e_1 + e_2\right) = 0 = \phi\left(e_3, \frac{1}{2}e_1 + e_2\right).$$

It implies that the procedure is now achieved.

Now the bilinear form ϕ can be represented concerning a new ordered basis $\left\{\frac{1}{2}e_1, e_3, \frac{1}{2}e_1 + e_2\right\}$ by the matrix $\mathbf{M}$, i.e.,

$$M = \begin{bmatrix} 0 & 1 & 0 \\ -1 & 0 & 0 \\ 0 & 0 & 0 \end{bmatrix},$$

which is in canonical form.

Now the replacement of basis from $\left\{\frac{1}{2}e_1, e_3, \frac{1}{2}e_1 + e_2\right\}$ to the standard ordered basis can be represented by the matrix

$$\mathbf{C} = \begin{bmatrix} \frac{1}{2} & 0 & \frac{1}{2} \\ 0 & 0 & 1 \\ 0 & 1 & 0 \end{bmatrix}.$$

It can be easily verified that $\mathbf{C}^T AC = \mathbf{M}$, the canonical form of the matrix $\mathbf{A}$, can be predicted by the proof of Theorem 9.8.

9.12 Application to the Reduction of Quadrics

Example 9.8:

Reduce the quadric $[x\ y\ z]\begin{bmatrix} 7 & -1 & -10 \\ -1 & 7 & 10 \\ -10 & 10 & -2 \end{bmatrix}\begin{bmatrix} x \\ y \\ z \end{bmatrix} = 36$ to its principal axes. Note that when this equation is written in full, it takes the form

$$7x^2 + 7y^2 - 2z^2 + 20yz - 20zx - 2xy = 36.$$

According to the preceding discussion, we need to reduce the matrix $\mathbf{A} = \begin{bmatrix} 7 & -1 & -10 \\ -1 & 7 & 10 \\ -10 & 10 & -2 \end{bmatrix}$ to the diagonal form.

The eigenvalues of $\mathbf{A}$ are $6, -12$, and 18. The corresponding eigenvectors are $(1,1,0)\,,(1,-1,2)$, and $(1,-1,-1)$.

Hence, the normalized eigenvectors are $\left(\frac{1}{\sqrt{2}}, \frac{1}{\sqrt{2}}, 0\right), \left(\frac{1}{\sqrt{6}}, -\frac{1}{\sqrt{6}}, \frac{2}{\sqrt{6}}\right)$, and $\left(\frac{1}{\sqrt{3}}, -\frac{1}{\sqrt{3}}, -\frac{1}{\sqrt{3}}\right)$.

These three vectors give the following orthogonal matrix:

$$H = \begin{bmatrix} \frac{1}{\sqrt{2}} & \frac{1}{\sqrt{6}} & \frac{1}{\sqrt{3}} \\ \frac{1}{\sqrt{2}} & -\frac{1}{\sqrt{6}} & -\frac{1}{\sqrt{3}} \\ 0 & \frac{2}{\sqrt{6}} & -\frac{1}{\sqrt{3}} \end{bmatrix}$$

Hence,

$$H^T AH = \begin{bmatrix} \frac{1}{\sqrt{2}} & \frac{1}{\sqrt{2}} & 0 \\ \frac{1}{\sqrt{6}} & -\frac{1}{\sqrt{6}} & -\frac{2}{\sqrt{6}} \\ \frac{1}{\sqrt{3}} & -\frac{1}{\sqrt{3}} & -\frac{1}{\sqrt{3}} \end{bmatrix} \begin{bmatrix} 7 & -1 & -10 \\ -1 & 7 & 10 \\ -10 & 10 & -2 \end{bmatrix} \begin{bmatrix} \frac{1}{\sqrt{2}} & \frac{1}{\sqrt{6}} & \frac{1}{\sqrt{3}} \\ \frac{1}{\sqrt{2}} & -\frac{1}{\sqrt{6}} & -\frac{1}{\sqrt{3}} \\ 0 & \frac{2}{\sqrt{6}} & -\frac{1}{\sqrt{3}} \end{bmatrix}$$

$$= \begin{bmatrix} 6 & 0 & 0 \\ 0 & -12 & 0 \\ 0 & 0 & 18 \end{bmatrix}$$

is the required diagonal matrix.

So, the equation of the quadric referred to its axes is

$$6x^2 - 12y^2 + 18z^2 = 36$$

or $\frac{x^2}{6} - \frac{y^2}{3} + \frac{z^2}{2} = 1.$

This is a hyperboloid of one sheet.

We shall conclude this chapter with another example where the matrix does not possess distinct eigenvalues. Though we have not developed the necessary theory for this, the following example shows that reducing diagonal form is possible even in such a case.

Example 9.9:

Reduce the matrix $\mathbf{A} = \begin{bmatrix} 1 & 2 & -2 \\ 2 & 1 & 2 \\ -2 & 2 & 1 \end{bmatrix}$ to the diagonal form and hence reduce the quadric

$$x^2 + y^2 + z^2 + 4yz - 4zx + 4xy = 27 \tag{9.5}$$

to its principal axes.

The eigenvalues of the matrix $\mathbf{A}$ are $3, 3$, and -3. Note that in this case, 3 is a repeated eigenvalue that depends on the fact that the matrix has distinct eigenvalues, and cannot be applied. However, an orthogonal matrix can be obtained by looking at the eigenvectors corresponding to 3 and -3.

The eigenvectors corresponding to the eigenvalue 3 are given by

$$\begin{bmatrix} -2 & 2 & -2 \\ 2 & -2 & 2 \\ -2 & 2 & -2 \end{bmatrix} \begin{bmatrix} x \\ y \\ z \end{bmatrix} = \begin{bmatrix} 0 \\ 0 \\ 0 \end{bmatrix}.$$

This gives only one equation: $x - y + z = 0$.

The solution set of this is the subspace $\{(y - z, y, z) : y, z \in R\}$ of V_3. We can choose two linearly independent vectors in this two-dimensional space by giving suitable values to y and z. Taking $y = 1$ and $z = 1$, we get $(0, 1, 1)$ and taking $y = 1$ and $z = 2$, we get $(-1, 1, 2)$. Since these two are linearly independent, we have only to orthogonalize them. Using the Gram–Schmidt process, we get two orthogonal vectors

$(0, 1, 1)$and $\left(-1, -{}^1\!/_2, {}^1\!/_2\right)$.

Normalizing them, we get

$\left(0, {}^1\!/_{\sqrt{2}}, {}^1\!/_{\sqrt{2}}\right)$and $\left(-{}^{\sqrt{2}}\!/_{\sqrt{3}}, -{}^1\!/_{\sqrt{6}}, {}^1\!/_{\sqrt{6}}\right)$.

The third eigenvector is the one that corresponds to the eigenvalue -3 and is $(1, -1, 1)$.

Normalizing this vector, we get $\left({}^1\!/_{\sqrt{3}}, -{}^1\!/_{\sqrt{3}}, {}^1\!/_{\sqrt{3}}\right)$. Without checking, we can say that this will be orthogonal to the two eigenvectors corresponding to 3 because the eigenvectors corresponding to distinct eigenvalues are orthogonal. Thus, we have the orthogonal matrix

$$H = \begin{bmatrix} 0 & -{}^{\sqrt{2}}\!/_{\sqrt{3}} & {}^1\!/_{\sqrt{3}} \\ {}^1\!/_{\sqrt{2}} & -{}^1\!/_{\sqrt{6}} & -{}^1\!/_{\sqrt{3}} \\ {}^1\!/_{\sqrt{2}} & {}^1\!/_{\sqrt{6}} & {}^1\!/_{\sqrt{3}} \end{bmatrix}.$$

Hence,

$$H^TAH$$
$$= \begin{bmatrix} 0 & 1/\sqrt{2} & 1/\sqrt{2} \\ -\sqrt{2}/\sqrt{3} & -1/\sqrt{6} & 1/\sqrt{6} \\ 1/\sqrt{3} & -1/\sqrt{3} & 1/\sqrt{3} \end{bmatrix} \begin{bmatrix} 1 & 2 & -2 \\ 2 & 1 & 2 \\ -2 & 2 & 1 \end{bmatrix} \begin{bmatrix} 0 & -\sqrt{2}/\sqrt{3} & 1/\sqrt{3} \\ 1/\sqrt{2} & -1/\sqrt{6} & -1/\sqrt{3} \\ 1/\sqrt{2} & 1/\sqrt{6} & 1/\sqrt{3} \end{bmatrix}.$$
$$= \begin{bmatrix} 3 & 0 & 0 \\ 0 & 3 & 0 \\ 0 & 0 & -3 \end{bmatrix}$$

Thus, the quadric (9.5) reduces to

$$3x^2 + 3y^2 - 3z^2 = 27$$

i.e., $\frac{x^2}{9} + \frac{y^2}{9} - \frac{z^2}{9} = 1$.
It may be noted here that eqn (9.5) can be written as

$$[x \; y \; z] \begin{bmatrix} 1 & 2 & -2 \\ 2 & 1 & 2 \\ -2 & 2 & 1 \end{bmatrix} \begin{bmatrix} x \\ y \\ z \end{bmatrix} = 27.$$

Exercises

1. Reduce the following matrices to diagonal form:

(a) $\begin{bmatrix} 5 & 1 & 1 \\ 1 & 5 & -1 \\ 1 & -1 & 5 \end{bmatrix}$

(b) $\begin{bmatrix} 6 & 4 & -2 \\ 4 & 12 & -4 \\ -2 & -4 & 13 \end{bmatrix}$

(c) $\begin{bmatrix} 4 & 3 & 3 \\ 3 & 0 & -1 \\ 3 & -1 & 3 \end{bmatrix}$.

2. Reduce the following conics to their principal axes:

(a) $7x^2 + 52xy - 32y^2 = 180$.
(b) $17x^2 + 312xy + 108y^2 = 900$.
(c) $145x^2 + 120xy + 180y^2 = 900$.

10

Canonical Forms

This chapter discusses the applications of special matrices such as block diagonal and triangular matrices, symmetric and Hermitian matrices, Schur's theorem, spectral theorem, Jordan and rational canonical form of special matrices. This chapter starts with some special types of matrices, including some essential theorems such as Schur's theorem, spectral theorem, and some crucial canonical and normal forms such as Jordan canonical form rational canonical form and minimum polynomial of Jordan normal form.

10.1 Triangularizable Matrices

Not every complex square matrix is *diagonalizable*, but it is always similar to an *upper triangular matrix*, and this type of consequence leads to many applications. For example, let us consider an $n \times n$ square matrix $\mathbf{A}$ that is defined over a field F. Then the square matrix $\mathbf{A}$ is *triangularizable* over the field F, if there is an invertible matrix C over F such that $C^{-1}AC = U$ is *upper triangular*. In other words, the invertible matrix C *triangularizes* the matrix $\mathbf{A}$.

Note: Upon using the fact, similar matrices have the same eigenvalues. It implies that the eigenvalues of the matrix $\mathbf{A}$ are the diagonal entries of the triangular matrix T. Hence, a necessary condition that a square matrix $\mathbf{A}$ is triangularizable is that the matrix $\mathbf{A}$ has n-eigenvalues in the field F. This condition is ever satisfied when $F = C^n$, which we are interested in discussing in the following section.

Theorem 10.1:

Every $n \times n$ square matrix $\mathbf{A}$ over C^n is *triangularizable.*

Proof:

Let us consider an $n \times n$ square complex matrix $\mathbf{A}$. Upon using the method of induction, we will show that the matrix $\mathbf{A}$ is triangularizable.

If $n = 1$, the matrix $\mathbf{A}$ is then upper triangular.

We shall employ the induction method for $n > 1$ and consider that the result is valid for square matrices with $n - 1$ rows.

As we know, a matrix $\mathbf{A}$ has at least one eigenvalue λ in the field of the complex number, having an associated eigenvector X. Since $X \neq 0$, it is possible to adjoin the vectors to X to produce a basis of C^n, i.e., $X = X_1, X_2, \cdots, X_n$.

Next, upon using the left multiplication of the matrix's vectors, it gives rise to a linear operator L on C^n. Now concerning the basis$\{X_1, X_2, \cdots, X_n\}$, the linear operator L will be represented by a matrix with the particular form as $B_1 = \begin{bmatrix} \lambda & \mathbf{A}_2 \\ 0 & \mathbf{A}_1 \end{bmatrix}$, where $\mathbf{A}_1$ and $\mathbf{A}_2$ are the complex matrices, $\mathbf{A}_1$ having $n - 1$ rows and columns. The logic for the specific form is that $L(X_1) = AX_1 = \lambda X_1$, since X_1 is an eigenvector of $\mathbf{A}$.

Here, the matrices $\mathbf{A}$ and B_1 are similar since they represent the same linear operator L; suppose that, indeed $B_1 = C_1{}^{-1}AC_1$, where C_1 is an invertible $n \times n$ matrix.

Upon using the induction axiom, there is an invertible matrix C_2 with $n - 1$ rows and columns, such that $B_2 = C_2{}^{-1}\mathbf{A}_1C_2$ is an upper triangular.

Write $C = C_1 \begin{bmatrix} 1 & 0 \\ 0 & C_2 \end{bmatrix}$.

Since the matrix C is the product of invertible matrices, C is invertible. Upon using the matrix computation, it shows that $C^{-1}AC$ equals

$$\begin{bmatrix} 1 & 0 \\ 0 & C_2^{-1} \end{bmatrix} (C_1{}^{-1}AC_1) \begin{bmatrix} 1 & 0 \\ 0 & C_1 \end{bmatrix} = \begin{bmatrix} 1 & 0 \\ 0 & C_2^{-1} \end{bmatrix} B_1 \begin{bmatrix} 1 & 0 \\ 0 & C_2 \end{bmatrix}.$$

Upon replacing the matrix B_1 by $\begin{bmatrix} \lambda & \mathbf{A}_2 \\ 0 & \mathbf{A}_1 \end{bmatrix}$ and multiplying the matrices together, we get

$$C^{-1}AC = \begin{bmatrix} \lambda & \mathbf{A}_2C_2 \\ 0 & C_2{}^{-1}\mathbf{A}_1C_2 \end{bmatrix} = \begin{bmatrix} \lambda & \mathbf{A}_2C_2 \\ 0 & B_2 \end{bmatrix}.$$

This shows that this matrix is upper triangular. Hence, the theorem is proved.

We can use the proof of the theorem as a procedure for triangularizing a matrix.

Example 10.1:

Triangularize the 2×2 matrix $\mathbf{A} = \begin{bmatrix} 1 & 1 \\ -1 & 3 \end{bmatrix}$.

Solution:

It can be observed that the characteristic equation of the matrix $\mathbf{A}$ is $|\mathbf{A} - \lambda I| = \lambda^2 - 4\lambda + 4 = 0$, which gives two eigenvalues λ_1, λ_2 and both the eigenvalues λ_1 and λ_2 are equal, i.e., $\lambda_1 = \lambda_2 = 2$.

For $\lambda = 2$, and upon solving $(\mathbf{A} - 2I_2)\,X = 0$, we find all the eigenvectors of $\mathbf{A}$ that are scalar multiples of $X_1 = \begin{bmatrix} 1 \\ 1 \end{bmatrix}$.

Thus, the matrix $\mathbf{A}$ is not *diagonalizable* since the eigenvalues are not distinct.

Let L be the linear operator defined on R^2 arising from left multiplication by $\mathbf{A}$.

Adjoin a vector X_2 to X_1 to get a basis

$\mathrm{B}_2 = \{X_1, X_2\}$ of R^2, say, $X_2 = \begin{bmatrix} 0 \\ 1 \end{bmatrix}$.

Denote by B_1 the standard basis of R^2. Then the change of basis $\mathrm{B}_1 \to \mathrm{B}_2$ is defined by the matrix

$$C_1 = \begin{bmatrix} 1 & 0 \\ -1 & 1 \end{bmatrix}.$$

Therefore, the matrix $\mathbf{A}$ that represents L concerning the basis B_2 is

$$C_1 A {C_1}^{-1} = \begin{bmatrix} 2 & 1 \\ 0 & 2 \end{bmatrix}.$$

Hence, $C = {C_1}^{-1} = \begin{bmatrix} 1 & 0 \\ 1 & 1 \end{bmatrix}$ triangularizes the matrix $\mathbf{A}$.

Jordan block:

Consider the following two r-square matrices, where $a \neq 0$:

$$J(\mu : r) = \begin{bmatrix} \mu & 1 & 0 & \vdots & 0 & 0 \\ 0 & \mu & 1 & \vdots & 0 & 0 \\ . & \cdots & . & \cdots & . & . \\ 0 & 0 & 0 & \vdots & \mu & 1 \\ 0 & 0 & 0 & \vdots & 0 & \mu \end{bmatrix} \quad \text{and}$$

$$\mathbf{A} = \begin{bmatrix} \mu & a & 0 & \vdots & 0 & 0 \\ 0 & \mu & a & \vdots & 0 & 0 \\ . & \cdots & . & \cdots & . & . \\ 0 & 0 & 0 & \vdots & \mu & a \\ 0 & 0 & 0 & \vdots & 0 & \mu \end{bmatrix}.$$

The matrix $J(\mu : r)$ is called a *Jordan block* having $\mu's$ on the diagonal, 1's on the super diagonal, and 0's elsewhere.

It can be shown that $\Delta(\lambda) = (\lambda - \mu)^r$ is both the *characteristic and minimal polynomial* of both the matrices $J(\mu : r)$ and $\mathbf{A}$.

Let us consider an arbitrary *monic polynomial*

$$f(\lambda) = \lambda^n + a_{n-1}\lambda^{n-1} + \cdots + a_1\lambda + a_0.$$

Let us consider an n-square matrix $C(f)$ with 1's on the sub-diagonal, the negative of the coefficients in the last column, and 0's elsewhere as follows:

$$C(f) = \begin{bmatrix} 0 & 0 & \vdots & 0 & -a_0 \\ 1 & 0 & \vdots & 0 & -a_1 \\ 0 & 1 & \vdots & 0 & -a_2 \\ . & \cdots & . & \cdots & . \\ 0 & 0 & . & 1 & -a_{n-1} \end{bmatrix}.$$

Then $C(f)$ is called the *companion matrix* of the polynomial $f(\lambda)$.

Moreover, the *minimal polynomial* $m(\lambda)$ and *characteristic polynomial* $\Delta(\lambda)$ of the *companion matrix* $C(f)$ are equal to the original polynomial $f(\lambda)$.

10.2 Block Triangular Matrices

Let us consider a block triangular matrix M, i.e., $M = \begin{bmatrix} \mathbf{A}_1 & B \\ 0 & \mathbf{A}_2 \end{bmatrix}$, where the diagonal elements $\mathbf{A}_1$ and $\mathbf{A}_2$ are square matrices.

Then the characteristic polynomials $M - \lambda I$ is also a block triangular matrix, where $\mathbf{A}_1 - \lambda I$ and $\mathbf{A}_2 - \lambda I$ are the diagonal blocks of $M - \lambda I$.

Thus,

$$\begin{aligned} |M - \lambda I| &= \begin{bmatrix} \mathbf{A}_1 - \lambda I & B \\ 0 & \mathbf{A}_2 - \lambda I \end{bmatrix}. \\ &= |\mathbf{A}_1 - \lambda I| \,.\, |\mathbf{A}_2 - \lambda I| \end{aligned}$$

This implies that the characteristic polynomial of the block triangular matrix M is the characteristic polynomial of the diagonal blocks $\mathbf{A}_1$ and $\mathbf{A}_2$.

Theorem 10.2:

Let M be a block triangular matrix with diagonal blocks $\mathbf{A}_1, \mathbf{A}_2, \cdots, \mathbf{A}_r$. Then the characteristic polynomial of the block triangular matrix M is the product of the characteristic polynomial of the diagonal blocks $\mathbf{A}_i^{\prime}s$, that is,

$$\Delta_m(\lambda) = \Delta_{\mathbf{A}_1}(\lambda)\,\Delta_{\mathbf{A}_2}(\lambda)\cdots\Delta_{\mathbf{A}_r}(\lambda).$$

Example 10.2:

Consider the block triangular matrix

$$M = \begin{bmatrix} 9 & -1 & \vdots & 5 & 7 \\ 8 & 3 & \vdots & 2 & -4 \\ \cdot & \cdot & \cdots & \cdot & \cdot \\ 0 & 0 & \vdots & 3 & 6 \\ 0 & 0 & \vdots & -1 & 8 \end{bmatrix}$$

having diagonal blocks

$$\mathbf{A}_1 = \begin{bmatrix} 9 & -1 \\ 8 & 3 \end{bmatrix} \text{ and } \mathbf{A}_2 = \begin{bmatrix} 3 & 6 \\ -1 & 8 \end{bmatrix}.$$

Here, $\operatorname{tr}(\mathbf{A}_1) = 9 + 3 = 12$, $|\mathbf{A}_1| = 27 + 8 = 35$ and so

$$\begin{aligned} \Delta_{\mathbf{A}_1}(\lambda) &= \lambda^2 - 12\lambda + 35 \\ &= (\lambda - 5)(\lambda - 7) \end{aligned}.$$

Similarly, $\operatorname{tr}(\mathbf{A}_2) = 3 + 8 = 11$ and $|\mathbf{A}_2| = 24 + 6 = 30$

and

$$\begin{aligned} \Delta_{\mathbf{A}_2}(t) &= \lambda^2 - 11\lambda + 30 \\ &= (\lambda - 5)(\lambda - 6) \end{aligned}.$$

Accordingly, the characteristic polynomial of the block triangular matrix M is the product

$$\begin{aligned} \Delta_m(\lambda) &= \Delta_{\mathbf{A}_1}(\lambda)\,\Delta_{\mathbf{A}_2}(\lambda) \\ &= (\lambda - 5)^2(\lambda - 6)(\lambda - 7) \end{aligned}.$$

10.3 Block Diagonalization

Theorem 10.3:

Let M be a *block diagonal matrix* with diagonal blocks $\mathbf{A}_1, \mathbf{A}_2, \cdots, \mathbf{A}_r$. Then the minimal polynomial of the *block diagonal matrix* M corresponds to the least common multiple (LCM) of the *minimal polynomial* of the diagonal blocks $\mathbf{A}_i$.

Remark: It can be noted that this theorem can be applied to *block diagonal matrix* M although the previous related theorem on characteristic polynomial can be applied to *block triangular matrices.*

Example 10.3:

Determine the characteristic polynomial $\Delta(\lambda)$ and the minimal polynomial $m(\lambda)$ of the *block diagonal matrix*

$$M = \begin{bmatrix} 2 & 5 & \vdots & 0 & 0 & \vdots & 0 \\ 0 & 2 & \vdots & 0 & 0 & \vdots & 0 \\ \cdots & . & \vdots & . & \cdots & \vdots & . \\ 0 & 0 & \vdots & 4 & 2 & \vdots & 0 \\ 0 & 0 & \vdots & 3 & 5 & \vdots & 0 \\ \cdots & . & \vdots & . & \cdots & \vdots & . \\ 0 & 0 & \vdots & 0 & 0 & \vdots & 7 \end{bmatrix} = diag\left(\mathbf{A}_1, \mathbf{A}_2, \mathbf{A}_3\right)$$

where $\mathbf{A}_1 = \begin{bmatrix} 2 & 5 \\ 0 & 2 \end{bmatrix}$, $\mathbf{A}_2 = \begin{bmatrix} 4 & 2 \\ 3 & 5 \end{bmatrix}$, and $\mathbf{A}_3 = [7]$.

Then the characteristic polynomials $\Delta(\lambda)$ can be expressed as the product of $\Delta_1(\lambda)$,$\Delta_2(\lambda)$, and $\Delta_3(\lambda)$, where $\Delta_1(\lambda)$, $\Delta_2(\lambda)$, and $\Delta_3(\lambda)$ are the characteristic polynomials $\mathbf{A}_1, \mathbf{A}_2,$ and $\mathbf{A}_3$.

It can be shown that

$$\begin{array}{l} \Delta_1(\lambda) = (\lambda - 2)^2 \\ \Delta_2(\lambda) = (\lambda - 2)(\lambda - 7) \\ \Delta_3(\lambda) = (\lambda - 7) \end{array} .$$

Thus, $\Delta(\lambda) = (\lambda - 2)^2 (\lambda - 7)^2$ (as expected, $\deg \Delta(\lambda) = 5$).

Similarly, the minimal polynomial $m_1(\lambda)$, $m_2(\lambda)$, and $m_3(\lambda)$ of the diagonal blocks $\mathbf{A}_1, \mathbf{A}_2,$ and $\mathbf{A}_3$ respectively correspond to the characteristic polynomials $\Delta_1(\lambda)$,$\Delta_2(\lambda)$, and $\Delta_3(\lambda)$, that is,

$$\begin{aligned} m_1(\lambda) &= (\lambda - 2)^2 \\ m_2(\lambda) &= (\lambda - 2)(\lambda - 7) \\ m_3(\lambda) &= (\lambda - 7) \end{aligned} .$$

But $m(\lambda)$ is equal to the *least common multiple* of the minimal polynomial $m_1(\lambda)$, $m_2(\lambda)$, and $m_3(\lambda)$.

Thus, the minimal polynomial $m(\lambda)$ of M can be written as

$$m(\lambda) = (\lambda - 2)^2 (\lambda - 7).$$

10.4 Hermitian Matrices

This section will discuss the orthogonality of Hermitian matrices with the help of eigenvalues and eigenvectors by keeping special regard to real symmetric matrices.

Definition 10.1:

An $n \times n$ square complex matrix $\mathbf{A}$ is said to be *Hermitian*, if $\mathbf{A} = \mathbf{A}^*$. Generally, the Hermitian matrices are the complex analogue of real symmetric matrices. The eigenvalues and eigenvectors of such Hermitian matrices have significant properties, which are usually not taken over by complex matrices. There is a noteworthy indication about Hermitian matrices' particular behavior because their eigenvalues are consistently real, while the eigenvectors become orthogonal.

Note 1: If $\mathbf{A}$ and B are $n \times n$ complex matrices, then $(AB)^* = B^*\mathbf{A}^*$.

Theorem 10.4:

Let us consider a *Hermitian* matrix $\mathbf{A}$. Then

(1) The eigenvalues of the *Hermitian* matrix $\mathbf{A}$ are all real.
(2) The eigenvectors of the *Hermitian* matrix $\mathbf{A}$ associated with distinct eigenvalues are orthogonal.

Proof:

Let us consider a *Hermitian* matrix $\mathbf{A}$. Let λ be an eigenvalue of the matrix $\mathbf{A}$ having its associated eigenvector be X such that

$$AX = \lambda X. \tag{10.1}$$

Upon proceeding with the complex transpose on both sides of eqn (10.1) and using the fact for complex matrices and from Note 1, i.e., $(AX)^* = (\lambda X)^*$, we obtain

$$X^*\mathbf{A} = \bar{\lambda}X^*, \text{ since} \mathbf{A} = \mathbf{A}^*. \tag{10.2}$$

Now on multiplying X on both sides of eqn (10.2), we obtain

$$X^*AX = \bar{\lambda}X^*X = \bar{\lambda}||X||^2, \text{ (since} X^*X = ||X||^2).$$

But **A**.

Thus, it implies that the scalar X^*AX corresponds to its complex conjugate, which means X^*AX is real, which indicates $\bar{\lambda}||X||^2$ is real. Now since the lengths are always real, it can be deduced that $\bar{\lambda}$ and hence λ is real.

Hence, part (i) of the theorem is proved.

Next, let us prove part (ii).

Let X_1 and X_2 be two eigenvectors of *Hermitian* matrices **A** associated with distinct eigenvalues λ_1 and λ_2.

Thus, $AX_1 = \lambda_1 X_1$ and $AX_2 = \lambda_2 X_2$,
which implies that

$$X_2{}^*AX_1 = X_2{}^* (\lambda_1 X_1) = \lambda_1 X_2{}^* X_1.$$

Similarly, it can be proved that

$$X_1{}^*AX_2 = \lambda_2 X_1{}^* X_2.$$

However, by Note 1 again,

$$(X_1{}^*AX_2)^* = X_2{}^*\mathbf{A}^*X_1 = X_2{}^*AX_1\,.$$

Therefore,

$$(\lambda_2 X_1{}^* X_2)^* = (X_1{}^*AX_2)^* = \lambda_1 X_2{}^* X_1$$

or $\lambda_2 X_2{}^* X_1 = \lambda_1 X_2{}^* X_1$ (since λ_2 is real).

$$\Rightarrow (\lambda_1 - \lambda_2)\, X_2{}^* X_1 = 0,$$

which implies $X_2{}^* X_1 = 0$, since $\lambda_1 \neq \lambda_2$.

Thus, X_1 and X_2 are orthogonal. □

10.5 Unitary Matrix

Let us consider an $n \times n$ unitary matrix **A** and $\{X_1, X_2, \cdots, X_m\}$ a set of m linearly independent eigenvectors of the matrix **A**, where m is chosen as large as desired.

Upon multiplying X_i by $\frac{1}{||X_i||}$, it produces a unit vector, which implies that each X_i is a unit vector.

By Theorem 10.4, $\{X_1, X_2, \cdots, X_m\}$ forms an orthonormal set. Now, if we write $U = \{X_1, X_2, \cdots, X_m\}$ as an $n \times m$ matrix, then the matrix U has the property

$$\begin{aligned} AU &= \mathbf{A}\left[X_1, X_2, \cdots, X_m\right] \\ &= \left[AX_1, AX_2, \cdots, AX_m\right] \\ &= \left[\lambda_1 X_1, \lambda_2 X_2, \cdots, \lambda_m X_m\right] \end{aligned},$$

where $\lambda_1, \lambda_2, \cdots, \lambda_m$ are the eigenvalues corresponding to the eigenvectors $X_1, X_2, \cdots, X_m$.

Hence,

$$\begin{aligned} AU &= \mathbf{A}\left[X_1, X_2, \cdots, X_m\right] \\ &= \left[AX_1, AX_2, \cdots, AX_m\right] \\ &= \left[\lambda_1 X_1, \lambda_2 X_2, \cdots, \lambda_m X_m\right] \\ &= \left[X_1, X_2, \cdots, X_m\right] \begin{bmatrix} \lambda_1 & 0 & \cdots & 0 \\ 0 & \lambda_2 & \cdots & 0 \\ \cdots & \cdots & \ddots & \cdots \\ 0 & 0 & \cdots & \lambda_m \end{bmatrix} \\ &= UD \end{aligned}.$$

Here, the matrix D is a diagonal matrix with its diagonal entries as $\lambda_1, \lambda_2, \cdots, \lambda_r$.

The columns of the unitary matrix U model is an orthonormal set, i.e., $U * U = I_r$.

So, for an $n \times n$ unitary matrix U, we shall have $U^{-1} = U^*$. (Primarily $r \leq n$, but here the case is for $r = n$.)

Thus, $U^* AU = D$ and hence the matrix $\mathbf{A}$ is diagonalized by the unitary matrix U. □

Remark: A matrix $\mathbf{A}$ can be diagonalized by a unitary matrix U, if there exist n mutually orthogonal eigenvectors of $\mathbf{A}$.

There is always a question in mind whether there are always many linearly independent eigenvectors for a matrix $\mathbf{A}$, which we will discuss in detail in the next section.

10.6 Schur's Theorem

Theorem 10.5 (Schur's theorem):

Let us consider an arbitrary $n \times n$ square complex matrix $\mathbf{A}$. Then there exists a unitary matrix U that $U^* AU$ is upper triangular.

Moreover, if a real symmetric matrix $\mathbf{A}$ of order n exists, then the unitary matrix U can be chosen as a real and orthogonal matrix.

Proof:

Let us consider an $n \times n$ square matrix $\mathbf{A}$.

The proof of the theorem can be obtained by the method of induction for n.

Indeed, if $n = 1$, then the matrix $\mathbf{A}$ is an upper triangular matrix.

Let us consider the case for $n > 1$. For $n > 1$, let λ_1 be an eigenvalue of the matrix $\mathbf{A}$ having the associated eigenvector as X_1, where the vector X_1 can be chosen to be a unit vector in C^n.

Now since X_1 is a linearly independent subset of R^n, we can adjoin vectors to the unit vector X_1 to form a basis of R^n. Then an orthonormal basis $X_1, X_2, \cdots, X_n$ of R^n is obtained by applying the Gram–Schmidt procedure, where X_1 will be on this basis.

Let U_0 denote the matrix $[X_1, X_2, \cdots, X_n]$; then U_0 is unitary since its column vectors form an orthonormal set.

Now $U_0^* AX_1 = U_0^* (\lambda_1 X_1) = \lambda_1 (U_0^* X_1)$.

Also $\begin{cases} X_i^* X_1 = 0 & \text{, if } i > 1 \\ \quad\quad = 1 & \text{, if } i = 1 \end{cases}$

Hence,

$$U_0^* AX_1 = \lambda_1 \begin{bmatrix} X_1^* X_1 \\ X_2^* X_2 \\ \vdots \\ X_n^* X_1 \end{bmatrix} = \begin{bmatrix} \lambda_1 \\ 0 \\ \vdots \\ 0 \end{bmatrix}.$$

Now since

$$\begin{aligned} U_0^* AU_0 &= U_0^* \mathbf{A} [X_1, X_2, \cdots, X_n] \\ &= [U_0^* AX_1, U_0^* AX_2, \cdots, U_0^* AX_n] \end{aligned}.$$

We analyze that

$$U_0^* AU_0 = \begin{bmatrix} \lambda_1 & B \\ 0 & \mathbf{A}_1 \end{bmatrix},$$

where the matrix $\mathbf{A}_1$ is a having $n - 1$ rows and columns, and the matrix B is an $(n - 1)$ row vector.

Similarly, upon applying the induction hypothesis on n, there exists a unitary matrix U_1 that $U_1^* \mathbf{A}_1 U_1 = T_1$ is upper triangular that results U_2, which is a unitary matrix.

Now write

$U_2 = \begin{bmatrix} 1 & 0 \\ 0 & U_1 \end{bmatrix}$, which is a unitary matrix.

Then put $U = U_0 U_2$. This is also unitary, since $U^* U = I$, i.e.,

$$\begin{aligned} U^* U &= (U_0 U_2)^* U_0 U_2 \\ &= U_2{}^* U_0{}^* U_0 U_2 \\ &= U_2^* (U_0^* U_0) U_2 = U_2^* U_2 = I \end{aligned}.$$

Finally,

$$\begin{aligned} U^* AU &= (U_0 U_2)^* \mathbf{A} (U_0 U_2) \\ &= U_2{}^* U_0{}^* A U_0 U_2 \\ &= U_2^* (U_0^* A U_0) U_2 \\ &= U_2^* \begin{bmatrix} \lambda_1 & B \\ 0 & \mathbf{A}_1 \end{bmatrix} U_2 = \begin{bmatrix} \lambda_1 & BU_1 \\ 0 & U_1^* \mathbf{A}_1 U_1 \end{bmatrix} \end{aligned},$$

which show that

$$U^* AU = \begin{bmatrix} \lambda_1 & BU_1 \\ 0 & T_1 \end{bmatrix},$$

which is an upper triangular matrix as needed. □

If $\mathbf{A}$ is a real and symmetric matrix, then there exists a real orthogonal matrix Q such that $Q^T AQ$ diagonal. Since here the eigenvalues of the matrix $\mathbf{A}$ are real, it implies that the matrix $\mathbf{A}$ has a real eigenvector.

Next, we will establish the spectral theorem based on the diagonalization of Hermitian matrices.

10.7 Spectral Theorem

Theorem 10.6 (spectral theorem):

Let $\mathbf{A}$ be a Hermitian matrix. Then there exists a unitary matrix U that $U^* AU$ is diagonal. If the matrix $\mathbf{A}$ is real and symmetric, then the unitary matrix U can be real and orthogonal.

Proof:

Let $\mathbf{A}$ be a Hermitian matrix, then by Theorem 10.5, there is a unitary matrix U such that $U^* AU = T$ is an upper triangular matrix.

Then $T^* = U^* \mathbf{A}^* U = U^* AU = T$. It implies that T is Hermitian.

But T is an upper triangular matrix and T^* is lower triangular. So the only way that the matrices T and T^* can be equal, if all the off-diagonal entries of the matrix T are zero, is if T is diagonal.

Thus, since $U^* AU = T$, $U^* AU$ is diagonal.

Similarly, the proof can be done for the case when the matrix $\mathbf{A}$ is real and symmetric.

Corollary 10.1:

Let us consider an $n \times n$ Hermitian matrix **A**. Then there exists an orthonormal basis of R^n that consists entirely of eigenvectors of the Hermitian matrix **A**. If the Hermitian matrix **A** is real, then there is an orthonormal basis of R^n consisting of all the eigenvectors **A**.

Proof:

Let **A** be a $n \times n$ Hermitian matrix. Then by Theorem 10.6, a unitary matrix U exists such that $U^*AU = D$ is diagonal. Here the diagonal entries of D are $d_{11}, d_{22}, \cdots, d_{nn}$ say.

If $X_1, X_2, \cdots, X_n$ are the columns of the unitary matrix U, then the equation $AU = UD$ implies that $AX_i = d_{ii}X_i$ for $i = 1, 2, \cdots, n$.

Thus, X_i are the eigenvectors of the Hermitian matrix, and as U is unitary, they form an orthonormal basis of R^n. □

The same altercation can be taken if the Hermitian matrix **A** is real.

Remark: An $n \times n$ Hermitian matrix always has ample eigenvectors to form an orthonormal basis of R^n.

Note: This case will also be applicable if all the eigenvalues of the matrix **A** are not distinct.

The following statement discusses an efficient method of diagonalizing an $n \times n$ Hermitian matrix **A** using a unitary matrix.

Procedure: For any eigenvalue λ, first, we find a basis for the analogous eigenspace. Then we apply the *Gram–Schmidt procedure* to obtain an orthonormal basis of each eigenspace. These bases are then united to form an orthonormal set $\{X_1, X_2, \cdots, X_n\}$. Thus, by Corollary 10.1, $\{X_1, X_2, \cdots, X_n\}$ will form a basis of R^n.

Similarly, if the unitary matrix U has column vectors $X_1, X_2, \cdots, X_n$, then the matrix U is Hermitian and U^*AU is diagonal, as shown in Theorem 10.5.

The same procedure can also be effective for the case of real symmetric matrices.

Example 10.4:

Determine a real orthogonal matrix that diagonalizes the real symmetric matrix

$$\mathbf{A} = \begin{bmatrix} 1 & 2 \\ 2 & 1 \end{bmatrix}.$$

Solution:

The characteristic equation of the matrix $\mathbf{A}$ is $|\mathbf{A} - \lambda I| = 0$, which gives the eigenvalues of the matrix $\mathbf{A}$ as 3 and -1. Hence, the corresponding eigenvectors are $\begin{bmatrix} 1 \\ 1 \end{bmatrix}$ and $\begin{bmatrix} -1 \\ 1 \end{bmatrix}$, respectively.

Since the eigenvectors $\begin{bmatrix} 1 \\ 1 \end{bmatrix}$ and $\begin{bmatrix} -1 \\ 1 \end{bmatrix}$ of $\mathbf{A}$ are orthogonal, an orthonormal basis of R^2 can be obtained by replacing it with the unit eigenvectors $\frac{1}{\sqrt{2}}\begin{bmatrix} 1 \\ 1 \end{bmatrix}$ and $\frac{1}{\sqrt{2}}\begin{bmatrix} -1 \\ 1 \end{bmatrix}$, respectively.

Thus, an orthogonal matrix $Q = \frac{1}{\sqrt{2}}\begin{bmatrix} 1 & -1 \\ 1 & 1 \end{bmatrix}$ can be obtained from which it predicts that $Q^T AQ = \begin{bmatrix} 3 & 0 \\ 0 & -1 \end{bmatrix}$.

(It can be easily verified by matrix multiplication.)

Hence, the matrix $\mathbf{A}$ is diagonalizable. □

Example 10.5:

Determine a unitary matrix U that diagonalizes the Hermitian matrix

$$\mathbf{A} = \begin{bmatrix} \frac{3}{2} & \frac{i}{2} & 0 \\ -\frac{i}{2} & \frac{3}{2} & 0 \\ 0 & 0 & 1 \end{bmatrix}, \text{ where } i = \sqrt{-1}.$$

Solution:

The characteristic equation of the matrix $\mathbf{A}$ is $|\mathbf{A} - \lambda I| = 0$, which gives the eigenvalues of the matrix $\mathbf{A}$ as $1, 2, 1$, and the associated unit eigenvectors of $\mathbf{A}$ can be found to be

$$\begin{bmatrix} -\frac{i}{\sqrt{2}} \\ \frac{1}{\sqrt{2}} \\ 0 \end{bmatrix}, \begin{bmatrix} \frac{1}{\sqrt{2}} \\ -\frac{i}{\sqrt{2}} \\ 0 \end{bmatrix}, \begin{bmatrix} 0 \\ 0 \\ 1 \end{bmatrix}.$$

Upon using the associated unit eigenvectors, a unitary matrix U can be obtained as

$$U = \frac{1}{\sqrt{2}}\begin{bmatrix} -i & 1 & 0 \\ 1 & -i & 0 \\ 0 & 0 & \sqrt{2} \end{bmatrix}.$$

Therefore,

$$U^*AU = \begin{bmatrix} 1 & 0 & 0 \\ 0 & 2 & 0 \\ 0 & 0 & 1 \end{bmatrix}.$$

Hence, the matrix **A** is diagonalizable. □

10.8 Normal Matrices

Every $n \times n$ Hermitian matrix **A** has the characteristic that there is an orthonormal basis of R^n consisting of the eigenvectors of the Hermitian matrix **A**. It is also noticed that this property directly leads to the matrix **A** being diagonalized by a unitary matrix U, namely the matrix whose columns are the vectors of the orthonormal basis.

Now we shall analyze what other matrices have this applicable property.

A square matrix **A** over R^n is said to be *normal*, if it commutes with its complex transpose

i.e., $\mathbf{A}^*\mathbf{A} = \mathbf{A}\mathbf{A}^*$.

Indeed, for a real matrix, this asserts that the square matrix **A** commutes with its transpose $\mathbf{A}^T$. Hermitian matrices are real and normal, for if $\mathbf{A} = \mathbf{A}^*$, then indeed, the matrix **A** commutes with the matrix $\mathbf{A}^*$.

The following section will find the conjunction between the normal matrices and discuss eigenvectors' existence that forms an orthonormal basis.

Theorem 10.7:

Let us consider an $n \times n$ complex matrix **A**. Then the matrix **A** is normal if and only if there is an orthonormal basis of R^n consisting of the eigenvectors of the matrix **A**.

Proof:

Let us consider that the eigenvectors of the matrix **A** form an orthonormal basis of R^n. Then by *spectral theorem*, there exists a unitary matrix U such that $U^*AU = D$ is diagonal.

This gives

$$\mathbf{A} = UDU^* (\because U^* = U^{-1}).$$

Next, we show that the matrix **A** commutes with its complex transpose through a direct computation, i.e.,

$$\begin{aligned} \mathbf{A}\mathbf{A}^* &= (UDU^*)\,(UDU^*)^* \\ &= UDU^*UD^*U^* \end{aligned}$$

$$
\begin{aligned}
&= UD\left(U^*U\right)D^*U^* \\
&= UDD^*U^*
\end{aligned}
$$

and in a similar manner

$$
\begin{aligned}
\mathbf{A}^*\mathbf{A} &= \left(UDU^*\right)^* UDU^* \\
&= UD^*U^*UDU^* \\
&= UD^*\left(U^*U\right)DU^* \\
&= UD^*DU^*
\end{aligned}
.
$$

But as we know that the diagonal matrices ever commute, so $DD^* = D^*D$.

It gives $\mathbf{A}^*\mathbf{A} = \mathbf{A}\mathbf{A}^*$. Thus, the matrix $\mathbf{A}$ is normal.

Conversely, if the matrix $\mathbf{A}$ is *normal*, it shows that there is an orthonormal basis consisting entirely of the matrix's eigenvectors $\mathbf{A}$.

If $\mathbf{A}$ is an $n \times n$ complex matrix, then from *Theorem 10.5 (Schur's theorem)*, we know a *unitary matrix* U that $U^*AU = T$ is upper triangular.

The following observation is to show that the upper triangular matrix T is also normal.

It can also be established by a direct computation, i.e.,

$$
\begin{aligned}
T^*T &= \left(U^*AU\right)^*\left(U^*AU\right) \\
&= U^*\mathbf{A}^*UU^*AU \\
&= U^*\mathbf{A}^*\left(UU\right)^* AU \\
&= U^*\left(\mathbf{A}^*\mathbf{A}\right)U
\end{aligned}
.
$$

Similarly, we can write $TT^* = U^*\left(\mathbf{A}\mathbf{A}^*\right)U$.

Now since the matrix $\mathbf{A}$ is normal, i.e., $\mathbf{A}^*\mathbf{A} = \mathbf{A}\mathbf{A}^*$. Thus, it follows that $T^*T = TT^*$.

Now let us equate the $(1,\ 1)$ entries of T^*T and TT^*, which yields the equation

$$
\left|t_{11}\right|^2 = \left|t_{11}\right|^2 + \left|t_{12}\right|^2 + \cdots + \left|t_{1n}\right|^2,
$$

which implies that $t_{12}, \cdots, t_{1n}$ are all zero.

By considering the $(2,2), (3,3), \cdots, (n,n)$ entries of T^*T and TT^*, we notice that all the other off-diagonal entries of the matrix T vanish. Thus, the upper triangular matrix T is diagonal.

Finally, as $AU = UT$, the columns of the unitary matrix U are the eigenvectors of the matrix $\mathbf{A}$, and they form an orthonormal basis of R^n where U is unitary.

Hence proved. □

Note: Complex unitary matrices or Hermitian are necessarily normal, as are real symmetric and real orthogonal matrices. A unitary matrix can therefore diagonalize any matrix of these types.

10.9 Nilpotent Operators

A linear operator $T : V \to V$ is said to be *nilpotent* if $T^n = 0_v$ for a positive integer n. If $T^k = 0_v$ but $T^{k-1} \neq 0_v$, then the *index of nilpotency* of the linear operator T is k.

Analogously: A square matrix $\mathbf{A}$ is said to be *nilpotent of an index* n, if $\mathbf{A}^n = 0$ for some positive integer n.

If $\mathbf{A}^k = 0$ but $\mathbf{A}^{k-1} \neq 0$, then $\mathbf{A}$ is said to be *nilpotent of an index* k.

Similarly, if we find the minimum polynomial of a nilpotent operator T (or matrix $\mathbf{A}$), then it is of an index k, i.e., $m(\lambda) = \lambda^k$ and 0 is its only eigenvalue.

Example 10.6:

Consider the following two r-square matrices:

$$N = N(r) = \begin{bmatrix} 0 & 1 & 0 & \dots & 0 & 0 \\ 0 & 0 & 1 & \dots & 0 & 0 \\ . & \dots & . & \dots & . & \dots \\ 0 & 0 & 0 & \dots & 0 & 1 \\ 0 & 0 & 0 & \dots & 0 & 0 \end{bmatrix} \text{ and}$$

$$J(\mu) = \begin{bmatrix} \mu & 1 & 0 & \dots & 0 & 0 \\ 0 & \mu & 1 & \dots & 0 & 0 \\ . & \dots & . & \dots & . & \dots \\ 0 & 0 & 0 & \dots & \mu & 1 \\ 0 & 0 & 0 & \dots & 0 & \mu \end{bmatrix}.$$

The first matrix $N = N(r)$ is called a *Jordan nilpotent block* that consists of 1's above the diagonal (called the *super diagonal*) and 0's elsewhere. Thus, $N = N(r)$ is said to be a *nilpotent matrix* of an index r.

Similarly, the second matrix $J(\mu)$ is called a *Jordan block* belonging to the eigenvalue λ that consists of$\mu's$ on the diagonal, 1's on the super diagonal, and 0's elsewhere.

It can be observed that $J(\lambda) = \mu I + N$.

Next, we will prove that any linear operator T can be decomposed into the sum of a scalar operator and a nilpotent operator. □

Theorem 10.8:

Let us consider a nilpotent operator $T : V \to V$ of an index k. Then the nilpotent operator T has a block diagonal matrix representation. Each diagonal entry is Jordan nilpotent block N, and there is at least one N of order k, and all others are of order $\leq k$. The nilpotent operator can uniquely determine the number N of each possible order T. The total number N of all orders is equal to the *nullity* of T.

The validation of the theorem demonstrates that the number of N order i is equal to $2m_i - m_{i+1} - m_{i-1}$, where m_i is the nullity of T^i.

10.10 Jordan Canonical Form

A linear operator $T : V \to V$ can be set in Jordan canonical form if its characteristic and minimal polynomial can be factored into products of linear polynomials. This is always true when the base field k is a complex C.

The base field k can always be extended to a field where the characteristic and the minimal polynomial do factor into linear factors.

The following theorem characterizes the Jordan canonical form J of a linear operator T.

Theorem 10.9:

Let $T : V \to V$ be a linear operator. The characteristic and the minimal polynomials of T are

$\Delta(\lambda) = (\lambda - \mu_1)^{n_1}(\lambda - \mu_2)^{n_2} \cdots (\lambda - \mu_r)^{n_r}$ and $m(\lambda) = (\lambda - \mu_1)^{m_1}(\lambda - \mu_2)^{m_2} \cdots (\lambda - \mu_r)^{m_r}$, respectively.

Here, the $\mu_i' s$ are different scalars. Then the operator T has a block diagonal matrix description J in which every diagonal entry is a Jordan block $J_{ij} = J(\mu_i)$.

For every μ_{ij}, the analogous J_{ij} has the subsequent properties.

There is at least one J_{ij} of order m_i; all other J_{ij} are of order $\leq m_i$.

The sum of the orders of J_{ij} is n_i.

The number of J_{ij} is equal to the geometric multiplicity of λ_i.

The number J_{ij} of each possible order is uniquely determined by T.

Example 10.7:

Let us consider that the characteristic and minimal polynomials of a linear operator T are respectively

$\Delta(\lambda) = (\lambda - 2)^4 (\lambda - 5)^3$ and $m(\lambda) = (\lambda - 2)^2 (\lambda - 5)^3$.

Then the Jordan canonical form of T is one of the following block diagonal matrices, i.e.,

$$diag\left(\begin{bmatrix} 2 & 1 \\ 0 & 2 \end{bmatrix}, \begin{bmatrix} 2 & 1 \\ 0 & 2 \end{bmatrix}, \begin{bmatrix} 5 & 1 & 0 \\ 0 & 5 & 1 \\ 0 & 0 & 5 \end{bmatrix}\right) \text{ or}$$

$$diag\left(\begin{bmatrix} 2 & 1 \\ 0 & 2 \end{bmatrix}, [2], [2], \begin{bmatrix} 5 & 1 & 0 \\ 0 & 5 & 1 \\ 0 & 0 & 5 \end{bmatrix}\right).$$

The first matrix occurs when the operator T has two independent eigenvectors that belong to the eigenvalue 2. In contrast, the second matrix appears when the operator T has three independent eigenvectors belonging to the eigenvalue 2.

10.11 Rational Canonical Form

Rational canonical form exists when the minimal polynomials cannot be factored into linear polynomials.

(**Note:** This is not the case for the Jordan canonical form.)

Theorem 10.10:

Let $T : V \to V$ be a linear operator including the minimal polynomial

$$m(\lambda) = (f_1(\lambda))^{m_1}, (f_2(\lambda))^{m_2}, \cdots, (f_s(\lambda))^{m_s},$$

where $f_i(\lambda)$ are discrete monic irreducible polynomials. Then the operator T has a unique block diagonal matrix description

$$M = diag(c_{11}, c_{12}, \cdots, c_{1r}, \cdots c_{s1}, c_{s1}, \cdots c_{sr}),$$

where $c'_{ij}s$ are the companion matrices of the polynomials $(f_i(\lambda))^{n_{ij}}$, and

$$m_1 = n_{11} \geq n_{12} \geq \cdots \geq n_{1r}, \cdots, m_s = n_{s1} \geq n_{s2} \geq \cdots \geq n_{sr}.$$

The above matrix representation T is called its rational canonical form, where the polynomials $(f_i(\lambda))^{n_{ij}}$ are called the elementary divisors T.

Example 10.8:

Let $T : V \to V$ be a vector space over the rational field Q and $\dim V = 8$. Let $m(\lambda)$ be the minimal polynomials of T defined by

$$\begin{aligned} m(\lambda) &= f_1(\lambda) . f_2(\lambda)^2 \\ &= (\lambda^4 - 4\lambda^3 + 6\lambda^2 - 4\lambda - 7)(\lambda - 3)^2. \end{aligned}$$

Since $\dim V = 8$, the characteristic polynomial of T is

$$\Delta(\lambda) = f_1(\lambda) . f_2(\lambda)^4 .$$

The rational canonical form M of the linear operator T is having one block of the companion matrix $f_1(\lambda)$ and the other block of the companion matrix $f_2(\lambda)^2$.

There are two possibilities:

$$\operatorname{diag}\left[c\left(\lambda^4 - 4\lambda^3 + 6\lambda^2 - 4\lambda - 7\right), c(\lambda-3)^2, c(\lambda-3)^2\right]$$
$$\operatorname{diag}\left[c\left(\lambda^4 - 4\lambda^3 + 6\lambda^2 - 4\lambda - 7\right), c(\lambda-3)^2, c(\lambda-3), c(\lambda-3)\right].$$

That is,

$$\text{(a) diag}\left(\begin{bmatrix} 0 & 0 & 0 & 7 \\ 1 & 0 & 0 & 4 \\ 0 & 1 & 0 & -6 \\ 0 & 0 & 1 & 4 \end{bmatrix}, \begin{bmatrix} 0 & -9 \\ 1 & 6 \end{bmatrix}, \begin{bmatrix} 0 & -9 \\ 1 & 6 \end{bmatrix}\right)$$

$$\text{(b) diag}\left(\begin{bmatrix} 0 & 0 & 0 & 7 \\ 1 & 0 & 0 & 4 \\ 0 & 1 & 0 & -6 \\ 0 & 0 & 1 & 4 \end{bmatrix}, \begin{bmatrix} 0 & -9 \\ 1 & 6 \end{bmatrix}, [3], [3]\right).$$

10.12 Minimum Polynomial and Jordan Canonical Form

This section discusses one of the most famous results in linear algebra, known as the *Jordan normal form of a matrix.* That is no less than a canonical form that applies to any squared complex matrix.

Jordan's normal form is usually presented as the climax of a series of difficult theorems; however, the approach adopted here, which is due to Filippov, is quite simple and depends on the only elementary fact about vector spaces.

We begin by establishing the essential concept of the minimum polynomial.

10.12.1 Jordan normal form

This section discusses the Jordan normal form for a square complex matrix. The essential components used in these specific complex matrices are called *Jordan blocks* of the matrix.

In general, an $n \times n$ Jordan block is a matrix of the form

$$J = \begin{bmatrix} \mu & 1 & 0 & 0 & \cdots & 0 & 0 \\ 0 & \mu & 1 & 0 & \cdots & 0 & 0 \\ 0 & 0 & \mu & 1 & \cdots & 0 & 0 \\ \cdots & \cdots & \cdots & \cdots & \cdots & \cdots & \cdots \\ 0 & 0 & 0 & 0 & \cdots & \mu & 1 \\ 0 & 0 & 0 & 0 & \cdots & 0 & \mu \end{bmatrix},$$

for any scalar μ.

Thus, J is an upper triangular $n \times n$ matrix with constant diagonal entries, a super diagonal of 1'*s* and zeros else, whereby the minimum and characteristic polynomial of J are $(\lambda - \mu)^n$ and $(\mu - \lambda)^n$, respectively.

10.12.2 Properties of Jordan matrix

Next, we will discuss some crucial property of the matrix J. Let $X_1, X_2, \cdots, X_n$ be the vectors of the standard basis of R^n. Then the matrix multiplication illustrates that

$JX_1 = \lambda X_1$ and $JX_i = \lambda X_i + X_i - 1$ where $1 < i \leq n$.

If there is any $n \times n$ complex matrix **A**, we shall call a sequence of vectors $X_1, X_2, \cdots, X_n$ in R^n a Jordan string **A**.

If the matrix **A** satisfies the equation $AX_1 = \lambda X_1$ and $AX_i = \lambda X_i + X_i - 1$, where λ is a scalar, and $1 < i \leq r$, then every Jordan block determines a Jordan string of length n.

Let us consider that there is a basis of $\mathbb{C}^n$ that consists of Jordan strings for the matrix **A** and group the basis elements in the same string. Then the linear operator on R^n given by $T(X) = AX$ representing the basis of Jordan strings by a Jordan matrix, which has Jordan blocks down the diagonal as:

$$\mathbf{N} = \begin{bmatrix} J_1 & 0 & \cdots & 0 \\ 0 & J_2 & \cdots & 0 \\ \cdots & \cdots & \cdots & \cdots \\ 0 & 0 & \cdots & J_k \end{bmatrix}.$$

Here J_i is a Jordan block with μ_i'*s* (say) as the diagonal elements.

Thus, we conclude that the matrix **A** is similar to a matrix N, called the *Jordan normal form* of **A** where the diagonal elements λ_i of N are the eigenvalues of **A**.

Next, we can show that every square matrix **A** has a Jordan normal form with the construction of the Jordan string.

Note: Every square complex matrix is similar to a matrix in Jordan normal form.

Remark: Every complex matrix is similar to an upper triangular matrix having zeros above the super diagonal.

Example 10.9:

Find the Jordan normal form of

$$\mathbf{A} = \begin{bmatrix} 3 & 1 & 0 \\ -1 & 1 & 0 \\ 0 & 0 & 2 \end{bmatrix}$$

Solution:

The eigenvalues of the matrix **A** are 2,2, and 2.

So, we can define

$$\mathbf{B} = \mathbf{A} - 2\mathbf{I} = \begin{bmatrix} 1 & 1 & 0 \\ -1 & -1 & 0 \\ 0 & 0 & 0 \end{bmatrix}$$

A single vector $\mathbf{X} = \begin{bmatrix} 1 \\ -1 \\ 0 \end{bmatrix}$ generates the column space C of the matrix B.

Since $AX = 2X$, X is a Jordan string of length 1 for the matrix **A**.

Also, the null space N of the matrix B is developed by X, and the vector is $\begin{bmatrix} 0 \\ 0 \\ 1 \end{bmatrix}$.

Thus, $D = C \bigcap N = C$ is developed by the vector X. The next step is to express X in the form BY, so that we can have $\mathbf{Y} = \begin{bmatrix} 1 \\ 0 \\ 0 \end{bmatrix}$. Therefore, the second basis element is Y.

Eventually, by putting $\mathbf{Z} = \begin{bmatrix} 0 \\ 0 \\ 1 \end{bmatrix}$, the set $\{X, Z\}$ forms a basis for N.

Thus, $BX = 0, BY = X$, and $BZ = 0$ and hence $AX = 2X, AY = 2Y + X$, and $AZ = 2Z$.

It can be now apparent that $\{X, Y, Z\}$ is a basis of $\mathbb{C}^3$ consisting of the two Jordan string X, Y, and Z. Thus, the Jordan form of **A** has two blocks and is

$$\mathbf{N} = \begin{bmatrix} 2 & 1 & \vdots & 0 \\ 0 & 2 & \vdots & 0 \\ \cdots & \cdots & \vdots & \cdots \\ 0 & 0 & \vdots & 2 \end{bmatrix}.$$

Next, we establish a network between a matrix and its transpose, an application of Jordan form.

Theorem 10.11:

Every $n \times n$ square complex matrix is similar to its transpose.

Proof:

Let us consider a square matrix $\mathbf{A}$ with complex entries and write N for the normal form of the matrix $\mathbf{A}$.

Thus, $C^{-1}AC = N$ for some invertible matrix C.

Now $N^T = C^T\mathbf{A}^T(C^T)^{-1}$ implies that N^T is similar to $\mathbf{A}^T$.

For the proof of the theorem, it is sufficient to prove that N and N^T are similar matrices. Here we can use the transitive property of similarity, i.e., if P is similar to Q, and Q is similar to R, then P is similar to R.

Due to the block decomposition of N, it is sufficient to prove that any Jordan block J is similar to its transpose. But it can be observed directly. Indeed, if the permutation matrix P has a line of 1's from top right to bottom left, then the matrix multiplication shows that $P^{-1}JP = J^T$.

Another use of the Jordan form is to check which matrices meet a given polynomial equation.

Example 10.10:

Determine up to similarity all $n \times n$ complex matrices $\mathbf{A}$ that satisfy the equation $\mathbf{A}^2 = I$.

Solution:

Let N be the Jordan normal form of the matrix $\mathbf{A}$ and $N = C^{-1}AC$. Then $N^2 = C^{-1}\mathbf{A}^2C$.

Thus, $\mathbf{A}^2 = I$, if and only if $N^2 = I$.

Since N consists of a string of Jordan blocks down the diagonal, we only have to decide which Jordan block J can satisfy $J^2 = I$. This is quickly done.

Indeed, the diagonal entries of J will have to be 1 or -1. Moreover, the matrix multiplication affirms that $J^2 \neq I$ if J has two or more rows.

Thus, the block J must be 1×1, which implies that N is a diagonal matrix with all its diagonal entries equal to $+1$ or -1.

After reordering the rows and columns, we obtain a matrix of the form

$$\mathbf{N} = \begin{bmatrix} I_r & 0 \\ 0 & -I_s \end{bmatrix}$$

where $r + s = n$.

Therefore, $\mathbf{A}^2 = I$ if and only if the matrix $\mathbf{A}$ is similar to a matrix with the form of N.

Note: We analyse the relationship between Jordan's normal form and the minimum and characteristic polynomials in the outcomes. The knowledge of the Jordan form will enable us to write down the minimum polynomial directly. Using the method of Example 10.10, one can obtain the Jordan form, which provides an organized way of computing the minimum polynomials.

10.13 Minimum Polynomial of Jordan Normal Form N

Let us consider a complex matrix $\mathbf{A}$ whose eigenvalues $\mu_1, \mu_2, \cdots, \mu_r$ are distinct. There is an analogous Jordan block in the Jordan normal form N for each μ_i, which have μ_i on their principal diagonals say $J_{i1}, J_{i2}, \cdots, J_{iL_i}$ and let n_{ij} be the number of rows of J_{ij}.

Since $\mathbf{A}$ and N are similar matrices, of course, they have the same minimum and characteristic polynomials.

Now J_{ij} is an $n_{ij} \times n_{ij}$ upper triangular matrix with μ_i on the principal diagonal; so its characteristic polynomial is $(\mu_i - \lambda)^{n_{ij}}$. Thus, the characteristic polynomial $\Delta(\lambda)$ of N is the product of all of these polynomials, i.e.,

$$\Delta(\lambda) = \prod_{i=1}^{r} (\mu_i - \lambda)^{m_i}, \text{ where } m_i = \sum_{j=1}^{L_i} n_{ij}.$$

Here the minimum polynomial is slightly harder to find. However, if there is any polynomial f, it is readily seen that $f(J_{ij})$ the matrix blocks of the matrix $f(N)$ down the principal diagonal and zeros elsewhere.

Thus, $f(N) = 0$ if and only if all the $f(J_{ij}) = 0$.

Hence, the minimum polynomial of N is the least common multiple of the minimum polynomials of the blocks J_{ij}. But we have seen that the minimum polynomial of the Jordan block J_{ij} is $(\lambda - \mu_i)^{n_{ij}}$.

So it observes that the minimum polynomial of N is $f = \prod_{i=1}^{n} (\lambda - \mu_i)^{k_i}$, where k_i is larger than the n_{ij} for $j = 1, 2, \cdots, L$.

These outcomes extend the method of computing *minimum polynomials* from *Jordan normal form.*

Theorem 10.12:

Let $\mathbf{A}$ be an $n \times n$ complex matrix and let $\lambda_1, \lambda_2, \cdots \lambda_r$ be the discrete eigenvalues of $\mathbf{A}$. Then the characteristic and minimum polynomial of the matrix $\mathbf{A}$ are $\prod_{i=1}^{n} (\mu_i - \lambda)^{m_i} 7$ and $\prod_{i=1}^{n} (\lambda_i - \mu)^{k_i}$ respectively, where m_i is the sum of the number of columns in Jordan blocks with eigenvalue μ_i and k_i being the number of columns in the larger such Jordan block.

Example 10.11:

Determine the *minimum polynomial* of the matrix

$$\mathbf{A} = \begin{bmatrix} 3 & 1 & 0 \\ -1 & 1 & 0 \\ 0 & 0 & 2 \end{bmatrix}.$$

Solution:

The Jordan form of the matrix $\mathbf{A}$ is

$$\mathbf{N} = \begin{bmatrix} 2 & 1 & \vdots & 0 \\ 0 & 2 & \vdots & 0 \\ \cdots & \cdots & \vdots & \cdots \\ 0 & 0 & \vdots & 2 \end{bmatrix}.$$

Here the only eigenvalue is 2, and two Jordan blocks have 2 and 1 columns. The minimum polynomial of $\mathbf{A}$ is, therefore $(\lambda - 2)^2$, and of course, the characteristic polynomial is $(2 - \lambda)^3$.

Exercises

1. If H is orthogonal, prove that $\det H = \pm 1$.
2. If U is unitary, prove that $|\det U| = 1$.
3. If U is unitary, show that $\bar{U}, U^T$, and U^k (k being a positive integer) are also unitary.
4. If H is orthogonal, show that H^T and H^k (a positive integer) are also orthogonal.

5. Show that $\begin{bmatrix} 0 & 1 \\ 1 & 0 \end{bmatrix}, \begin{bmatrix} 0 & -i \\ i & 0 \end{bmatrix} \begin{bmatrix} 1 & 0 \\ 0 & -1 \end{bmatrix}$ are unitary, involutory, as well as Hermitian.
6. Prove that if U is unitary and U^*AU and U^*BU are both diagonal matrices, then $AB = BA$. Is this result true if U is replaced by a real orthogonal matrix H?
7. Let T be the linear operator on V_3 defined by $T(x, y, z) = (x', y', z')$, where $x' = x\cos\phi - y\sin\phi$, $y' = x\sin\phi + y\cos\phi$, $z' = z$, concerning a Cartesian coordinate system. Prove that T is given by an orthogonal matrix.

11

Least Square Problems

This chapter discusses the approximation of functions, least square problems having their vital application in numerical analysis.

The chapter closes with a discussion of the approximation of functions and the use of pseudo inverse to determine least square curves for given data.

11.1 Approximation of Functions

Many problems in the physical sciences and engineering involve approximating a given function by polynomials or trigonometric functions.

For example:

It may be necessary to approximate $f(x) = e^x$ by a linear function $g(x) = a + bx$, over the interval $[0, 1]$ or by a trigonometric function of the form $h(x) = a + b\sin x + c\cos x$ over the interval $[-\pi, \pi]$.

Furthermore, the approximating functions by polynomials are central to software development since computers can only evaluate polynomials. Other functions are evaluated through polynomial approximation.

We now introduce the technique for approximating functions.

Let $C[a, b]$ be the inner product space of continuous functions over the interval $[a, b]$ with an inner product $\langle f, g\rangle = \int_a^b f(x)g(x)dx$ and geometrically defined by the inner product.

Let $\mathbf{W}$ be the subspace of $C[a, b]$. Suppose f is in $C[a, b]$, but outside $\mathbf{W}$, and we want to find the best approximation that lies in $\mathbf{W}$.

We define the "best" approximation to be the function g in $\mathbf{W}$ such that the distance $||f - g||$ between f and g is minimum.

Definition 11.1:

Let $C[a, b]$ be the vector space of continuous functions defined over the interval $[a, b]$. Let f be an element of $C[a, b]$ and $\mathbf{W}$ be a subspace of $C[a, b]$. Then the function g in $\mathbf{W}$ such that $\int_a^b [f(x) - g(x)]^2 dx$ is a *minimum* is called the *least square approximation* to f.

This approximation is called *the least square approximation* since this distance formula is based on squares.

We now give a method for finding the least squares approximation $g(x)$. We can appreciate extending geometrical structure and results from R^n to more abstract surrounding.

We know that if x is a point in R^n and u is a subspace of R^n, then the element of u that is closest to X is $\mathrm{Proj_u X}$. Therefore, it is beneficial also to imagine functions as geometrical vectors. For example, Figure 11.1 with this picture in mind, the analogous result for the function space is as follows.

The least squares approximation to f in the subspace $\mathbf{W}$ is $g = \mathrm{Proj_w f}$.

We build on the definition of $\mathrm{Proj_u X}$ to get an expression for $\mathrm{Proj_u X}$. If $u_1, u_2, \ldots, u_m$ is an orthonormal basis for U, then we know that

$$\mathrm{Proj}_u\mathbf{X} = (\mathbf{X} \cdot u_1)\, u_1 + (\mathbf{X} \cdot u_2)\, u_2 + \ldots + (\mathbf{X} \cdot u_m)\, u_m$$

Let $\{g_1, g_2, \ldots, g_n\}$ be an orthonormal basis for $\mathbf{W}$. Replacing the dot product of $\mathbb{R}^n$ by the inner product of the function space, we get

$$\mathrm{Proj}_w f = (f \cdot g_1)\, g_1 + (f \cdot g_2)\, g_2 + \ldots + (f \cdot g_n)\, g_n.$$

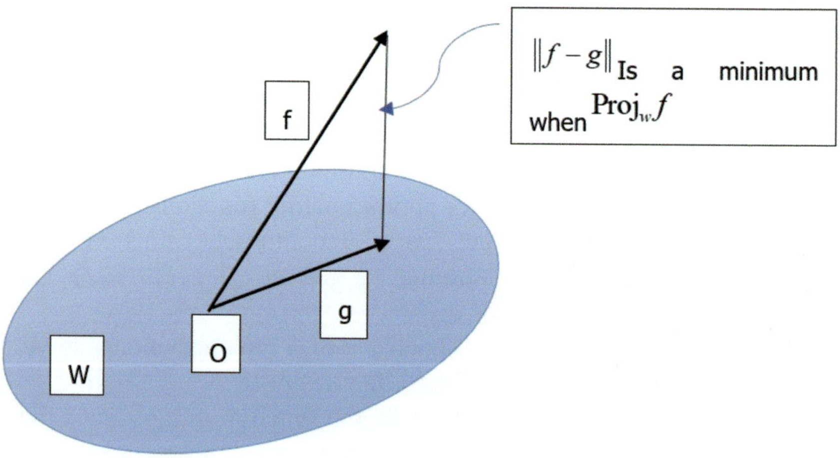

Figure 11.1 Least squares approximation to f in the subspace W.

Example 11.1:

Find the least square linear approximation to $f(x) = e^x$ over the interval $[-1, 1]$.

Solution:

Let the linear approximation $g(x) = a + bx$ be an element of $C[-1, 1]$, and $g(x)$ be an element of the subspace $P_1[-1, 1]$, where $P_1[-1, 1]$ is the set of polynomials of degree less than or equal to 1 over $[-1, 1]$. Then the set $\{1, x\}$ is a basis for $P_1[-1, 1]$.

We get $\langle 1, x\rangle = \int_{-1}^{1}(1 \cdot x)dx = 0$.

Thus, the functions are orthogonal.

The magnitudes of these vectors are given by

$$\|1\|^2 = \int_{-1}^{1}(1 \cdot x)dx = 2 \text{ and } \|x\|^2 = \int_{-1}^{1}(x \cdot x)dx = \frac{2}{3}$$

Thus, the set $\left\{\frac{1}{\sqrt{2}}, \sqrt{\frac{3}{2}}x\right\}$ is an orthonormal basis for $P_1[-1, 1]$.

We now get

$$\begin{aligned}\text{Proj}_w f &= (f \cdot g_1)\, g_1 + (f \cdot g_2)\, g_2 \\ &= \int_{-1}^{1}\left(e^x\sqrt{\frac{1}{2}}\right)dx\sqrt{\frac{1}{2}} + \int_{-1}^{1}\left(e^x\sqrt{\frac{3}{2}}x\right)dx\sqrt{\frac{3}{2}}x \\ &= \frac{1}{2}\left(e - e^{-1}\right) + 3e^{-1}x\end{aligned}$$

The least square linear approximation to $f(x) = e^x$ over the interval $[-1, 1]$ is

$$g(x) = \frac{1}{2}\left(e - e^{-1}\right) + 3e^{-1}x.$$

This gives $g(x) = 1.18 + 1.1x$ to two decimal places.

In this example, we have found the linear approximation to f in $P_1[-1, 1]$, where the higher degree polynomial approximation can be found in the space $P_n[-1, 1]$ of polynomials of degree less than or equal to n.

An orthogonal basis has to be constructed in $P_n[-1, 1]$ by applying the Gram–Schmidt orthogonalization process to arrive at the approximation. The orthogonal functions found in this manner are called Legendre polynomials.

The first six Legendre polynomials are

$$1, x, x^2 - \frac{1}{3}, x^3 - \frac{3}{5}x, x^4 - \frac{6}{7}x^2 + \frac{3}{35}, x^5 - \frac{10}{9}x^3 + \frac{5}{21}x.$$

We next look at approximations of functions in terms of trigonometric functions. Such approximations are widely used in heat conduction, electromagnetism, electric circuits, and mechanical vibrations. The initial work in

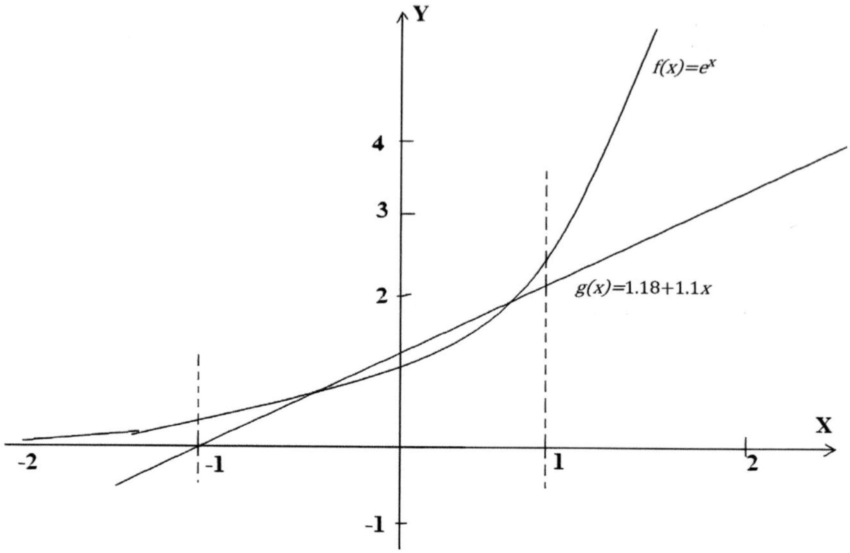

Figure 11.2 Least square linear approximation to $f(x) = ex$.

this area was undertaken by *Jean Baptiste Fourier*, a French mathematician who developed the methods to analyze conduction in an insulated bar. In addition, the approximation is often used in discussing solutions to partial differential equations that describe physical situations.

11.2 Fourier Approximation

Let f be a function in $C[-\pi, \pi]$ with an inner product

$\langle f, g \rangle = \int_{-\pi}^{\pi} f(x)g(x)dx$, and geometry defined by this inner product.

Let us find the least square approximation of f in the space $T[-\pi, \pi]$ of trigonometric polynomials spanned by the sets

$\{1, \cos x, \sin x, \ldots, \cos nx, \sin nx\}$, where n is a positive integer.

It can be shown that the vectors $1, \cos x, \sin x, \ldots, \cos nx, \sin nx$ are mutually orthogonal in this space.

The magnitude of these vectors is given by

$$\|1\|^2 = \int_{-\pi}^{\pi} (1.1)dx = 2\pi$$

$$\|\cos nx\|^2 = \int_{-\pi}^{\pi} (\cos nx \cdot \cos nx)dx = \pi$$

$$\|\sin nx\|^2 = \int_{-\pi}^{\pi} (\sin nx \cdot \sin nx)dx = \pi.$$

Thus, the following set is an orthonormal basis for $T[-\pi, \pi]$.

$$\{g_0, g_1, \ldots, g_{2n}\}$$
$$= \left\{ \frac{1}{\sqrt{2\pi}}, \frac{1}{\sqrt{\pi}} \cos x, \frac{1}{\sqrt{\pi}} \sin x, \ldots, \frac{1}{\sqrt{\pi}} \cos nx, \frac{1}{\sqrt{\pi}} \sin nx \right\}.$$

Let us use this orthonormal basis in the following formula to find the least square approximation g to f.

$$g(x) = \mathrm{Proj}_T f$$
$$= \langle f, g_0 \rangle g_0 + \langle f, g_1 \rangle g_1 + \cdots + \langle f, g_{2n} \rangle g_{2n}.$$

We get

$$g(x) = \left\langle f, \frac{1}{\sqrt{2\pi}} \right\rangle \frac{1}{\sqrt{2\pi}} + \left\langle f, \frac{1}{\sqrt{\pi}} \cos x \right\rangle \frac{1}{\sqrt{\pi}} \cos x$$
$$+ \ldots + \left\langle f, \frac{1}{\sqrt{\pi}} \sin nx \right\rangle \frac{1}{\sqrt{\pi}} \sin nx.$$

Let us introduce the following convenient notation:

$$a_0 = \left\langle f, \frac{1}{\sqrt{2\pi}} \right\rangle \frac{1}{\sqrt{2\pi}}$$
$$= \int_{-\pi}^{\pi} \left(f(x) \frac{1}{\sqrt{2\pi}} \right) dx \frac{1}{\sqrt{2\pi}}$$
$$= \frac{1}{2\pi} \int_{-\pi}^{\pi} f(x) dx$$

$$a_k = \left\langle f, \frac{1}{\sqrt{\pi}} \cos kx \right\rangle \frac{1}{\sqrt{\pi}}$$
$$= \int_{-\pi}^{\pi} \left(f(x) \frac{1}{\sqrt{\pi}} \cos kx \right) dx \frac{1}{\sqrt{\pi}}$$
$$= \frac{1}{\pi} \int_{-\pi}^{\pi} f(x) \cos kx dx$$

$$
\begin{aligned}
b_k &= \left\langle f, \frac{1}{\sqrt{\pi}} \sin kx \right\rangle \frac{1}{\sqrt{\pi}} \\
&= \int_{-\pi}^{\pi} \left(f(x) \frac{1}{\sqrt{\pi}} \sin kx \right) dx \frac{1}{\sqrt{\pi}}. \\
&= \frac{1}{\pi} \int_{-\pi}^{\pi} f(x) \sin kx dx
\end{aligned}
$$

The trigonometric approximation of $f(x)$ can now be written as

$$
g(x) = a_0 + \sum_{k=1}^{n} (a_k \cos kx + b_k \sin kx),
$$

where $g(x)$ is called the nth-order Fourier approximation of $f(x)$. The coefficients $a_0, a_1, b_1, \ldots, a_n, b_n$ are called Fourier coefficients.

As n increases, this approximation naturally becomes an increasingly better approximation in the sense that $\|f - g\|$ gets smaller.

The infinite sum $g(x) = a_0 + \sum_{k=1}^{\infty} (a_k \cos k x + b_k \sin k x)$ is known as the *Fourier series* of f on the interval $[-\pi, \pi]$.

Example 11.2:

Find the fourth-order Fourier approximation to $f(x) = x$ over the interval $[-\pi, \pi]$.

Solution:

Using the above Fourier coefficient with $f(x) = x$ and using integration by parts, we get

$$
\begin{aligned}
a_0 &= \frac{1}{2\pi} \int_{-\pi}^{\pi} f(x) dx \\
&= \frac{1}{2\pi} \int_{-\pi}^{\pi} x \cdot dx \\
&= \frac{1}{2\pi} \left[\frac{x^2}{2} \right]_{-\pi}^{\pi} = 0 \\
a_k &= \frac{1}{\pi} \int_{-\pi}^{\pi} f(x) \cdot \cos kx dx \\
&= \frac{1}{\pi} \int_{-\pi}^{\pi} (x \cdot \cos k\mathrm{x}) dx \\
&= \frac{1}{\pi} \left[\frac{x}{k} \sin kx + \frac{1}{k^2} \cos kx \right]_{-\pi}^{\pi} = 0
\end{aligned}
$$

$$\begin{aligned} b_k &= \frac{1}{\pi}\int_{-\pi}^{\pi} f(x)\cdot \sin kx dx \\ &= \frac{1}{\pi}\int_{-\pi}^{\pi} (x\cdot \sin kx) dx \\ &= \frac{1}{\pi}\left[-\frac{x}{k}\cos kx + \frac{1}{k^2}\sin kx\right]_{-\pi}^{\pi} \\ &= \frac{2(-1)^{k+1}}{k} \end{aligned}$$

The Fourier approximation of f is

$$g(x) = \sum_{k=1}^{n} \frac{2(-1)^{k+1}}{k} \sin kx.$$

Taking k =1,2,3,4, we get the fourth-order approximation

$$g(x) = 2\left(\sin x - \frac{1}{2}\sin 2x + \frac{1}{3}\sin 3x - \frac{1}{4}\sin 4x\right)$$

11.3 Least Square Solutions

Here we derive the method of finding a polynomial that best fits the given data points, which is extremely important to the natural sciences, social sciences, and engineering.

We have seen that a system $AX = Y$ of $n-$equations in $n-$variables, where **A** is invertible, has a unique solution $X = \mathbf{A}^{-1}\mathrm{Y}$; however, $AX = Y$ is a system of $n-$equations in $m-$variables, $n > m$. The system does not, in general, have a solution and is said to be *overdetermined*. Here the matrix **A** is not square, and for such a system, $\mathbf{A}^{-1}$ does not exist.

We shall introduce a matrix called the *pseudo inverse* of **A**, denoted $Pinv(\mathbf{A})$, leading to a least square solution $X = Pinv(\mathbf{A})$ for an *overdetermined system*. Of course, this is not a good solution but is, in some sense, the closest we can get to a reasonable explanation for the system.

We shall see an application of overdetermined systems in finding curves that "best" fit data.

Definition 11.2:

Let **A** be a matrix. The matrix $\left(\mathbf{A}^{\mathrm{T}}\mathbf{A}\right)^{-1}\mathbf{A}^{\mathrm{T}}$ is called the *pseudo inverse* of **A** and is denoted as $Pinv(\mathbf{A})Y$.

Remark:

We have seen that not every matrix has an inverse. Similarly, not every matrix has *a pseudo inverse*. The matrix $\mathbf{A}$ has a *pseudo inverse*, if $\left(\mathbf{A}^{\mathrm{T}}\mathbf{A}\right)^{-1}$ exists.

Example 11.3:

Find the pseudo inverse of

$$\mathbf{A} = \begin{bmatrix} 1 & 2 \\ -1 & 3 \\ 2 & 4 \end{bmatrix}$$

Solution:

We compute the pseudo inverse of $\mathbf{A}$ in stages:

$$\begin{aligned}
\mathbf{A}^{\mathrm{T}}\mathbf{A} &= \begin{bmatrix} 1 & -1 & 2 \\ 2 & 3 & 4 \end{bmatrix} \begin{bmatrix} 1 & 2 \\ -1 & 3 \\ 2 & 4 \end{bmatrix} \\
&= \begin{bmatrix} 6 & 7 \\ 7 & 29 \end{bmatrix} \\
\left(\mathbf{A}^{\mathrm{T}}\mathbf{A}\right)^{-1} &= \frac{1}{|\mathbf{A}^{\mathrm{T}}\mathbf{A}|}\mathrm{Adj}\left(\mathbf{A}^{\mathrm{T}}\mathbf{A}\right) \\
&= \frac{1}{125}\begin{bmatrix} 29 & -7 \\ -7 & 6 \end{bmatrix} \\
\mathrm{P}inv(\mathbf{A}) &= \left(\mathbf{A}^{\mathrm{T}}\mathbf{A}\right)^{-1}\ \mathbf{A}^{\mathrm{T}} \\
&= \frac{1}{125}\begin{bmatrix} 29 & -7 \\ -7 & 6 \end{bmatrix}\begin{bmatrix} 1 & -1 & 2 \\ 2 & 3 & 4 \end{bmatrix}. \\
&= \frac{1}{25}\begin{bmatrix} 3 & -10 & 6 \\ 1 & 5 & 2 \end{bmatrix}
\end{aligned}$$

Next, we use the concept of *pseudo inverse* further to extend our understanding of systems of linear equations.

Let $AX = Y$ be a system of $n-$linear equations in $m-$variables with $n > m$, where $\mathbf{A}$ is of rank m.

Multiply each side of this matrix equation by $\mathbf{A}^T$ to get $\mathbf{A}^T AX = \mathbf{A}^T Y$.

The matrix $\mathbf{A}^T\mathbf{A}$ can be shown to be invertible for such a system.

Multiply each side of this equation by $\left(\mathbf{A}^{\mathrm{T}}\mathbf{A}\right)^{-1}$ to get

$$\mathrm{X} = \left[\left(\mathbf{A}^{\mathrm{T}}\mathbf{A}\right)^{-1}\ \mathbf{A}^{\mathrm{T}}\right]\mathrm{Y} = \mathrm{P}inv(\mathbf{A})\mathrm{Y}$$

This value of X is called the *least square solution* to the system of equations.

AX=Y	X=$Pinv$(**A**)**Y**
System	Least square solution

Let $AX = Y$ be a system of $n-$linear equations in $m-$variables with $n > m$, where **A** is of rank m. This system has a *least square solution.*

If the system has a unique solution, the *least square solution* is that unique solution. If the system is *overdetermined*, the *least square solution* is the closest solution to get a valid solution. The system cannot have a matrix solution.

Example 11.4:

Find the least square solution of the following overdetermined system of equations:

$$\begin{aligned} x + y &= 6 \\ -x + y &= 3 \quad \text{and sketch.} \\ 2x + 3y &= 9 \end{aligned}$$

Solution:

The matrix of coefficients is

$$\mathbf{A} = \begin{bmatrix} 1 & 1 \\ -1 & 1 \\ 2 & 3 \end{bmatrix} \text{ and } Y = \begin{bmatrix} 6 \\ 3 \\ 9 \end{bmatrix}$$

The column vectors of **A** are linearly independent. Thus, the rank of **A** is 2.

This system has a least square solution.

We compute $Pinv(\mathbf{A})$.

$$\begin{aligned} \mathbf{A}^{\mathrm{T}}\mathbf{A} &= \begin{bmatrix} 1 & -1 & 2 \\ 1 & 1 & 3 \end{bmatrix} \begin{bmatrix} 1 & 1 \\ -1 & 1 \\ 2 & 3 \end{bmatrix} \\ &= \begin{bmatrix} 6 & 6 \\ 6 & 11 \end{bmatrix} \\ \left(\mathbf{A}^{\mathrm{T}}\mathbf{A}\right)^{-1} &= \frac{1}{|\mathbf{A}^{\mathrm{T}}\mathbf{A}|} \operatorname{Adj}\left(\mathbf{A}^{\mathrm{T}}\mathbf{A}\right) \\ &= \frac{1}{30} \begin{bmatrix} 11 & -6 \\ -6 & 6 \end{bmatrix} \end{aligned}$$

$$\begin{aligned}\text{Pinv}(\mathbf{A}) &= \left(\mathbf{A}^{\text{T}}\mathbf{A}\right)^{-1}\mathbf{A}^{\text{T}} \\ &= \frac{1}{30}\begin{bmatrix} 11 & -6 \\ -6 & 6 \end{bmatrix}\begin{bmatrix} 1 & -1 & 2 \\ 2 & 3 & 4 \end{bmatrix}. \\ &= \frac{1}{30}\begin{bmatrix} 5 & -17 & 4 \\ 0 & 12 & 6 \end{bmatrix}\end{aligned}$$

The least square solution is

$$\begin{aligned}\text{Pinv}(\mathbf{A})Y &= \frac{1}{30}\begin{bmatrix} 5 & -17 & 4 \\ 0 & 12 & 6 \end{bmatrix}\begin{bmatrix} 6 \\ 3 \\ 9 \end{bmatrix}. \\ &= \begin{bmatrix} 1/2 \\ 3 \end{bmatrix}\end{aligned}$$

The least square solution is the point $p\left(\frac{1}{2}, 3\right)$ as in Figure 11.3.

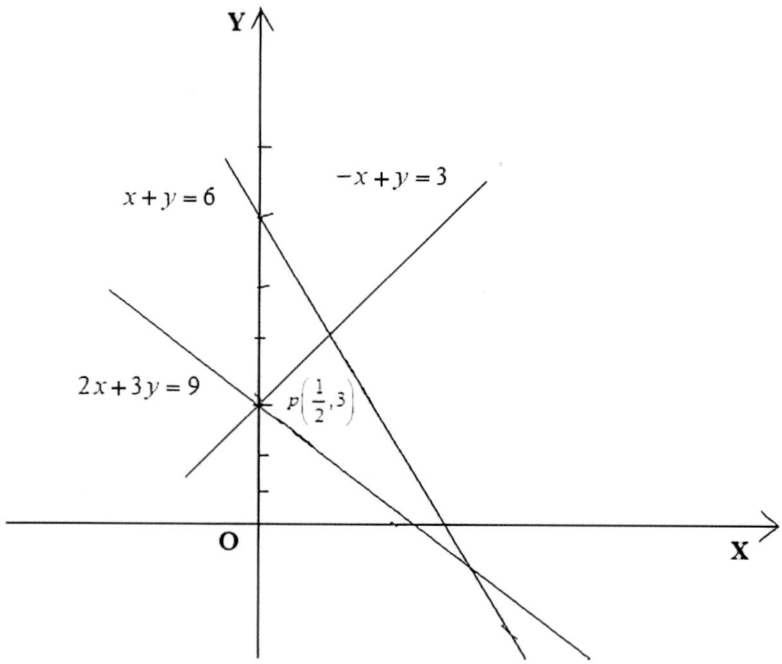

Figure 11.3 Least square solution of the following overdetermined system.

11.4 Least Square Curves

Many branches of science and business use equation based on data that has been determined from experimental results. It is more straightforward in finding a unique polynomial of degree two that passes through three data points. However, too much data in many applications leads to an equation that can exactly fit all the data. One then uses the equation of a line or curve that, in some sense, "best" fit the data.

For example:

Suppose the data consists of the points $(x_1, y_1), (x_2, y_2), \ldots, (x_n, y_n)$ as shown in Figure 11.4 (a). These points lie approximately on a line. We would want the equation of the line that best fit these points. On the other hand, the points might closely fit a parabola, as shown in Figure 11.4 (b). We would then want to find the parabola that most closely fits these points.

Many criteria can be used for the "best" fit in such cases. The one that has generally been found to be most satisfactory is called the least square line or curve found by solving an overdetermined system of equations.

The least squares line and curve is such that

$$d_1^2 + d_2^2 + \cdots + d_n^2 \quad \text{in Figure 11.4} \quad \text{is minimum.}$$

In the previous section, we discussed the least square approximations to functions. This discussion is the discrete analogue of that problem. In the

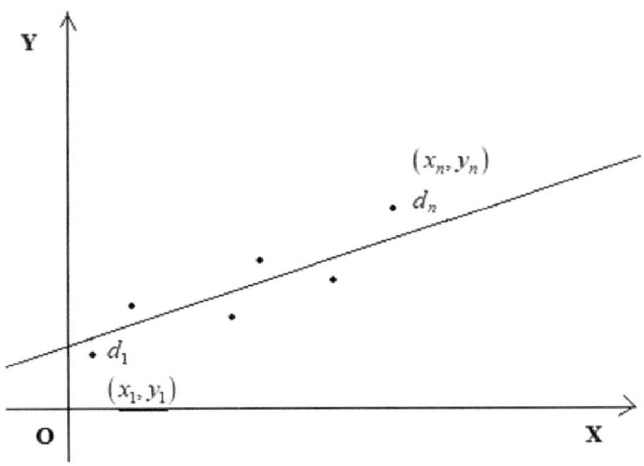

Figure 11.4(a)

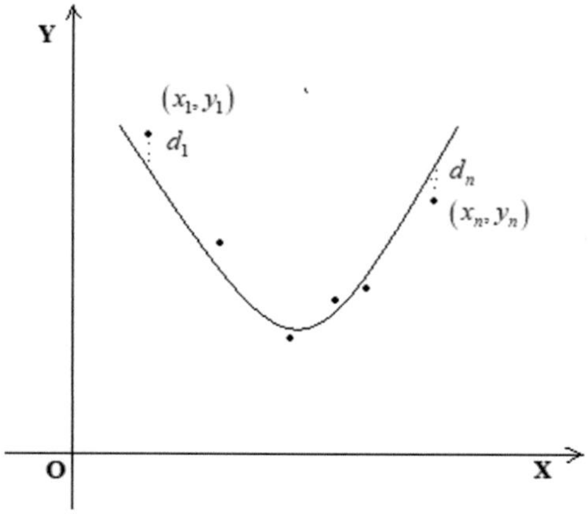

Figure 11.4(b)

Figure 11.4 Least square approximations to functions.

function situation, we have to find a curve of a specific type that best fitted a continuous set of data points over a given interval (the graph of the function). Here we want the "best" fit to a discrete set of data points over a given interval. Although the techniques that we develop to arrive at results are different, look for certain similarities in concepts in the two situations. In both cases, the best results are the one obtained by minimizing certain squares. Hence the term "least square."

We now illustrate how to fit a least squares polynomial to given data. The method involves constructing a system of linear equations. The least squares solution to this system of equations gives the coefficients of the polynomial. We shall justify the technique after seeing how it works.

Example 11.5:

Find the least square line for the following data points (1,1), (2,2,4), (3,3,6), (4,4).

Solution:

Let the equation of the line be $y = a + bx$. Substituting for these points into the equation of the line, we get the overdetermined system

$$a + b = 1$$

$$a + 2b = 2.4$$

$$a + 3b = 3.6$$

$$a + 4b = 4$$

We find the least square solution. The matrix of coefficient **A** and column vector Y as follows:

$$\mathbf{A} = \begin{bmatrix} 1 & 1 \\ 1 & 2 \\ 1 & 3 \\ 1 & 4 \end{bmatrix} \text{ and } Y = \begin{bmatrix} 1 \\ 2.4 \\ 3.6 \\ 4 \end{bmatrix}.$$

It can be shown that

$$\begin{aligned} \text{Pinv}(\mathbf{A}) &= \left(\mathbf{A}^{\mathrm{T}}\mathbf{A}\right)^{-1}\mathbf{A}^{\mathrm{T}} \\ &= \frac{1}{20}\begin{bmatrix} 20 & 10 & 0 & -10 \\ -6 & -2 & 2 & 6 \end{bmatrix}. \end{aligned}$$

The least square solution is

$$\begin{aligned} \left[\left(\mathbf{A}^{\mathrm{T}}\mathbf{A}\right)^{-1}\mathbf{A}^{\mathrm{T}}\right]Y &= \frac{1}{20}\begin{bmatrix} 20 & 10 & 0 & -10 \\ -6 & -2 & 2 & 6 \end{bmatrix}\begin{bmatrix} 1 \\ 2.4 \\ 3.6 \\ 4 \end{bmatrix} \\ &= \begin{bmatrix} 0.2 \\ 1.02 \end{bmatrix} \end{aligned}$$

Thus,

$$a = 0.2, b = 1.02.$$

The equation of the least square line for this data is $y = 0.2 + 1.02x$.

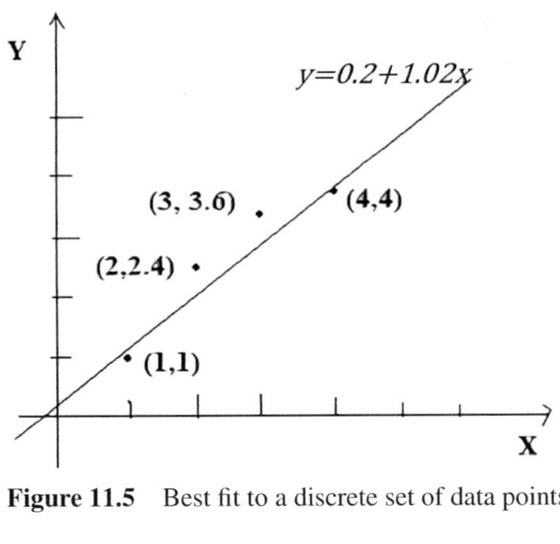

Figure 11.5 Best fit to a discrete set of data points.

This is the line that is generally considered to be the line of best fit for these points.

Example 11.6:

Find the least squares parabola for the following data points (1,7), (2,2), (3,1), (4,3).

Solution:

Let the sequence of the parabola be

$$y = a + bx + cx^2$$

Substituting these points into the equation of the parabola, we get the system

$$\begin{aligned}
a + b + c &= 7 \\
a + 2b + 4c &= 2 \\
a + 3b + 9c &= 1 \\
a + 4b + 16c &= 3
\end{aligned}$$

We find the least square solution. The matrix of coefficients **A** and column vector Y are as follows:

$$\mathbf{A} = \begin{bmatrix} 1 & 1 & 1 \\ 1 & 2 & 4 \\ 1 & 3 & 9 \\ 1 & 4 & 16 \end{bmatrix} \text{ and } Y = \begin{bmatrix} 7 \\ 2 \\ 1 \\ 3 \end{bmatrix}.$$

It can be shown that

$$\begin{aligned}
\text{Pinv}(\mathbf{A}) &= \left(\mathbf{A}^{\mathrm{T}}\mathbf{A}\right)^{-1}\mathbf{A}^{\mathrm{T}} \\
&= \frac{1}{20}\begin{bmatrix} 45 & -15 & -25 & 15 \\ -31 & 23 & 27 & -19 \\ 5 & -5 & -5 & 5 \end{bmatrix}
\end{aligned}$$

The least square solution is

$$\begin{aligned}
\left[\left(\mathbf{A}^{\mathrm{T}}\mathbf{A}\right)^{-1}\mathbf{A}^{\mathrm{T}}\right]Y &= \frac{1}{20}\begin{bmatrix} 45 & -15 & -25 & 15 \\ -31 & 23 & 27 & -19 \\ 5 & -5 & -5 & 5 \end{bmatrix}\begin{bmatrix} 7 \\ 2 \\ 1 \\ 3 \end{bmatrix}. \\
&= \begin{bmatrix} 15.25 \\ -10.05 \\ 1.75 \end{bmatrix}
\end{aligned}$$

Thus, $a = 15.25, b = -10.05, c = 1.75$.
The equation of the least square parabola for these data points is

$$y = 15.25 - 10.05x + 1.75x^2.$$

We illustrate this parabola in Figure 11.6.

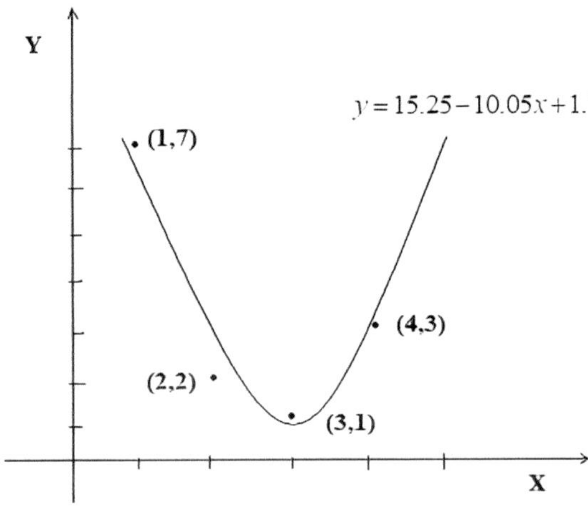

Figure 11.6 Least square line.

11.5 Eigenvalues by Iteration and Connectivity of Networks

Numerical technique exists for evaluating certain eigenvalues and eigenvectors of various types of matrices. Here we present an iterative method called the power method. It can be used to determine the eigenvalue with the most significant absolute value (if it exists) and a corresponding eigenvector for specific matrices.

In many applications, one is only interested in the dominant eigenvalue. Applications in geography and history that illustrates the importance of the dominant eigenvalue will be given.

Definition 11.3:

Let $\mathbf{A}$ be a square matrix with eigenvalues $\lambda_1, \lambda_2, \ldots, \lambda_n$. The eigenvalue λ_i is said to be a dominant eigenvalue if

$$|\lambda_i| > |\lambda_k|, k \neq i.$$

The eigenvectors corresponding to the dominant eigenvalue are called the dominant eigenvectors of $\mathbf{A}$.

Example 11.7:

Let $\mathbf{A}$ be a square matrix with eigenvalues $-5, -2, 1 \,\&\, 3$. Then -5 is the dominant eigenvalue since

$$|-5| > |-2|, |-5| > |1| \text{ and } |-5| > |3|.$$

Let B be a square matrix with eigenvalues $-4, -2, 1 \,\&\, 4$. There is no dominant eigenvalues

Since $|-4| = |4|$.

The power method for finding a dominant eigenvector is based on the following theorem.

Theorem 11.1:

Let $\mathbf{A}$ be an $n \times n$ matrix having $n-$ linearly independent eigenvectors and a dominant eigenvalue. Furthermore, let $x_\circ$ be any non-zero column vector in $\mathbb{R}^n$ having a non-zero component in the direction of a dominant eigenvector.

Then the sequence of vectors

$x_1 = \mathbf{A}x_0, x_2 = \mathbf{A}x_1, \ldots, x_k = \mathbf{A}x_{k-1}, \ldots,$ will approach a dominant eigenvector of $\mathbf{A}$.

Proof:

Let the eigenvalues of the matrix $\mathbf{A}$ be $\lambda_1, \lambda_2, \ldots, \lambda_n$ with λ_1 being the dominant eigenvalue.

Let $y_1, y_2, \ldots, y_n$ be corresponding linearly independent eigenvectors. These eigenvectors will form a basis for R^n.

Thus, there exist scalars $a_1, a_2, \ldots, a_n$ such that

$$x_0 = a_1y_1 + a_2\mathrm{y}_2 + \ldots. + a_n\mathrm{y}_n, \text{ where } a_1 \neq 0$$

We get

$$\begin{aligned} x_k &= \mathbf{A}x_{k-1} = \mathbf{A}^2x_{k-2} = \cdots = \mathbf{A}^{\mathrm{k}}x_0 \\ &= \mathbf{A}^{\mathrm{k}}\left[a_1y_1 + a_2\mathrm{y}_2 + \cdots + a_n\mathrm{y}_n\right] \\ &= \left[a_1\mathbf{A}^{\mathrm{k}}y_1 + a_2\mathbf{A}^{\mathrm{k}}\mathrm{y}_2 + \cdots + a_n\mathbf{A}^{\mathrm{k}}\mathrm{y}_n\right] \\ &= \left[a_1\left(\lambda_1\right)^k y_1 + a_2\left(\lambda_2\right)^k \mathrm{y}_2 + \cdots + a_n\left(\lambda_n\right)^k \mathrm{y}_n\right] \\ &= \left(\lambda_1\right)^k\left[a_1y_1 + a_2\left(\frac{\lambda_2}{\lambda_1}\right)^k \mathrm{y}_2 + \cdots + a_n\left(\frac{\lambda_n}{\lambda_1}\right)^k \mathrm{y}_n\right] \end{aligned}$$

Since $\left|\frac{\lambda_i}{\lambda_1}\right| < 1$ for $i = 2, 3, \ldots$, then $\left(\frac{\lambda_i}{\lambda_1}\right)^k$ will approach 0 as k increases and the vector x_k will approach $(\lambda_1)^k\, a_1 y_1$, a dominant eigenvector.

The following theorem tells us how to determine the eigenvalue corresponding to a given eigenvector. Once a dominant eigenvector has been found, the dominant eigenvalue can be determined by applying these results.

Theorem 11.2:

Let X be an eigenvector of a matrix $\mathbf{A}$. The corresponding eigenvalue is given by

$$\lambda = \frac{AX.X}{X.X}.$$

This quotient is called the Rayleigh quotient.

Proof:

Since λ is the eigenvalue corresponding to X, $AX = \lambda X$,

$$\frac{AX \cdot X}{X \cdot X} = \frac{\lambda X \cdot X}{X \cdot X} = \frac{\lambda (X \cdot X)}{X \cdot X} = \lambda.$$

If Theorem 11.2 as it now stands is used to compute a dominant eigenvector, the components of the vectors may become very large, causing significant round-off errors to occur. This problem is overcome by dividing each element by the absolute value of its most potent component and then using it (a vector in the same direction as X_i) in the following iteration. We refer to this procedure as scaling the vector.

We now summarize the power method.

11.5.1 The power method for an $n \times n$ matrix

Select an arbitrary non-zero column vector $\mathbf{X}_0$ having $n-$components.

Iteration I:

Compute AX_0.

Scale AX_0 to get X_1.

Compute $\frac{AX_1 \cdot X_1}{X_1 \cdot X_1}$.

Iteration II:

Compute AX_1.

Scale AX_1 to get X_2.

Compute $\frac{AX_2 \cdot X_2}{X_2 \cdot X_2}$ and so on.

Then $X_0, X_1, X_2, \ldots$ converges to a dominant eigenvector, and $\frac{AX_1 \cdot X_1}{X_1 \cdot X_1}, \frac{AX_2 \cdot X_2}{X_2 \cdot X_2}, \ldots$ converges to the dominant eigenvalue if the matrix **A** has $n-$linearly independent eigenvectors and X_0 has a non-zero component in the direction of a dominant eigenvector.

Remark:

Since $n \times n$ symmetric matrices have $n-$linearly independent eigenvectors, the method can thus be applied to find the dominant eigenvalue of a symmetric matrix.

Example 11.8:

Find the dominant eigenvalue and a dominant eigenvector of the following symmetric matrix:

$$\begin{bmatrix} 5 & 4 & 2 \\ 4 & 5 & 2 \\ 2 & 2 & 2 \end{bmatrix}.$$

Solution:

Let $X_0 = \begin{bmatrix} 1 \\ -2 \\ 3 \end{bmatrix}$ be an arbitrary column vector with three components.

Computation is carried out to nine decimal places. We display the results, rounded to three decimal places for clarity of viewing.

Table 11.1 Dominant eigenvalue and a dominant eigenvector.

Iteration	AX	Scaled vectors	$\frac{\langle X.AX \rangle}{\langle X.X \rangle}$
1	$\begin{bmatrix} 3 \\ 0 \\ 4 \end{bmatrix}$	$X_1 = \begin{bmatrix} 0.75 \\ 0 \\ 1 \end{bmatrix}$	5
2	$\begin{bmatrix} 5.75 \\ 5 \\ 3.5 \end{bmatrix}$	$X_2 = \begin{bmatrix} 1 \\ 0.870 \\ 0.609 \end{bmatrix}$	9.889
3	$\begin{bmatrix} 9.696 \\ 9.565 \\ 4.957 \end{bmatrix}$	$X_3 = \begin{bmatrix} 1 \\ 0.987 \\ 0.511 \end{bmatrix}$	9.999
4	$\begin{bmatrix} 0.997 \\ 9.955 \\ 4.996 \end{bmatrix}$	$X_4 = \begin{bmatrix} 1 \\ 0.999 \\ 0.501 \end{bmatrix}$	10
5	$\begin{bmatrix} 0.997 \\ 0.996 \\ 5 \end{bmatrix}$	$X_5 = \begin{bmatrix} 1 \\ 1 \\ 0.5 \end{bmatrix}$	10

Thus, after five iterations, we arrived at the dominant eigenvalue of 10 with a corresponding eigenvector $\begin{bmatrix} 1 \\ 1 \\ 0.5 \end{bmatrix}$. These results agree with the previous discussion, where we computed the eigenvalues and eigenvectors of this matrix using determinants.

This method has the same advantage as the Gauss–Seidel iterative method. Any error in the computation only means that a new arbitrary vector has been introduced at that stage. Thus, the procedure is very accurate. The only round-off errors that occur are those that arise in the final iteration. However, the method has the disadvantage that it may converge only very slowly for large matrices.

11.6 Difficulties in the Solution of the System of Equations

In earlier chapters, we have discussed different approaches such as Gauss–Jordan, Gaussian, and $\mathbf{LU}$ decomposition for finding the solution of the system of linear equations and its applications to various real-world problems.

The matrix of coefficients $\mathbf{A}$ and the matrix of constants B in a system of the equation $AX = B$ often derive from measurements and are unknown exactly. As a result, minor flaws in the elements of these matrices can matter significant errors in the solution, dominant to very imprecise results.

This section will review ways of appraising such effects and ways of reducing them.

11.6.1 The condition number $c(A)$ of a matrix

Non-singular matrices have shown a significant aspect in our discussion in this course. Here we introduce the notion of a condition number of a non-singular matrix $\mathbf{A}$. This number is described in terms of a norm (or magnitude) $||\mathbf{A}||$ of the matrix.

Definition 11.4:

Let $\mathbf{A}$ be a non-singular matrix. The condition number of a matrix $\mathbf{A}$ is denoted and defined as $c(\mathbf{A})$. If $c(\mathbf{A})$ is small, then the matrix $\mathbf{A}$ is said to be well-conditioned. If $c(\mathbf{A})$ is large, then the matrix $\mathbf{A}$ is ill-conditioned.

Let us now examine how $c(\mathbf{A})$ can be used to demonstrate the accuracy of the solution of a system of equation $AX = B$.

Suppose the matrix equation $AX = B$ illustrates a given experiment where the elements of **A** and B derive from the measurement. Such data depends on the efficiency of instruments and are hardly exact.

Let the slight error in **A** be described by a matrix E and the corresponding errors in X expressed by e.

Thus, $(\mathbf{A} + E)(X + e) = B$.

If we prefer proper norms for the vectors and matrices, then it can be demonstrated that

$$\frac{\|e\|}{\|X + e\|} \leq c(\mathbf{A})\frac{\|E\|}{\|\mathbf{A}\|}.$$

Thus, if $c(\mathbf{A})$ is small, errors can only be minor errors, and the results are exact. Therefore, the system of equations is said to be *well-behaved.*

Contrarily, if $c(\mathbf{A})$ is large, then there is the prospect that minor errors in **A** can outcome significant errors in X leading to unreliable results. Thus, such a system of equations is ill-conditioned.

Note:

A significant value of $c(\mathbf{A})$ is an indication, not an assurance of a substantial error in the solution.

If there is an identity matrix I, then

$$c(I) = 1 \quad \text{and} \quad c(\mathbf{A}) \geq 1$$

.

Thus, 1 is a lower bound for condition numbers.

We can intuitively assume the system $IX = B$ as being a well-behaved system.

The smaller $c(\mathbf{A})$, then in some impression, the closer **A** is to I, and the better behaved the system $AX = B$ becomes. But, on the other end of the scale, the larger $c(\mathbf{A})$, the closer **A** is to be singular, and the problem can arise.

The definite value of $c(\mathbf{A})$ will, indeed, depend upon the norm used for **A**. A norm that is generally used is the so-called I$-$norm

$$\|\mathbf{A}\| = \max\{|a_{1j}| + |a_{2j}| + \cdots + |a_{nj}|\} \quad \text{for } j = 1, 2, \ldots, n.$$

This norm is the most significant number resulting from adding up the absolute values of elements in each column.

Other norms are more reliable but less efficient.

This norm is a good compromise of reliability and efficiency.

Example 11.9:

A natural question to ask is "How large is large for a condition number?" If $c(\mathbf{A})$ is written in the form $c(\mathbf{A}) \approx 0.d \times 10^k$, then the above inequality implies that

$$\frac{\|e\|}{\|X+e\|} \leq 10^k \frac{\|E\|}{\|\mathbf{A}\|}.$$

This inequality gives the following rule of thumb.

If $c(\mathbf{A}) \approx 0.d \times 10^k$, then the components of X can usually be expected to have k a fewer significant digit of accuracy than the element of $\mathbf{A}$.

Thus, for example:

If $c(\mathbf{A}) \approx 100 = (0.1 \times 10^3)$, and the components of $\mathbf{A}$ are known to five-digit precision, the components of X may have only two-digit accuracy. Much, therefore, depends upon the accuracy of the measurements and the desired accuracy of the solution. $c(\mathbf{A}) = 100$ would, however, be considered significant by any standard.

A similar relationship involving $c(\mathbf{A})$ exists between errors in the matrix of constants B and the resulting errors in the solution X. Clarify these ideas in the following example.

Example 11.10:

The following system of equations describes a specific experiment. First, let us show that this system is sensitive to changes in the coefficients and the constants on the right.

$$\begin{aligned} 34.9x_1 + 23.6x_2 &= 234 \\ 22.9x_1 + 15.6x_2 &= 154 \end{aligned}$$

The exact solution is $x_1 = 4$ and $x_2 = 4$.

Solution:

Let us compute the condition number.

The matrix of coefficient and its inverse are

$$\mathbf{A} = \begin{bmatrix} 34.9 & 23.6 \\ 22.9 & 15.6 \end{bmatrix} \text{ and } \mathbf{A}^{-1} = \begin{bmatrix} 3.9 & -5.9 \\ -5.725 & 8.725 \end{bmatrix}$$

We get

$$\begin{aligned} \|\mathbf{A}\| &= \max\{(34.9 + 22.9), (23.6 + 15.6)\} \\ &= \max\{57.8, 39.2\} = 57.8 \\ \|\mathbf{A}^{-1}\| &= \max\{(3.9 + 5.725), (5.9 + 8.725)\} \\ &= \max\{9.625, 14.625\} = 14.625 \end{aligned}$$

Thus,

$$c(\mathbf{A}) = \|\mathbf{A}\| \, \|\mathbf{A}^{-1}\|$$
$$= 57.8 \times 14.625 = 845.325$$

The number is enormous.

The system is ill-conditioned, and the solution may not be reliable.

Let us change the coefficient of x_1 in the first equation from 34.9 to 34.8. The new system has a solution $x_1 = 6.5574$, $x_2 = 0.2459$. The system is also sensitive to small changes in constant terms. For example: changing the first constant term from 234 to 235 gives a new explanation of $x_1 = 7.9$, $x_2 = -1.725$.

Note:

What does one do in a case like this to get meaningful results? If possible, the problem should be reformulated in terms of a well-behaved system of equations. If this is not possible, then one should attempt to derive more accurate data with more precise data. Finally, the system should be solved on a computer using double-precision arithmetic, applying techniques that minimize errors during computational called round-off errors.

Example 11.11:

Find the condition number of the following matrices. Then, decide whether a system of linear equations defined by such a matrix of the coefficient is well-behaved.

$$\text{(a) } \mathbf{A} = \begin{bmatrix} 1 & 1 & -1 \\ 4 & 0 & 1 \\ 0 & 4 & 1 \end{bmatrix} \text{ (b) } B = \begin{bmatrix} 1 & 1 & 1 \\ 1 & 2 & 4 \\ 1 & 3 & 9 \end{bmatrix}.$$

Solution:

(a) The inverse is found to be

$$\mathbf{A}^{-1} = \begin{bmatrix} \frac{1}{6} & \frac{5}{24} & -\frac{1}{24} \\ \frac{1}{6} & -\frac{1}{24} & \frac{5}{24} \\ -\frac{2}{3} & \frac{1}{6} & \frac{1}{6} \end{bmatrix}.$$

We get

$$\|\mathbf{A}\| = \max\{(1+4+0), (1+0+4), (1+1+1)\} = 5$$
$$\|\mathbf{A}^{-1}\| = \max\left\{\left(\frac{1}{6}+\frac{1}{6}+\frac{2}{3}\right), \left(\frac{5}{24}+\frac{1}{24}+\frac{1}{6}\right), \left(\frac{1}{24}+\frac{5}{24}+\frac{1}{6}\right)\right\}$$
$$= 1$$

Thus, $c(\mathbf{A}) = \|\mathbf{A}\| \left\|\mathbf{A}^{-1}\right\| = 5.$
This value is considered small, and the system is thus well-behaved.

(b) This example is of the ill-conditioned system and illustrates the possibility of constructing an alternative system that is not ill-conditioned. The inverse of **A** is found.

$$\mathbf{A}^{-1} = \begin{bmatrix} 3 & -3 & 1 \\ -2.5 & 4 & -1.5 \\ 0.5 & -1 & 0.5. \end{bmatrix}$$

We get

$$\begin{aligned} \|\mathbf{A}\| &= \max\{(1+1+1), (1+2+3), (1+4+9)\} = 14 \\ \left\|\mathbf{A}^{-1}\right\| &= \max\{(3+2.5+0.5), (3+4+1), (1+1.5+0.5)\} = 8 \end{aligned}$$

Thus, $c(\mathbf{A}) = \|\mathbf{A}\| \left\|\mathbf{A}^{-1}\right\| = 112.$
This is a large number. The system is ill-conditioned.

11.7 Coding Theory

Messages are sent electronically as sequences of 0's and 1's. Errors can occur in messages due to noise or interference. For example, the message $(1,1,0,1,1,0,1)$ could be sent, and the message $(1,1,0,1,1,0,1)$ might be received. An error has occurred in the fifth entry. We shall now look at methods that are used to detect and correct such errors.

The scalars for a vector space can be sets other than real or complex numbers. The requirement is that the scalars form an algebraic field. A field is a set of elements with two operations that satisfy certain axioms. Readers who go on to take a course in modern algebra will study fields. In this example, we shall use the field $\{0,1\}$ of scalars having only these two elements with operations of addition and multiplication defined as follows:

$$\begin{aligned} &0+0=0, \quad 0+1=1, \quad 1+0=1, 1+1=0 \\ &0.0=0, \quad 0.1=0, \quad 1.0=0, \quad 1.1=1 \end{aligned}$$

Let V_7 be the vector space of seven-tuples of 0's and 1's over this field of scalars where addition and scalar multiplication are defined in the usual component-wise manner. For example

$$\begin{aligned} &(1,0,0,1,1,0,1) + (0,1,1,1,0,0,1) \\ &\qquad = (1+0, 0+1, 0+1, 1+1, 1+0, 0+0, 1+1) \\ &\qquad = (1,1,1,0,1,0,0) \end{aligned}$$

$$0(1,0,0,1,1,0,1) = (0.1,0.0,0.0,0.1,0.1,0.0,0.1) = (0,0,0,0,0,0,0)$$
$$1(1,0,0,1,1,0,1) = (1.1,1.0,1.0,1.1,1.1,1.0,1.1) = (1,0,0,1,1,0,1)$$

Since each vector in V_7 has seven components and each of these components can be either 0 or 1, there are 2^7 vectors in this space. The four-dimensional subspace of V_7 having basis

$$\{(1,0,0,1,1,0,1),(0,1,0,0,1,0,1),(0,0,1,0,1,1,0),(0,0,0,1,1,1,1)\}$$

is called a *Hamming code* and is denoted $C_{7,4}$. The vector in $C_{7,4}$ can be used as a message. Each vector in $C_{7,4}$ can be written as

$$v_i = a_1(1,0,0,0,0,1,1) + a_2(0,1,0,0,1,0,1) + a_3(0,0,1,0,1,0,1) + a_4(0,0,0,1,1,1,1).$$

Since each of the four scalars a_1, a_2, a_3, a_4 can take one of the values 1 or 0, we have 2^4 that is 16 vectors in $C_{7,4}$. The hamming code can thus be used to send sixteen different messages $v_1, v_2, \ldots, v_{16}$. The reader is asked to list these vectors in the following exercises.

When an error occurs in one location of a transmitted message, the resulting incorrect vector lies in V_7 outside the subspace $C_{7,4}$. It can prove that there is precisely one vector in $C_{7,4}$ that differs from this incorrect vector in one location. Thus, the error can be detected and corrected. For example, suppose the received vector is $(1,0,1,1,0,1)$. This vector cannot be expressed as a linear combination of the above base vectors; it is not in $C_{7,4}$. There is a single vector in $C_{7,4}$ that differs from this vector in one entry, namely $(1,0,1,0,1,0,1)$. The corrected message is $(1,0,1,0,1,0,1)$. The Hamming code is called an *error-correcting code* because of this facility to detect and correct errors.

Let us look at the geometry underlying this code. The distance between two vectors u and w in V_7 is denoted $d(u,w)$ and is defined to be the number of components that differ in u and w. Thus, for example,

$$d((1,0,0,1,0,1,1)(1,1,0,0,1,1,1)) = 3.$$

The second, fourth, and fifth components of these vectors differ. A sphere of radius 1 about a vector in V_7 contains all those vectors that differ in one component from the vector. It can be shown that the spheres of radii 1 centered at the vectors of $C_{7,4}$ will be disjoint and that every element of V_7 lies in one such sphere (Figure 11.7).

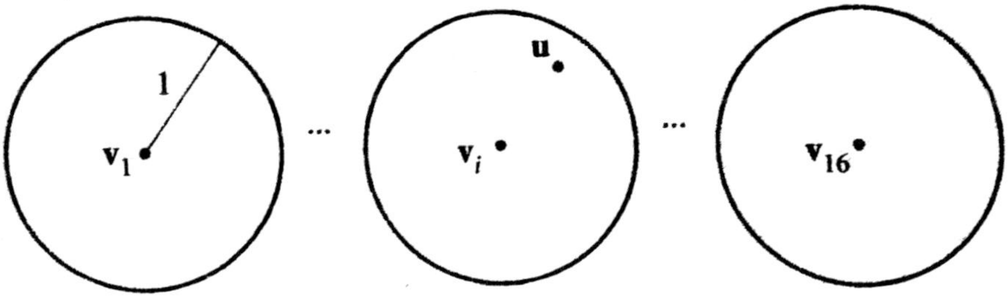

Hamming Code $\{v_1, v_2, \ldots, v_{16}\}$
Incorrect message u; correct message v_i

Figure 11.7 Least square parabola.

Thus, if a vector u is received that is not in $C_{7,4}$, it will be in a single sphere having a center v_i – the correct message. In practice, electrical circuits called gates are used to test whether the received message is in $C_{7,4}$ and if it is not, to determine the center of the sphere in which it lies, giving the correct message.

Other error detecting codes exist. The *Golay code*, for example, is a 12-dimensional subspace of V_{23} that is denoted $C_{23,12}$. The code space has 2^{12} elements that can be used to represent 4096 messages. This code can detect and correct errors in one, two, or three locations. It can represent that the spheres of radii 3 centered at the vectors of $C^{23,12}$ are disjoint and that every element of V_{23} lies in one such sphere. If a received vector u has an error in one, two, or three locations, it will lie in a sphere centered at the correct message v_i and v_i.

A good introduction to coding theory can be found in *A Common-sense Approach to the Theory of Error-correcting Codes* by Benjamin Arazi, The MIT Press, 1988.

Index

About the Author

Ramakanta Meher currently works as an Associate Professor at Sardar Vallabhbhai National Institute of Technology, Surat, Gujarat, India. His current research interests include differential and integral equations, fractional differential equations, linear algebra, approximation theory, numerical analysis, and fuzzy set theory. He has supervised 7 Ph.D. students and 30 M.Sc. students, and oversees 6 additional Ph.D. students. He has published over 80 international research publications in various SCI-indexed Q1 and Q2 international journals and solely published 7 textbooks. He has coordinated many prestigious programmes such as Mathematics Training Talent Search Programme (MTTS)-2013 (Level-0,1 and 2) sponsored by National Board of Higher Mathematics (NBHM), two STTPs by receiving grants from the Indian Society of Technical Education, New Delhi and some STTPs by receiving grants from TEQIP and SVNIT, Surat. He has delivered a series of invited talks at several scientific events held in India, Thailand, Saudi Arabia, Dubai, Mongolia, and South Africa. He has received several international travel grants such as the CIMPA-KAUST Travel Grant to visit King Abdullah University of Science and Technology, Thuwal, Saudi Arabia in 2012, CIMPA-UNESCO-MESR-MICINN-APSA-SOUTH AFRICA Travel Grant to visit African Institute of Mathematical Sciences, Muizenberg, Cape town, South Africa in 2011, CIMPA-UNESCO-MICINN-THAILAND Travel Grant to visit Chulalongkorn University, Bangkok, Thailand in 2011 and CIMPA-UNESCO-SCHOOL MONGOLIA Travel Grant to visit Mongolian State University, Mongolia in 2013.

For Product Safety Concerns and Information please contact our
EU representative GPSR@taylorandfrancis.com Taylor & Francis
Verlag GmbH, Kaufingerstraße 24, 80331 München, Germany